# Studies in Computational Intelligence

Volume 731

**Series editor**

Janusz Kacprzyk, Polish Academy of Sciences, Warsaw, Poland
e-mail: kacprzyk@ibspan.waw.pl

*About this Series*

The series "Studies in Computational Intelligence" (SCI) publishes new developments and advances in the various areas of computational intelligence—quickly and with a high quality. The intent is to cover the theory, applications, and design methods of computational intelligence, as embedded in the fields of engineering, computer science, physics and life sciences, as well as the methodologies behind them. The series contains monographs, lecture notes and edited volumes in computational intelligence spanning the areas of neural networks, connectionist systems, genetic algorithms, evolutionary computation, artificial intelligence, cellular automata, self-organizing systems, soft computing, fuzzy systems, and hybrid intelligent systems. Of particular value to both the contributors and the readership are the short publication timeframe and the worldwide distribution, which enable both wide and rapid dissemination of research output.

More information about this series at http://www.springer.com/series/7092

Yazmin Maldonado · Leonardo Trujillo
Oliver Schütze · Annalisa Riccardi
Massimiliano Vasile
Editors

# NEO 2016

## Results of the Numerical and Evolutionary Optimization Workshop NEO 2016 and the NEO Cities 2016 Workshop Held on September 20–24, 2016 in Tlalnepantla, Mexico

*Editors*
Yazmin Maldonado
Instituto Tecnológico de Tijuana
Tijuana
Mexico

Leonardo Trujillo
Instituto Tecnológico de Tijuana
Tijuana
Mexico

Oliver Schütze
Departamento de Computación
CINVESTAV-IPN
Mexico City
Mexico

Annalisa Riccardi
University of Strathclyde
Glasgow
UK

Massimiliano Vasile
University of Strathclyde
Glasgow
UK

ISSN 1860-949X ISSN 1860-9503 (electronic)
Studies in Computational Intelligence
ISBN 978-3-319-87712-9 ISBN 978-3-319-64063-1 (eBook)
https://doi.org/10.1007/978-3-319-64063-1

© Springer International Publishing AG 2018
Softcover reprint of the hardcover 1st edition 2017
This work is subject to copyright. All rights are reserved by the Publisher, whether the whole or part of the material is concerned, specifically the rights of translation, reprinting, reuse of illustrations, recitation, broadcasting, reproduction on microfilms or in any other physical way, and transmission or information storage and retrieval, electronic adaptation, computer software, or by similar or dissimilar methodology now known or hereafter developed.
The use of general descriptive names, registered names, trademarks, service marks, etc. in this publication does not imply, even in the absence of a specific statement, that such names are exempt from the relevant protective laws and regulations and therefore free for general use.
The publisher, the authors and the editors are safe to assume that the advice and information in this book are believed to be true and accurate at the date of publication. Neither the publisher nor the authors or the editors give a warranty, express or implied, with respect to the material contained herein or for any errors or omissions that may have been made. The publisher remains neutral with regard to jurisdictional claims in published maps and institutional affiliations.

Printed on acid-free paper

This Springer imprint is published by Springer Nature
The registered company is Springer International Publishing AG
The registered company address is: Gewerbestrasse 11, 6330 Cham, Switzerland

# Preface

Computer science continues to provide an integral part of today's real-world systems, which are built on top of continuously more sophisticated digital systems. In particular, search and optimization techniques are more widely used than ever, in both traditional domains and modern data-centered systems. In this context, the goal of the Numerical and Evolutionary Optimization (NEO) workshop series is to bring researchers together that work in complimentary areas of search and optimization, namely mathematical programming techniques as well as heuristic and meta-heuristic approaches. The NEO vision is that successful real-world systems will need to integrate both types of techniques to address some of the most complex and ambitious tasks.

The NEO has been founded as an international workshop series with possible venues mainly in Mexico but also other locations all over the world. The first two editions of this series, NEO 2013 and NEO 2014, were held in Tlalnepantla, State of Mexico, Mexico, hosted by the Cinvestav-IPN. NEO 2015 was held in Tijuana, Baja California, Mexico, hosted by the Instituto Tecnolgico de Tijuana (ITT). This book represents the second post-proceedings of the NEO series, from the NEO 2016 event which was held once again in Tlalnepantla from 20 to 24 of September. For more, and up to date information, please visit http://neo.cinvestav.mx for more details.

Moreover, the NEO 2016 event included a spin-off event called NEO Cities, focusing how optimization can, and should, be used in the development of future and smart cities, with funding provided by the Newton Fund of the British Council. Cities are very dynamic environments, they are breeding places for innovation, opportunities and research, but also sources of great challenges. The term Smart City is intended to describe a city where multiple computer technologies are integrated to manage monitor and improve city assets for a better living experience. Search and optimization techniques, as well as modern data analytics, can help address some of the open questions and shape these cities of the future.

This volume comprises a selection of extended works that have mainly been presented at NEO 2016, but speakers from previous NEO editions have also been invited to contribute. The book contains 11 chapters, organized into three parts, these are: (I) Smart Cities; (II) Search, Optimization and Hybrid Algorithms; and (III) Electronic and Embedded Systems.

Part I presents four chapters dealing with the application of search and optimization techniques to problems in the Smart Cities of today and tomorrow. In particular, the chapters deal with issues related to self driving cars, ubiquitous computing, transportation scheduling, and cloud computing. Part II contains chapters related to the core of the NEO series, search and optimization. Finally, Part III also contains application chapters, in this case related to an important component of today's technological revolution, and tomorrow's Smart Cities, electrical and embedded systems. In particular, the chapters deal with microelectromechanical systems, radio frequency communication and electronic amplifiers. These works present novel theoretical and experimental studies, that might be applicable to the domains from the first two parts of the book. In particular, the chapters deal with integer programming, multi-objective optimization and multidimensional assignment problems.

To conclude, we would like to express our gratitude to all invited speakers and attendees of the NEO 2016 event, they make this series possible! Finally, we thank all authors for their valuable contributed chapters and the reviewers who allowed us to meet the high quality standards of our scientific community.

We believe that this volume presents a valuable contribution to the fields of numerical and evolutionary optimization, which we hope you the reader will enjoy.

Tijuana, Mexico — Yazmin Maldonado
Tijuana, Mexico — Leonardo Trujillo
Mexico city, Mexico — Oliver Schütze
Glasgow, UK — Annalisa Riccardi
Glasgow, UK — Massimiliano Vasile
May 2017

# Acknowledgements

The editors of the NEO 2016 post-proceedings book would like to thank all of the authors for their unique and excellent contributions to the present book. Special thanks are also given to all that participated at the NEO 2016 workshop, particularly the student local organizers, without whom the event would not have been a success.

We also thank the Instituto Tecnológico de Tijuana and the Departamento de Ingeniería Eléctrica y Electrónica and the Posgrado en Ciencias de la Ingeniería, for their support and encouragement.

Additionally, we acknowledge the funding provided by Newton Fund Researcher Links Workshop Grant No. 216435254; CINVESTAV-IPN, CONACYT Basic Science Research Project No. 178323; CONACYT project FC-2015-2/944 "Aprendizaje evolutivo a gran escala"; CONACYT PRODECYT-DADC Project No. 263101; TecNM (México) Project No. 6350.17-P; FP7-PEOPLE-2013-IRSES project ACOBSEC financed by the European Commission with contract No. 612689.

# Contents

# Contributors

**R. Acosta-Bermejo** Instituto Politécnico Nacional Centro de Investigación en Computación, Mexico City, Mexico

**E. Aguirre-Anaya** Instituto Politécnico Nacional Centro de Investigación en Computación, Mexico City, Mexico

**E. Allende-Chávez** Tecnológico Nacional de México, Instituto Tecnológico de Tijuana, Tijuana, B.C., Mexico

**Elhadj Benkhelifa** School of Computing and Digital Tech, Staffordshire University, Staffordshire, UK

**Rodrigo Alexander Castro Campos** Graduate Program in Optimization, Universidad Autónoma Metropolitana Azcapotzalco, Mexico City, Mexico

**Heriberto Cruz Hernández** Computer Science Department, Cinvestav, Mexico City, México

**J.R. Cárdenas-Valdez** Tecnológico Nacional de México, Instituto Tecnológico de Tijuana, Tijuana, B.C., Mexico

**Michael Dellnitz** Department of Mathematics, Paderborn University, Paderborn, Germany

**Victor H. Díaz-Ramírez** CITEDI-IPN, Tijuana, Baja California, Mexico

**P.J. Escamilla-Ambrosio** Instituto Politécnico Nacional Centro de Investigación en Computación, Mexico City, Mexico

**Lizbeth Escobedo** UCSD, La Jolla, CA, USA

**Michael Farnsworth** Manufacturing Informatics Centre, Cranfield University, Cranfield, UK

**J.A. Galaviz-Aguilar** Instituto Politécnico Nacional, CITEDI, Tijuana, B.C., Mexico

**Marco Antonio Heredia Velasco** Departamento de Sistemas, Universidad Autónoma Metropolitana Unidad Azcapotzalco, Mexico City, Mexico

**S.A. Juárez-Cázares** Instituto Politécnico Nacional, CITEDI, Tijuana, B.C., Mexico

**Xiang Li** Department of Mechanics, Tianjin University, Tianjin, China; College of Sciences, Northeastern University, Shenyang, China

**Víctor R. López-López** Posgrado en Ciencias de la Ingeniería, Departamento de Ingeniería Eléctrica y Electrónica, Tijuana, Baja California, Mexico

**J.C. Nuñez-Pérez** Instituto Politécnico Nacional, CITEDI, Tijuana, B.C., Mexico

**Sebastian Peitz** Department of Mathematics, Paderborn University, Paderborn, Germany

**Sergio Luis Pérez Pérez** Graduate Program in Optimization, Universidad Autónoma Metropolitana Azcapotzalco, Mexico City, Mexico

**A. Rodríguez-Mota** Instituto Politécnico Nacional Escuela Superior de Ingeniería Mecánica y Eléctrica, Mexico City, Mexico

**M. Salinas-Rosales** Instituto Politécnico Nacional Centro de Investigación en Computación, Mexico City, Mexico

**Adriana C. Sanabria-Borbón** Department of Electrical and Computer Engineering, Texas A&M University, College Station, TX, USA

**Y. Sandoval-Ibarra** Tecnológico Nacional de México, Instituto Tecnológico de Tijuana, Tijuana, B.C., Mexico

**Jian-Qiao Sun** School of Engineering, University of California, Merced, CA, USA

**Ashutosh Tiwari** Manufacturing Informatics Centre, Cranfield University, Cranfield, UK

**Esteban Tlelo-Cuautle** Computer Science Department, Cinvestav, Mexico City, Mexico; Department of Electronics, INAOE, Tonantzintla, Puebla, Mexico

**Leonardo Trujillo** Tecnológico Nacional de México, Instituto Tecnológico de Tijuana, Tijuana, B.C., Mexico

**Carlos E. Valencia** Departamento de Matemáticas, Centro de Investigación y de Estudios Avanzados del IPN, Mexico City, Mexico

**Gualberto Vazquez Casas** Posgrado en Optimización, Universidad Autónoma Metropolitana Unidad Azcapotzalco, Mexico City, Mexico

**Francisco Javier Zaragoza Martínez** Departamento de Sistemas, Area de Optimización Combinatoria, Universidad Autónoma Metropolitana Azcapotzalco, Mexico City, Mexico

**Meiling Zhu** College of Engineering, Mathematics and Physical Sciences, University of Exeter, Exeter, UK

**Luis Gerardo de la Fraga** Computer Science Department, Cinvestav, Mexico City, Mexico

# Part I
# Smart Cities

# Defensive Driving Strategy and Control for Autonomous Ground Vehicle in Mixed Traffic

**Xiang Li and Jian-Qiao Sun**

**Abstract** One of the challenges of autonomous ground vehicles (AGVs) is to interact with human driven vehicles in the traffic. This paper develops defensive driving strategies and controls for AGVs to avoid problematic vehicles in the mixed traffic. The multi-objective optimization algorithms for local trajectory planning and adaptive cruise control are proposed. The dynamic predictive control is used to derive optimal trajectories in a rolling horizon. The intelligent driver model and lane-changing rules are employed to predict the movement of the vehicles. Multiple performance objectives are optimized simultaneously, including traffic safety, transportation efficiency, driving comfort, tracking error and path consistency. The multi-objective optimization problems are solved with the cell mapping method. Different scenarios are created to test the effectiveness of the defensive driving strategies and adaptive cruise control. Extensive experimental simulations show that the proposed defensive driving strategy and PID-form control are promising and may provide a new tool for designing the intelligent navigation system that helps autonomous vehicles to drive safely in the mixed traffic.

**Keywords** Defensive driving · Motion planning · Trajectory planning · Multi-objective optimization · Adaptive cruise control

## 1 Introduction

Intelligent autonomous vehicles have great potential to improve human mobility, enhance traffic safety and achieve fuel economy [5, 43]. However, to realize this potential, the autonomous driving technologies such as sensing, decision making and motion planning must learn how to interact with the vehicles driven by human. The

---

X. Li
Department of Mechanics, Tianjin University, Tianjin 300072, China

J.-Q. Sun (✉)
School of Engineering, University of California, Merced, CA 95343, USA
e-mail: jqsun@ucmerced.edu

© Springer International Publishing AG 2018

Y. Maldonado et al. (eds.), *NEO 2016*, Studies in Computational Intelligence 731,
https://doi.org/10.1007/978-3-319-64063-1_1

mixed traffic of human-driven and autonomous vehicles has been drawing increasing attention in recent years. This chapter proposes a novel multi-objective optimization algorithm for designing *defensive driving strategies* and *adaptive cruise controls* (ACC) for autonomous ground vehicles (AGVs) to avoid problematic human-driven vehicles in the mixed traffic. The research findings are obtained from extensive computer simulations. Since the relationship between the performances and control parameters of the traffic network is highly nonlinear, complex and random, it is very common to use computer simulations in the traffic research.

Driving autonomous vehicles is usually planned in four levels [55], i.e. route planning, path planning, maneuver choice and trajectory planning. Route planning is concerned with finding the best global reference path from a given origin to a destination. Path, maneuver and trajectory planning, which are often combined as one, provide the AGV with a safe and efficient local trajectory along the reference path considering vehicle dynamics, maneuver capabilities and road structure in the presence of other traffic [24]. *A reference path is assumed to be obtained as prior information from the high-level planner in this study.* We develop a feasible multi-objective optimal local trajectory for the AGV to follow the reference path in a safe manner. In the low-level, a PID-form control is proposed for the AGV to track the leader.

## 1.1 Motion Planning

Recently, the model predictive control (MPC) which is an optimal control method applied in a rolling horizon framework, appears to be very promising and has been popularly used to solve trajectory planning problems [67]. It is robust to uncertainty, disturbance and model mismatch. However, solving the optimization problem in real-time requires remarkable computational effort. To deal with the difficulties, a lot of researches have been carried out on sampling-based trajectory planning approaches. Two classes of trajectory generations are studied, i.e. control-space sampling and state-space sampling [18].

The control-space sampling method aims to generate a feasible control in the parameterized control space. The trajectories can be generated through forward simulation of the differential equations. However, the road environment must be considered in the sampling process, which may lead to discrepancy between two consecutive plans, and overshoot and oscillation in trajectory tracking process [18, 37].

On the other hand, the state-space sampling method selects a set of terminal states and computes the trajectories connecting initial and terminal states. The information of the environment can be well exploited. System-compliant trajectories can be obtained via forward simulations using the vehicle system model [25, 47]. The state-space sampling method is used in this study for trajectory generation.

In the planning of autonomous vehicle trajectory, static obstacles are often considered to avoid collision. Li et al. have proposed an integrated local trajectory planning and tracking control framework for AGVs driving along a reference path with obstacle avoidance [34]. An MPC-based path generation algorithm is applied to produce a set of smooth and kinematically-feasible paths connecting the initial state with the sampled terminal states. The algorithm proposed in [14] generates a collision-free path that considers vehicle dynamic constraints, road structure and different obstacles inside the horizon of view. Yoon et al. have developed a model-predictive approach for trajectory generation of AGVs combined with a tire model [62]. Information on static obstacles is incorporated online in the nonlinear model-predictive framework as they are sensed within a limited sensing range.

On public roads, the influence of surrounding traffic on AGVs is significant. Trajectory planning techniques need to anticipate the behaviors of the surrounding traffic. Some trajectory planning is based on the assumption that the surrounding traffic will keep constant speed [38, 67]. Shim and colleagues have proposed a motion planner that can compute a collision-free reference trajectory [48]. When the path of a moving obstacle is estimated and a possible collision in the future is detected, a new path is computed to avoid the collision. A real-time autonomous driving motion planner with trajectory optimization has been proposed by Xu and colleagues [60]. Wei et al. have proposed a motion planning algorithm considering social cooperation between the autonomous vehicle and surrounding cars [56]. Kinematically and dynamically feasible paths are generated first assuming no obstacle on the road. A behavioral planner then takes static and dynamic obstacles into account. An optimization-based path planner is presented by Hardy et al. [15] that is capable of planning multiple contingent paths to account for uncertainties in the future trajectories of dynamic obstacles. The problem of probabilistic collision avoidance is addressed for AGVs that are required to safely interact with other vehicles with unknown intentions.

A human driver makes decisions on lane-changing, overtaking *etc*. based on his/her experience and judgement of the behavior of the surrounding vehicles. The ability to judge the surrounding vehicle behavior including the driving pattern, aggressiveness and intentions and to make adjust his/her own driving to avoid problematic vehicles is an important part of knowledge an experienced driver has. This work intents to develop a multi-objective optimization algorithm to allow AGVs to possess such an ability. Specifically, car-following and lane-changing behaviors of AGVs in straight road segments are optimized to avoid problematic vehicles. The objectives include transportation efficiency, traffic safety, driving comfort, path consistency *etc*. The multi-objective optimization problem (MOP) of trajectory planning is formulated where multiple driving performance objectives are optimized simultaneously. The cell mapping method [19], which is a deterministic searching algorithm compared with stochastic algorithms such as genetic algorithm, is adopted to solve the MOP. The solutions of MOP form a set in the design space called the *Pareto set*, and the corresponding objective evaluations are the *Pareto front* [44]. This work may provide a useful tool for intelligent navigation of AGVs in the mixed traffic.

## 1.2 Adaptive Cruise Control

Generally, an ACC-equipped vehicle collects the information such as the range and range rate to the preceding vehicle with a sensor (radar), in order to keep a desired distance with its leader. Moreover, to obtain richer travelling information of the platoons on road, cooperative adaptive cruise control (CACC) is developed with wireless communication with neighboring vehicles (vehicle location, acceleration etc.) and surrounding infrastructures (traffic light states etc.) such as the dedicated short range communication (DSRC). The CACC equipped vehicle in this way can be fast responsive to condition changes and achieve tracking performance more accurate than ACC. However, DSRC devices are unable to transmit information instantaneously, which introduces time delay in communication that should be taken into consideration.

ACC provides better driving experience and driver convenience in comparison with traditional cruise control (CC) [41, 42, 45]. Delay effects and stability of ACC vehicle platoon were studied by Orosz and colleagues [12, 65]. Generally, space headway and velocity difference with the leader can be selected as input signals for ACC [42]. And moreover, delayed acceleration feedback signal is suggested to be taken into account for the fact that it can help stabilize the vehicle platoons even in cases when human drivers are incapable due to reaction time [12]. However, the ACC systems available at present only put emphasis on tracking performance with limited attention on other control objectives in reality [28], e.g., driving comfort and fuel economy. Reduction in fuel consumption of vehicles with the help of ACC was suggested by Tsugawa [54] and Ioannou et al. [3]. Focusing on energy economy, a nonlinear filter-based PI controller was designed by Zhang et al. [64] that set limits on acceleration, and a dynamic programming based off-line control method was proposed by Jonsson [23] to lower the fuel consumption level. Multi-objectives problem (MOP) on ACC system has been hardly solved based on the current literature review. The goals associated are often conflicting with each other, e.g., high accuracy of tracking with the preceding vehicle usually leads to uncomfortable driving experience and increasing fuel consumption, and fuel economy can be achieved at the expense of more tracking error. Hence, it's quite challenging to consider various objectives simultaneously for ACC system. This chapter proposes a proportion-integral-derivative (PID) form controller to solve the MOP in ACC, the simulation results of which are shown effective with respect to several objectives. Meanwhile, the stability problem of the system with the proposed controller is solved.

Model predictive control has been popularly used in solving MOP in ACC at present, but with a weak point in its considerable computational burden. A systematic way for the design and tuning of ACC controller was presented by Naus et al. [40]. An explicit MPC approach was adopted to design the control off-line incorporating multiple objectives and constraints, and generate an analytical solution. Spacing-control laws for transitional maneuvers of ACC vehicles are computed by Bageshwar and colleagues [1] using MPC where the acceleration is restricted to achieve driving comfort and safety. A rather complete set of control objectives of tracking error, fuel economy, driving comfort and driver desired traveling characteristics are considered by Li et al. [28] to build a MPC framework for ACC. Detailed simulations have proved the effectiveness of MPC in the optimization of the various design objectives. In this study, MPC is used in ACC system as a comparison to show the effectiveness of the proposed method.

So far, many studies have been available for MOP control design for linear systems, but only a handful references can be used with respect to nonlinear system. Different from single objective optimization problem (SOP), the solutions of the MOP consist of a set of optimal ones, which is called Pareto set, instead of a unique point in the design space. And the corresponding objective function values (Pareto front) can be all considered to be optimal. Many algorithms are able to obtain the desired Pareto set and Pareto front [8, 27]. Among them the cell mapping method appears to be very promising and is adopted in this study to solve the MOP in ACC.

The remainder of this chapter starts with a brief description of the autonomous vehicle planning scheme in Sect. 2. Section 3 presents the motion and trajectory planning algorithms. The traffic prediction models and multi-objective optimization method are provided. The PID-form adaptive cruise control algorithm is proposed in Sect. 4, together with stability proof, optimal designs and multi-objective optimization formulation. The proposed methods are tested in extensive different scenarios to verify the effectiveness. We close the chapter with conclusions in Sect. 5.

## 2 Autonomous Driving Architecture

Figure 1 shows the architecture of the automated driving control. In the following, we discuss the architecture and the assumptions of this study.

The control architecture includes a dynamic route planning module, which makes use of on-board sensors. The sensors are assumed to be able to provide real-time information of the surrounding vehicles such as positions, velocities, careless-driving, aggressiveness, *etc.*. The motion planning algorithm for defensive driving to avoid careless or aggressive drivers selects one of ten possible motions in Fig. 2. This approach significantly reduces the online computational time in the dynamic optimization of trajectories.

In the dynamic trajectory planning, the Pareto optimal set of trajectories can be obtained. Multiple performance objectives such as safety, transportation efficiency, driving comfort *etc.* are optimized simultaneously with the cell mapping method. All the optimal trajectories must respect the longitudinal and lateral vehicle constraints to ensure the safety, stability and controllability. The multi-objective optimization problems are solved online in a rolling horizon way to derive optimal designs for the next control horizon. The time-domain updating scheme of the control is presented in Fig. 3. The prediction horizon $N_p$ and control horizon $N_c$ are assumed to be constant. The final output is a desired trajectory representing the spatiotemporal longitudinal and lateral position, velocity and acceleration of the AGV.

The prediction horizon is supposed to be much longer than the control horizon, allowing the controller to tolerate delays. It also increases the reliability of the system, since the lower-level controller always has a relatively long trajectory to execute, even if the higher-level planner stops working for some time. At the next control step, the prediction horizon is shifted one step forward, and the optimization is carried out again with the updated traffic information. The control system benefits from the rolling horizon framework with feedbacks from the real traffic periodically, which makes the proposed control robust to uncertainties and disturbances.

As stated earlier, the AGV is assumed to be collecting real-time information of the neighboring vehicles within the sensor range $l_{sense}$, including the longitudinal and lateral positions, velocities and accelerations. In the time step $t_{ctrl,p} = t_{ctrl} - t_{delay}$ where $t_{delay}$ is the delayed time step reserved for online computing, the multi-objective optimization problem starts to be solved based on the current information. The optimal controller $\mathbf{k}(t_{ctrl})$ is assumed to be derived at control step $t_{ctrl}$ where $t_{ctrl} = n_{int} \times N_c$ and $n_{int}$ is an integer. The prediction horizon is the following $N_p$ steps from $t_{ctrl,p}$. The future traffic with a wide variety of control designs is forecasted, and the design leading to the optimal driving performance is used at the control step.

Finally, we assume that the lower level cruise control of the AGV can track the desired trajectory with high accuracy if it is not in the adaptive cruise control mode. When it is in the ACC mode, the proposed PID-form controller will be implemented, that will be introduced in the following sections.

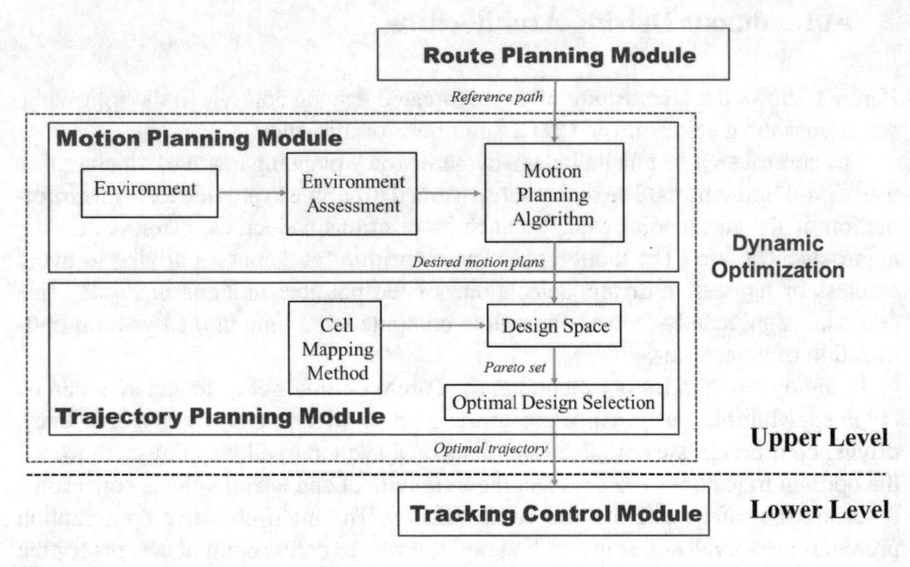


**Fig. 1** Autonomous driving architecture

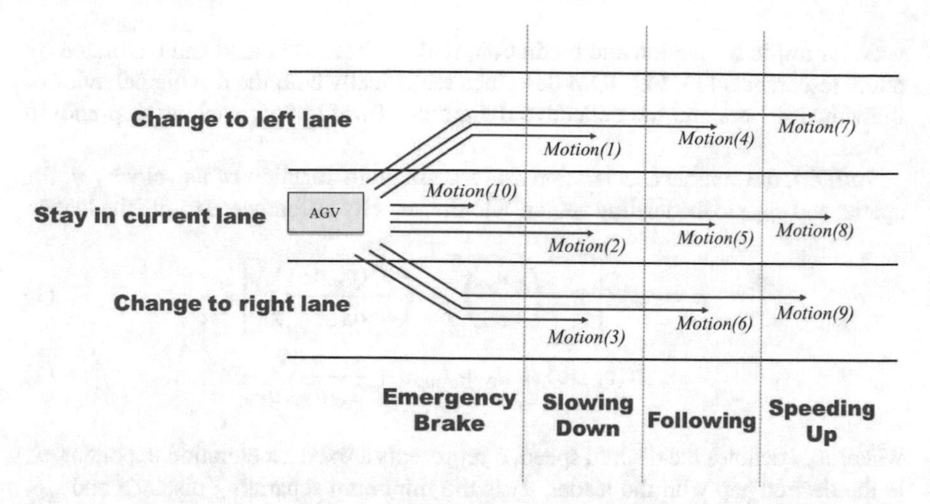


**Fig. 2** Options for local motion planning

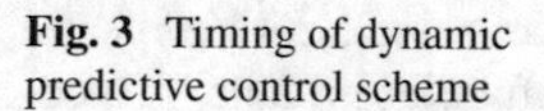
**Fig. 3** Timing of dynamic predictive control scheme

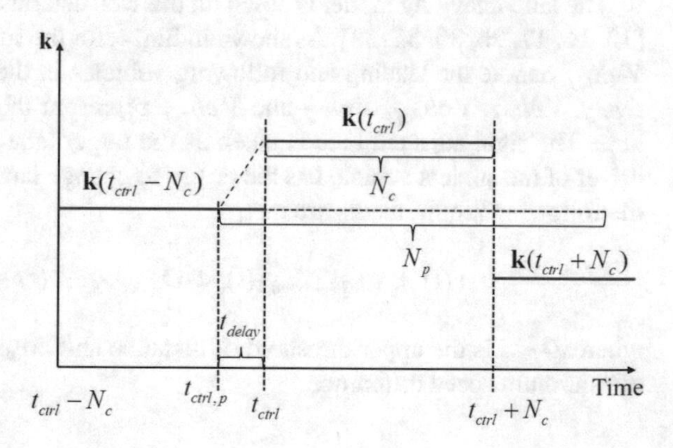


# 3 Motion Planning

## 3.1 *Traffic Prediction Models*

The models for traffic simulation and prediction are widely available in the literature and are briefly reviewed here. Only the straight road is considered in this study. Vehicles are assumed to travel longitudinally in a single lane and to move laterally when changing lane.

Car-following models that describe how drivers follow the leading vehicles have been studied for more than half a century [4, 53]. Popular models include intelligent driver model [35, 52], Gazis-Herman-Rothery model [7, 29], safety distance or collision avoidance model, Helly linear model, fuzzy logic-based model and optimal velocity model [2]. In this chapter, the popular intelligent driver model (IDM) is used for traffic simulation and prediction, that has been validated and calibrated by many researchers [35, 52]. IDM describes realistically both the driving behavior of individual drivers and the collective dynamics of traffic flow such as stop-and-go waves.

In IDM, the vehicle acceleration $a$ is a continuous function of its velocity $v$, the spatial gap $d_{lead}$ to the leading vehicle and the velocity difference $\Delta v$ with the leader.

$$a = a_{\max}\left[1 - \left(\frac{v}{v_{des}}\right)^{\delta} - \left(\frac{d^*(v, \Delta v)}{d_{lead}}\right)^2\right], \tag{1}$$

$$d^*(v, \Delta v) = d_0 + t_{gap}v + \frac{v\Delta v}{2\sqrt{a_{\max}a_{\min}}}, \tag{2}$$

where $v_{des}$ denotes the desired speed, $\delta$ represents a fixed acceleration exponent, $d^*$ is the desired gap with the leader, $d_0$ is the minimum separating distance and $t_{gap}$ denotes a preferred time gap.

The lane-changing model is based on the well established models in the literature [13, 16, 17, 26, 30, 32, 33]. As shown in Fig. 4, for the subject vehicle, let $Veh_{l,l}$ and $Veh_{l,f}$ denote the leading and following vehicles in the left adjacent lane, respectively. $Veh_{c,l}$, $Veh_{c,f}$, $Veh_{r,l}$ and $Veh_{r,f}$ represent those in the current and right lane. The right adjacent lane is taken as the target lane for instance. At time $t$, the driver of the subject vehicle has the desire to change lane for speed advantage when the following requirements are met,

$$v(t) < v_{des},\quad d_{lead}(t) < D_{free},\quad \Delta v(t) < \Delta v_{dif,\min}, \tag{3}$$

where $D_{free}$ is the upper threshold of distance and $\Delta v_{dif,\min}$ denotes a preset value of minimum speed difference.

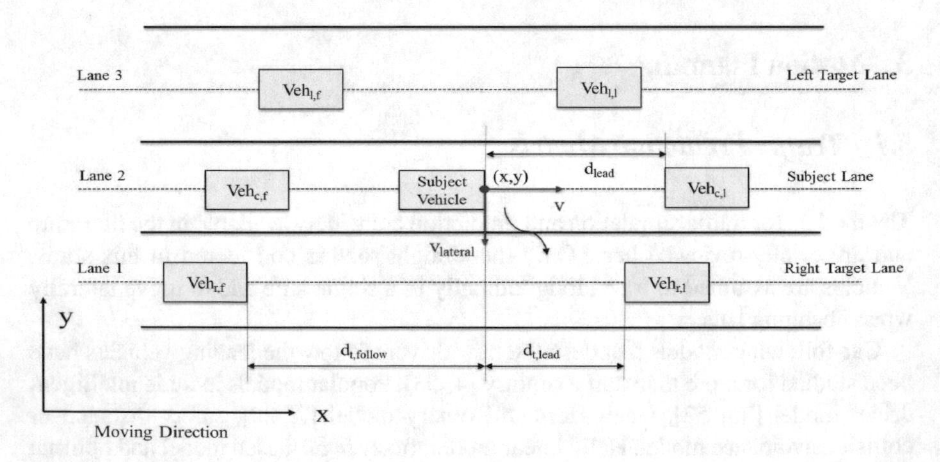


**Fig. 4** A three-lane traffic scenario

Feasibility and safety must be checked before changing lane, such that the following inequalities must hold,

$$d_{t,lead}(t) > d_{\min}, \quad d_{t,follow}(t) > d_{\min}, \tag{4}$$

where $d_{t,lead}$ and $d_{t,follow}$ denote the longitudinal gaps between the subject vehicle and the leader and follower in the target lane, and $d_{\min}$ is the minimum acceptable gap.

Time-to-collision (TTC) is a safety indicator that represents the time for two vehicles to collide if they remain at the current longitudinal speed without lane-changing. A minimum time-to-collision $t_{safe}$ is required for safety to change lane such that,

$$t_{t,lead}(t) > t_{safe}, \quad t_{t,follow}(t) > t_{safe}, \tag{5}$$

where $t_{t,lead}$ and $t_{t,follow}$ denote the longitudinal TTCs between the subject vehicle and the leader and follower in the target lane. In addition, the target lane has to provide enough space for the subject vehicle to accelerate to higher speed. The following condition must hold,

$$d_{t,lead}(t) > v(t)\, t_{extra}, \tag{6}$$

where $t_{extra}$ is the required minimum time headway. We also impose a minimum time $t_{\min}$ between two consecutive lane-changings for one vehicle in order to prevent frequent lane-changings.

The duration of lane-changing sometimes is ignored and the lane-changing process is regarded as an instantaneous movement in some models [13, 26, 31]. However, the influence of lane-changing duration on the traffic is significant and cannot be disregarded [51]. In this study, each vehicle $Veh_i$ is assumed to have a pre-determined lane-changing time $t_{i,lc}$ that is a random number uniformly distributed in an interval $[t_{lc,\min}, t_{lc,\max}]$ where $t_{lc,\max}$ and $t_{lc,\min}$ denote the maximum and minimum lane-changing time. In the lane-changing process, the lateral velocity is assumed to be constant while the longitudinal movement follows the IDM. For the two leading vehicles in the current and target lanes, the one with shorter longitudinal distance from the subject vehicle is considered the preceding vehicle in the car-following model.

### 3.2 *Defensive Strategy*

The defensive motion planning provides strategies for the AGV to deal with the surrounding traffic. As Fig. 2 shows, ten motion patterns are proposed to reduce the online computational time in the trajectory optimization. In the longitudinal direction, three regimes are defined, i.e. following, slowing down and speeding up. Equation (7) shows nine motions as combinations of three longitudinal moves and three lane maneuvers (changing to the left lane, staying in the current lane and changing to the right lane). The tenth move is an emergency braking motion to avoid collision.

$$Motion(i) = \begin{cases} \text{Change to the left lane and slow down}, & i = 1, \\ \text{Stay in the current lane and slow down}, & i = 2, \\ \text{Change to the right lane and slow down}, & i = 3, \\ \text{Change to the left lane and follow}, & i = 4, \\ \text{Stay in the current lane and follow}, & i = 5, \\ \text{Change to the right lane and follow}, & i = 6, \\ \text{Change to the left lane and speed up}, & i = 7, \\ \text{Stay in the current lane and speed up}, & i = 8, \\ \text{Change to the right lane and speed up}, & i = 9, \\ \text{Emergency brake}, & i = 10. \end{cases} \quad (7)$$

The defensive motion planning algorithm is presented in Table 1 to avoid careless and aggressive drivers. Note that the emergency braking occurs if the gap between the AGV and its leader is shorter than a safe distance $d_{safe}$.

Careless-driving has significant influence on traffic [49], which includes vehicle waving, close cut-in, frequent lane-changings *etc*. In this study, only the vehicle waving behavior is considered. In the defensive driving, the AGV is supposed to avoid the waving vehicle $Veh_{wav}$ in the neighborhood if the waving amplitude $Veh_{wav}.Amp$ exceeds a critical value $d_{wave}$. We assume that $Veh_{wav}$ can be detected by the onboard sensors of the AGV.

The defensive planner also suggests not to follow the leader closely if the leader is aggressive. If the follower shows remarkable aggressiveness, the defensive planner suggests to change lane to provide courtesy and avoid the aggressive driver. The vehicle aggressiveness $Veh.Agg$ is quantified by the detected maximum absolute acceleration. Let $Veh_{agg}$ represent the more aggressive driver between the leader and follower, when $Veh_{agg}.Agg$ exceeds a critical value $c_{agg}$, a defensive-driving motion will be followed through.

If the speed advantage is demanded (see Eq. (3)) and lane-changing feasibilities are satisfied (see Eqs. (4)–(6)), the AGV will attempt to change lane to overtake.

The AGV either speeds up if no leading vehicle is ahead, or follows the leader. The minimum time $t_{\min}$ between two consecutive lane-changings is also imposed to prevent frequent lane-changings.

When lane-changing is attempted but not feasible, the longitudinal motion will follow through.

### 3.3 Trajectory Formulation

Once a defensive driving move is decided, short-term trajectories will be generated dynamically. A novel multi-objective optimization algorithm is proposed to find the optimal trajectory.

The local trajectory is generated for each prediction horizon. Let $t_1$ and $t_2$ denote the initial and terminal time in one prediction horizon, and $x_0$, $v_0$ and $a_0$ denote the initial longitudinal position, velocity and acceleration when the AGV is at $(x, y)$ as shown in Fig. 4. $v_1$ and $a_1$ represent the terminal velocity and acceleration. Hence,

$$\begin{aligned} x(t_1) = x_0,\ \dot{x}(t_1) = v_0,\ \ddot{x}(t_1) = a_0, \\ \dot{x}(t_2) = v_1,\ \ddot{x}(t_2) = a_1. \end{aligned} \tag{8}$$

Note that $x(t_2)$ is unspecified and the velocity and acceleration $v_1$ and $a_1$ can fully specify the local trajectory. For this reason, we take $v_1$ and $a_1$ as the design parameters of the local trajectory.

In the lateral direction, when lane-changing is considered, the AGV is assumed to complete the lane-changing in a pre-determined time $t_{lc}$. Let $y_0$ and $y_1$ denote the lateral positions of the center axises of the current and target lanes respectively.

$$\begin{aligned} y(t_1) = y_0,\ \dot{y}(t_1) = 0,\ \ddot{y}(t_1) = 0, \\ y(t_1 + t_{lc}) = y_1,\ \dot{y}(t_1 + t_{lc}) = 0,\ \ddot{y}(t_1 + t_{lc}) = 0. \end{aligned} \tag{9}$$

The above initial and terminal conditions indicate that the trajectory $x(t)$ can be fourth-order polynomials of time in the interval $[t_1, t_2]$ [63]. A fifth-order polynomial of time for $y(t)$ is needed to describe the lateral movement in the time interval $[t_1, t_3]$ where $t_3 = t_1 + t_{lc}$.

$$x(t) = \sum_{i=0}^{4} a_i t^i,\ y(t) = \sum_{i=0}^{5} b_i t^i. \tag{10}$$

Hence, we have

$$\begin{aligned} [x_0, v_0, a_0, v_1, a_1]^T &= \mathbf{M}_x \mathbf{A}^T, \\ [y_0, 0, 0, y_1, 0, 0]^T &= \mathbf{M}_y \mathbf{B}^T, \end{aligned} \tag{11}$$

where

$$\begin{aligned} \mathbf{A}^T &= [a_4, a_3, a_2, a_1, a_0], \\ \mathbf{B}^T &= [b_5, b_4, b_3, b_2, b_1, b_0], \end{aligned} \tag{12}$$

$$\mathbf{M}_x = \begin{bmatrix} t_1^4 & t_1^3 & t_1^2 & t_1^1 & 1 \\ 4t_1^3 & 3t_1^2 & 2t_1^1 & 1 & 0 \\ 12t_1^2 & 6t_1^1 & 2 & 0 & 0 \\ 4t_2^3 & 3t_2^2 & 2t_2^1 & 1 & 0 \\ 12t_2^2 & 6t_2^1 & 2 & 0 & 0 \end{bmatrix},\ \mathbf{M}_y = \begin{bmatrix} t_1^5 & t_1^4 & t_1^3 & t_1^2 & t_1^1 & 1 \\ 5t_1^4 & 4t_1^3 & 3t_1^2 & 2t_1^1 & 1 & 0 \\ 20t_1^3 & 12t_1^2 & 6t_1^1 & 2 & 0 & 0 \\ t_3^5 & t_3^4 & t_3^3 & t_3^2 & t_3^1 & 1 \\ 5t_3^4 & 4t_3^3 & 3t_3^2 & 2t_3^1 & 1 & 0 \\ 20t_3^3 & 12t_3^2 & 6t_3^1 & 2 & 0 & 0 \end{bmatrix}. \tag{13}$$

The trajectory in one prediction horizon can be created by solving Eqs. (11) and (12).

### 3.4 Multi-objective Optimization

As noted before, we consider the design vector $\mathbf{k} = [v_1, a_1]^T$ for optimizing the local trajectories. The following constraints on the design parameters are imposed.

$$\begin{aligned} 0 \leq \dot{x}(t) \leq v_{\max}, \\ a_{\min} \leq \ddot{x}(t) \leq a_{\max}, \end{aligned} \tag{14}$$

where $\forall t \in [t_1, t_2]$, $a_{\max}$ and $a_{\min} < 0$ denote the longitudinal maximum acceleration and deceleration, and $v_{\max}$ represents the maximum speed of the AGV subject to the legal speed limit. When lane-changing is attempted, the lateral trajectory can be uniquely determined based on the lane structure and corresponding constraints. The control objectives of the AGV in the defensive driving include (1) Safety, (2) Transportation efficiency, (3) Driving comfort and fuel economy, and (4) Path consistency. The solutions of a multi-objective optimization problem (MOP) form a set in the design space called the *Pareto set*, and the corresponding objective evaluations are the *Pareto front* [44].

The safety indicator is defined as the sum of the time headway $t_{th}(t)$ and time-to-collision (TTC) $t_{ttc}(t)$. Time headway is the elapsed time between the front of the leading vehicle passing a point on the roadway and the front of the following vehicle passing the same point. TTC represents the time before two consecutive vehicles collide if they remain their current longitudinal speed. Let $G_{sf}$ denote the safety,

$$G_{sf} = \int_{t_1}^{t_2} (\alpha_{th} t_{th}(t) + \alpha_{ttc} t_{ttc}(t)) dt, \tag{15}$$

$$t_{th}(t) = \frac{v(t)}{d_{lead}(t)}, \tag{16}$$

$$t_{ttc}(t) = \begin{cases} \dfrac{v(t) - v_{leader}(t)}{d_{lead}(t)}, & v(t) > v_{leader}(t) \\ 0, & v(t) \leq v_{leader}(t) \end{cases} \tag{17}$$

where $v_{leader}$ is the speed of the leader, and $\alpha_{th}$ and $\alpha_{ttc}$ represent the weights of the two items.

When safety is ensured, the AGV is suggested to travel longer distance in one horizon. This distance is taken as the measure of the transportation efficiency denoted as $G_{te0}$,

$$G_{te0} = x(t_2). \tag{18}$$

The objective $G_{dc}$ of driving comfort and fuel consumption is measured by the longitudinal acceleration and jerk of the vehicle [36, 54, 64], given by

$$G_{dc} = \int_{t_1}^{t_2} (\omega_a a^2(t) + \omega_j \dot{a}^2(t))dt. \tag{19}$$

where $\omega_a$ and $\omega_j$ denote the weighting factors of $a(t)$ and $\dot{a}(t)$.

The inconsistency between the consecutive plans can result in sharp steerings, control overshoots or even instability [34]. The path consistency in the trajectory re-planning process is imposed in this study. The 2-norm objective $G_{con}$ penalizes the inconsistency between the current and previous longitudinal trajectories, defined as,

$$G_{con} = \int_{t_1}^{t_2} \left(x(t) - x_{pre}(t)\right)^2 dt. \tag{20}$$

The multi-objective optimal local trajectory problem is formulated as,

$$\min_{\mathbf{k}\in Q}\{G_{sf}, G_{con}, G_{dc}, G_{te}\}, \tag{21}$$

where $Q$ is a bounded domain of the design parameters, $G_{te} = G_{te,\max} - G_{te0}$ and $G_{te,\max} > 0$ is a large number such that $G_{te}$ is positive. Furthermore, we impose the constraints on $t_{th}(t)$ and $t_{ttc}(t)$ to keep the driving performance within an acceptable range.

$$\begin{aligned} \max t_{th}(t) &\le t_{th,\lim}, \\ \max t_{ttc}(t) &\le t_{ttc,\lim}, \end{aligned} \tag{22}$$

where $t_{th,\lim}$ and $t_{ttc,\lim}$ are the pre-selected upper limits.

## 3.5 *Cell Mapping Method*

The simple cell mapping (SCM) method, proposed by Hsu [19] and further developed by Sun and his group [58], is used to solve the MOP. The cell mapping method is well suited to search for the solutions of the MOP [58, 59]. The method proposes to discretize the continuum design space into a collection of boxes or cells and to describe the searching process by cell-to-cell mappings in a finite region of interest in the design space.

The SCM accepts one image cell for a given pre-image cell. The SCM can be symbolically expressed as $z_{k+1} = C[z_k]$ where $k$ is the iteration step, $z_k$ is an integer representing the cell where the system resides at the $k^{th}$ step, and $C[\cdot]$ is the integer mapping constructed from an optimization search strategy. The region out of the domain of the computational interest is called the sink cell. If the image of a cell is out of the domain of interest, we say that it is mapped to the sink cell. The sink cell always maps to itself.

The SCM can be implemented initially with relatively large cells. The cyclic groups of the cells in the SCMs form a set covering the Pareto solution of the MOP. The cells in the cyclic groups can be sub-divided to improve the accuracy of the MOP solutions [58].

### 3.6 Design Optimization

The bounded design space is defined as

$$Q = \left\{ \mathbf{k} \in \mathbf{R}^2 | \; [v_{lb}, a_{lb}] \le \mathbf{k} \le [v_{ub}, a_{ub}] \right\}. \tag{23}$$

where the lower and upper bounds $[v_{lb}, a_{lb}]$ and $[v_{ub}, a_{ub}]$ are specified for four different driving patterns of the AGV under consideration.

1. Following

$$\begin{aligned} a_{lb} &= -a_f, \;\; a_{ub} = a_f, \\ v_{lb} &= \max(v_{lead,p}(t_2) - v_f, 0), \\ v_{ub} &= \min(v_{lead,p}(t_2) + v_f, v_{des}), \end{aligned} \tag{24}$$

2. Slowing down

$$\begin{aligned} a_{lb} &= -a_c, \;\; a_{ub} = 0, \\ v_{lb} &= \max(v(t_1) - a_c N_p, 0), \;\; v_{ub} = v(t_1), \end{aligned} \tag{25}$$

3. Speeding up

$$\begin{aligned} a_{lb} &= 0, \;\; a_{ub} = a_c, \\ v_{lb} &= v(t_1), \;\; v_{ub} = \min(v(t_1) + a_c N_p, v_{des}), \end{aligned} \tag{26}$$

4. Emergency braking

$$\begin{aligned} a_{lb} &= a_{\min}, \;\; a_{ub} = 0, \\ v_{lb} &= \max(v(t_1) + a_{\min} N_p, 0), \;\; v_{ub} = v(t_1), \end{aligned} \tag{27}$$

where $a_f$ and $a_c$ denote the pre-determined acceleration fluctuating amplitudes, $v_f$ represents the velocity fluctuating amplitude, and $v_{lead,p}(t_2)$ is the predicted speed of the leading vehicle at time $t_2$. Appropriate values of $a_f$, $a_c$ and $v_f$ are determined based on the experimental results.

**Table 1** Motion planning module of autonomous driving process

```
Motion Planning Algorithm
Input: Surrounding vehicle information, Output: Desired motion plans M_d,
1: if d_lead < d_safe
2:    M_d ← Motion(10)
3: elseif Veh_wav.Amp ≥ d_wave
4:    if Veh_wav = Veh_{l,l}
5:       M_d ← Motion(2) AND Motion(6)
6:    else if Veh_wav = Veh_{l,f}
7:       M_d ← Motion(8) AND Motion(6)
8:    else if Veh_wav = Veh_{c,l}
9:       M_d ← Motion(2) AND Motion(4) AND Motion(6)
10:   else if Veh_wav = Veh_{c,f}
11:      M_d ← Motion(8) AND Motion(4) AND Motion(6)
12:   else if Veh_wav = Veh_{r,l}
13:      M_d ← Motion(2) AND Motion(4)
14:   else if Veh_wav = Veh_{r,f}
15:      M_d ← Motion(4) AND Motion(6)
16    end
17: elseif Veh_agg.Agg ≥ c_agg
18:      if Veh_agg = Veh_{c,l}
19:         M_d ← Motion(2)
20:      else if Veh_agg = Veh_{c,f}
21:         M_d ← Motion(4) AND Motion(6)
22:      end
23: elseif speed demand is satisfied
24:    M_d ← Motion(7) AND Motion(9)
25: else
26:    M_d ← Motion(5)
27: end
```

The local motion planning algorithm is listed in Table 1. At each step, we decide if the AGV to change to the left lane, or to stay in the current lane, or to change to the right lane. For each decision, an MOP is solved. The Pareto sets of the three MOPs are combined to form a single Pareto set. The optimal trajectory is chosen from this Pareto set. In the following numerical studies, we have used a $5 \times 5 \times 5$ coarse partition of the design space. One refinement of $3 \times 3 \times 3$ is applied.

## 3.7 Selection of Optimal Designs

To facilitate the user to pick up an optimal design from the Pareto set to implement, we propose an algorithm that operates on the Pareto front. Let $f_{i,\min}$ denote the minimum of the $i$th objective in the Pareto front and $f_{i,\max}$ be the corresponding maximum. Define a vector,

$$\mathbf{F}_{ideal} = [f_{1,\min}, f_{2,\min}, ..., f_{n_{obj},\min}]. \tag{28}$$

where $\mathbf{F}_{ideal}$ is considered to be an ideal point in the objective space, whose objective evaluation is minimum in each dimension. Let $n_s$ denote the number of Pareto solutions. To eliminate the effect of different scale of the objectives, the Pareto front is normalized as,

$$\bar{f}_{i,j} = \frac{f_{i,j} - f_{i,\min}}{f_{i,\max} - f_{i,\min}}, \quad 1 \le i \le n_s, 1 \le j \le n_{obj}, \tag{29}$$

where $f_{i,j}$ is the $i$th objective of the $j$th optimal solution, and $\bar{f}_{i,j}$ represents the normalized value of $f_{i,j}$. The normalized objective vector $\bar{\mathbf{F}}_i$ reads,

$$\bar{\mathbf{F}}_i = [\bar{f}_{1,i}, \bar{f}_{2,i}, ..., \bar{f}_{n_{obj},i}], \tag{30}$$

and its norm is denoted as $r_i = \left\|\bar{\mathbf{F}}_i\right\|$. Let $r_{\max}$ and $r_{\min}$ denote the maximum and minimum of the norm of the points in the Pareto front.

Let $n_{per}$ denote the percentage of control designs in the Pareto set such that their corresponding norms in the Pareto front are among the $n_{per}$ percent smallest ones. We refer these designs as the top $n_{per}$ percent. A special sub-set of top designs consists of the so-called *knee points* [46]. They are defined as,

$$\mathbf{k}_{knee} = \left\{ \mathbf{k}_i \mid i = \min_{1 \le i \le n_s} r_i \right\}. \tag{31}$$

It should be noted that there are different ways to normalize the Pareto front and the knee point can be defined differently in the literature [11, 46], but they share similar control selection purpose. Furthermore, the normalization affects the classification of the designs. The normalization used in this chapter is intuitive and simple to implement.

**Table 2** Parameters used in this paper

| Parameter | Value | Parameter | Value | Parameter | Value |
|---|---|---|---|---|---|
| $t_{lc}$ | 4 s | $l_{sense}$ | 150 m | $d_{\min}$ | 2 m |
| $a_{\max}$ | $5\,m^2/s$ | $v_f$ | 2m/s | $t_{safe}$ | 2.5 s |
| $a_{\min}$ | $-8\,m^2/s$ | $t_{gap}$ | 0.8 s | $t_{extra}$ | 1.5 s |
| $N_p$ | 6 s | $D_{free}$ | 100 m | $t_{\min}$ | 5 s |
| $N_c$ | 1 s | $\Delta v_{dif,\min}$ | 2 m/s | $G_{te,\max}$ | 3000 m |
| $t_{delay}$ | 0.5 s | $\omega_a$ | 1 | $t_{th,\lim}$ | 1 s |
| $t_{lc,\max}$ | 5 | $\omega_j$ | 1 | $t_{ttc,\lim}$ | 0.7 s |
| $t_{lc,\min}$ | 3 | $\alpha_{th}$ | 0.1 | $c_{agg}$ | $3\,m^2/s$ |
| $a_f$ | $1\,m^2/s$ | $\alpha_{ttc}$ | 1 | $d_{safe}$ | 15 m |
| $a_c$ | $2\,m^2/s$ | $D$ | 8 veh/km | $d_{wave}$ | 0.4 m |

### 3.8 Experimental Results

To verify the effectiveness of the proposed multi-objective optimization algorithm for local trajectory planning, we have tested the algorithm in different scenarios. The simulations in this study are implemented in Matlab. The sampling time step is 0.1 s. The simulation parameters are listed in Table 2. The simulations in this study are carried out in Matlab 2015b on a PC.

The best way to demonstrate the defensive driving strategies is to use the animation of the AGV with the mixed traffic in the presence of human-driven vehicles. Due to the limitation of the printed publication, we present a few special cases to show how the AGV responds to different situations.

**Overtaking** We first show how the AGV performs an overtaking task. As Fig. 5 shows, the AGV $Veh_A$ is approaching the leader $Veh_1$ on the same lane. However, the desired speed of $Veh_A$ is higher than that of $Veh_1$. Therefore, the AGV is supposed to overtake the leader to gain speed advantage. The longitudinal position and velocity of the AGV are presented in Fig. 6. It is observed that the AGV slows down first to cooperate with $Veh_1$. In order to achieve the desired speed again, $Veh_A$ tends to change lane to overtake the leader. The applied speed profile is optimal in consideration of safety, driving comfort, traffic efficiency and path consistency simultaneously. The lateral trajectory is smooth. More numerical results and comparisons with other non-optimal designs are provided in the following examples.

**Avoiding** Next, we demonstrate the ability of the AGV to avoid waving vehicles. Figure 7 shows that vehicle $Veh_3$ is waving. At the time indicated by number 1, the AGV detects that the waving amplitude of $Veh_3$ is larger than the safety level, and attempts to change to the right lane to avoid it. The movement profile is optimized considering the interactions with all other vehicles nearby, which are not shown in the figure. The optimal trajectory is obtained with the best compromise over the four performance metrics.

**Road Test** In the following, a realistic road section is taken to test the performance of the proposed algorithm. The numerical experiment is carried out in a one-way three-lane road section of 3000 meters with a periodic boundary as shown in Fig. 4. We denote the vehicle density, i.e. vehicles per kilometer per lane, as $D$. A random number of vehicles with different length and desired speed are considered. 20% of all the vehicles are assumed to be waving, and the maximum waving amplitudes of them are randomly determined in the range (0, 0.8] meters. The simulations are carried out over 1000 s including a 500 s to let the transient effect die out. Other parameters are presented in Table 2.

Due to the complex interactions with the surrounding vehicles, different motion decisions are made in different situations throughout the simulation. Figure 8 shows the Pareto front obtained in the longitudinal car-following scenario ($Motion(5)$)

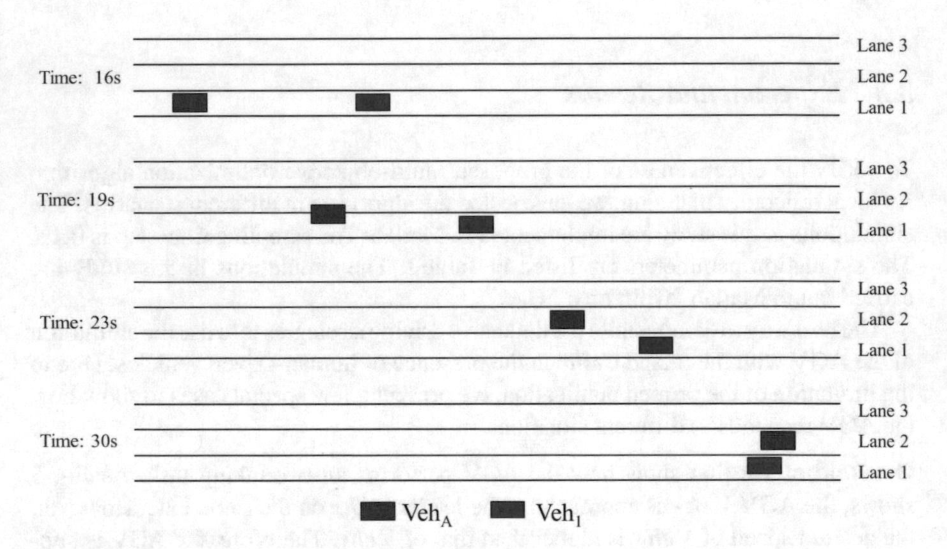


**Fig. 5** The vehicle trajectories in the overtaking scenario

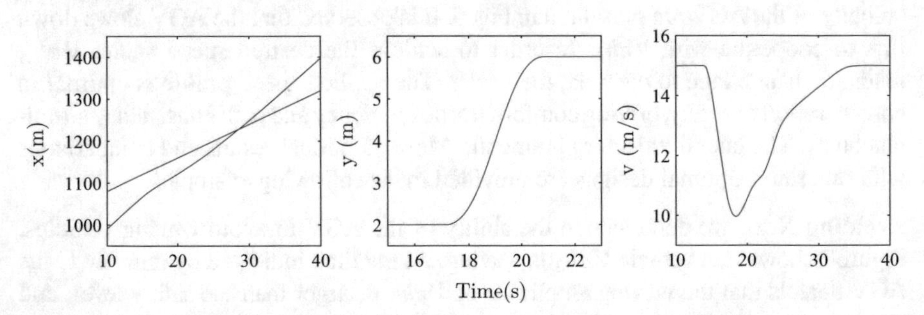


**Fig. 6** The longitudinal position, lateral position and velocity of the autonomous vehicle in the overtaking scenario

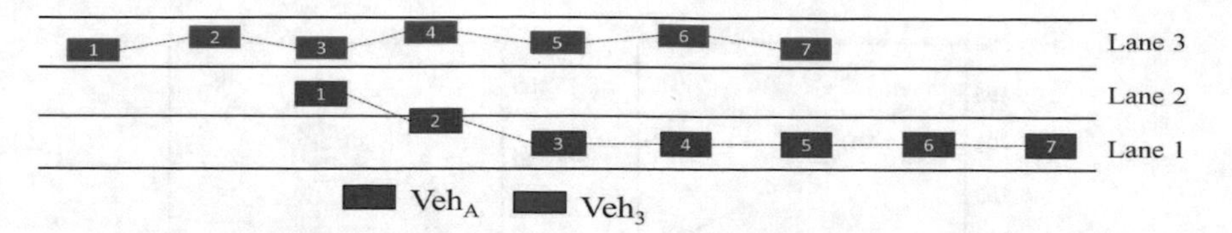


**Fig. 7** The vehicle trajectories when the AGV is avoiding a waving vehicle. The numbers in the vehicle blocks denote the time sequence. Other surrounding vehicles are not shown for clarity

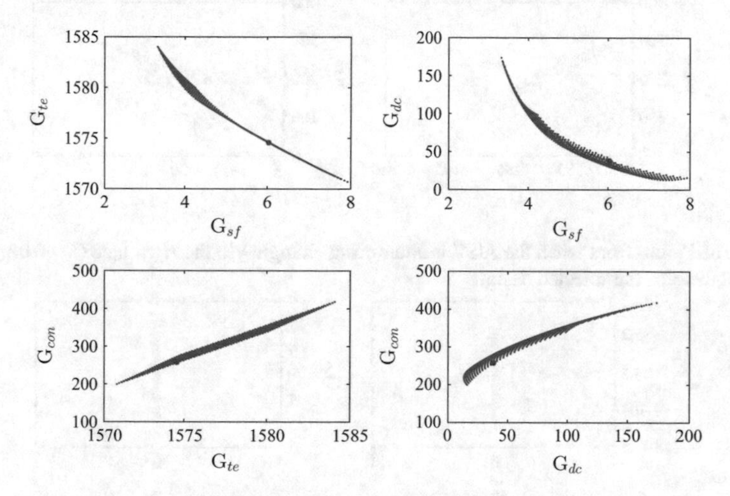


**Fig. 8** The Pareto front in a longitudinal following scenario ($Motion(5)$). The *red* point denotes the selected design

at time 804 s. The conflicting nature between different objectives is observed. The design with the balanced compromises among the four objectives is selected.

As the simulation continues to time 932 s, the following vehicle of the AGV in the current lane is detected to be very aggressive. The planner attempts to change lane to provide courtesy ($Motion(4)$ and $Motion(6)$). It is not feasible to change to the left lane at that time, and the right lane is available. The Pareto front is presented in Fig. 9 and the optimal trajectory is obtained.

At time 952 s in the simulation, the follower of the AGV in the current lane is sensed to be waving. Based on the motion planning algorithm, $Motion(8)$, $Motion(4)$ and $Motion(6)$ are considered. The right lane is not available, and the Pareto fronts for the other two motions are presented in Fig. 10. It can be observed that the two Pareto fronts for the two motions are similar to each other, with small difference in traffic efficiency ($G_{te}$). Applying the selection algorithm for the optimal design, $Motion(8)$ is implemented in this case.

When the simulation reaches time 969 s, the following vehicle of the AGV in the left lane is detected to be waving. The planner considers speeding up in the current lane ($Motion(8)$) and changing to the right lane ($Motion(6)$) at the same time to avoid. Figure 11 shows the combined Pareto fronts with respect to the two

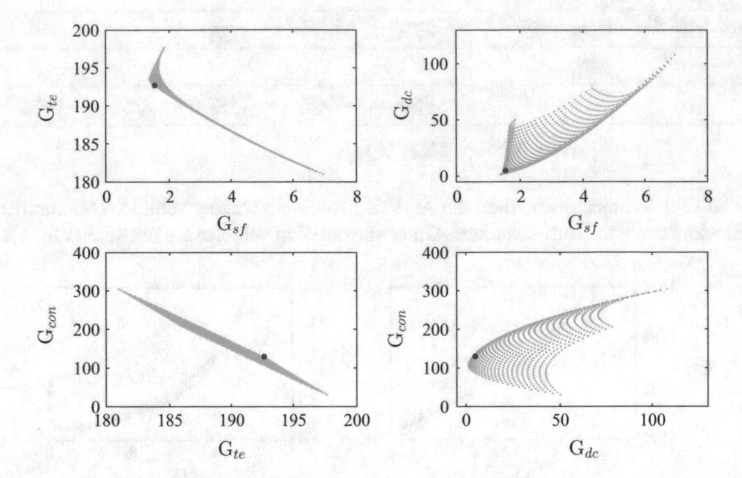


**Fig. 9** The Pareto front when the AGV is attempting changing to the *right* lane ($Motion(6)$). The *red* point denotes the selected design

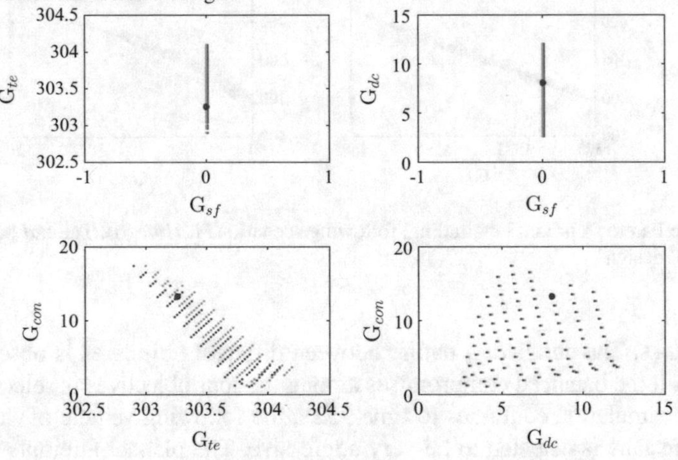


**Fig. 10** The Pareto fronts when the AGV is considering speeding up in the current lane ($Motion(8)$ and denoted by the *blue* points) and changing to the *left* lane ($Motion(4)$ and denoted by the *orange* points). The *red* point denotes the selected design

motions. It is seen that the blue Pareto front lies on the left side of the orange Pareto front in the two upper figures. This suggests that a higher safety can be achieved by staying in the current lane than lane-changing. No significant difference in the other objectives is observed with the two motions. Consequently, the optimal trajectory design of staying in the current lane is implemented. Similarly at time $980s$, the leading vehicle in the left lane is observed to be waving, and the planner has to make a decision between $Motion(2)$ and $Motion(6)$. The Pareto fronts are shown in Fig. 12. While safety is compromised significantly for staying in the current lane, the optimal motion becomes changing to the right lane.

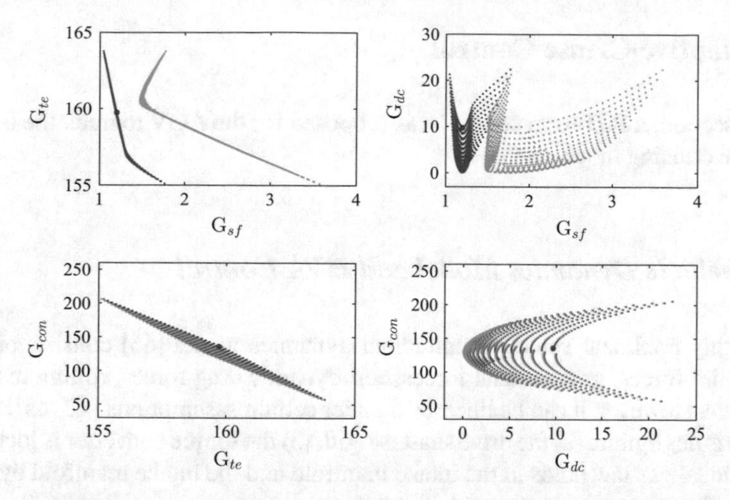


**Fig. 11** The Pareto fronts when the AGV is considering speeding up in the current lane ($Motion(8)$ and denoted by the *blue* points) and changing to the *right* lane ($Motion(6)$ and denoted by the *orange* points). The *red* point denotes the selected design

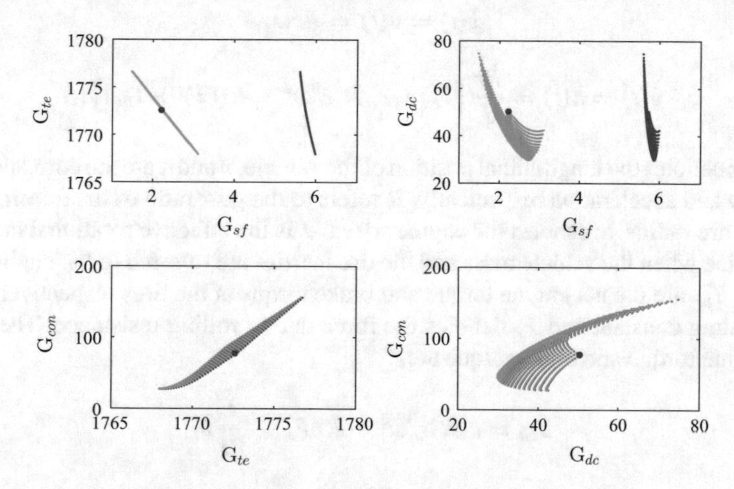


**Fig. 12** The Pareto fronts when the AGV is considering slowing down in the current lane ($Motion(2)$ and denoted by the *blue* points) and changing to the *right* lane ($Motion(6)$ and denoted by the *orange* points). The *red* point denotes the selected design

In the experiments, only the straight roads are considered for simplicity. However, it should be noted that the proposed motion planning algorithm can be applied in various road scenarios with curves.

# 4 Adaptive Cruise Control

In this section, a PID-form controller is proposed for the AGV to track the leader in adaptive cruising mode.

## 4.1 Vehicle Dynamics Model and HVA Control

The highly nonlinear vehicle longitudinal dynamics model [66] consists of longitudinal tire forces, gravitational forces, aerodynamic drag forces, rolling resistance forces *etc*. However, it can be linearized under certain assumptions [42, 45]. (1) The tire slip is negligible, (2) the drive shaft is rigid, (3) the torque converter is locked and (4) the ideal gas law holds in the intake manifold and the intake manifold dynamics are very fast compared to the vehicle dynamics.

Based on the assumptions, the following equations are obtained.

$$\dot{x}(t) = v(t) = Rr\omega, \tag{32}$$

$$\dot{v}(t) = a(t) = \frac{Rr}{J}\left[T_{en} - c_a R^3 r^3 \omega^2 - R(rF_r + T_{br})\right], \tag{33}$$

where $x$ denotes the longitudinal position of the vehicle, $v$ and $a$ are the corresponding velocity and acceleration respectively, $R$ refers to the gear ratio of the transmission, $r$ is the tire radius, $\omega$ denotes the engine speed, $J$ is the effective rotational inertia of the engine when the vehicle mass and the tire inertias are referred to the engine side, $T_{en}$ and $T_{br}$ are the net engine torque and brake torque at the tires respectively, $c_a$ is the air drag constant and $F_r$ denotes the force due to rolling resistance. Therefore, the engine torque and brake torque are,

$$T_{en} = c_a R^3 r^3 \omega^2 + RrF_r + \frac{J}{Rr}\dot{v}, \tag{34}$$

when the braking maneuver is deactivated, and

$$T_{br} = -\frac{c_a R^3 r^3 \omega^2 + RrF_r}{R} - \frac{J}{R^2 r}\dot{v}, \tag{35}$$

when the braking maneuver is activated.

Considering the nonlinearities in the engine model, the linearization is developed [10, 61]. The control law $u$ is employed as,

$$T_{des} = c_a R^3 r^3 \omega^2 + RrF_r + \frac{J}{Rr}u, \tag{36}$$

where $T_{des}$ denotes the desired net engine torque. Based on Eqs. (34) and (35),

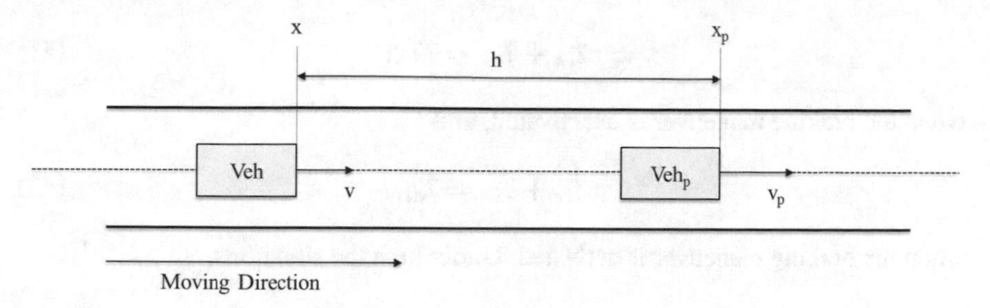


**Fig. 13** The scheme of ACC vehicles showing positions, velocities and space headway

$$\dot{v}(t) = u(t). \tag{37}$$

If the brake maneuver is activated, Eq. (37) can still be obtained. Therefore, the acceleration or deceleration of the AGV is commanded and controlled in ACC process.

Two consecutive identical vehicles are considered to be on a single lane as represented in Fig. 13 where $x_p$ and $v_p$ denote the longitudinal position and velocity of the preceding vehicle respectively, and $h = x_p - x$ means the space headway between neighbors which includes the length of the vehicle.

The constant time headway (CTH) range policy, which is widely accepted as a safety practice for drivers and popularly used in ACC designs, is adopted in this chapter. The desired space gap between the two vehicles can be calculated as a function of the current speed of the follower.

$$h_{des} = t_{pth}v + h_0,\, v \in [0, v_{\max}]\,, \tag{38}$$

where $h_0$ is the minimum separation distance between vehicles at standstill within which they may collide and that is unacceptable for safety consideration, $t_{pth}$ denotes the prefered constant time headway and $v_{\max}$ denotes the legal speed limit of the road. The dynamics of the ACC vehicle can be expressed as,

$$\frac{d}{dt}h\,(t) = -v\,(t) + v_p\,(t)\,, \tag{39}$$

$$\frac{d}{dt}h_{des}\,(t) = t_{pth}a\,(t)\,. \tag{40}$$

In mechanical and control systems, the time lags are inevitable [57]. Therefore, the actual net engine torque is not able to track the desired engine torque immediately when the control signal is computed and transferred from the ACC controller to the actuators [21]. A first-order system is used to model the time lag. Let $\tau$ refer to the time constant due to the lag.

$$\tau \dot{T}_{en} + T_{en} = T_{des}, \tag{41}$$

when the braking maneuver is deactivated, and

$$\tau \dot{T}_{br} + T_{br} = T_{des}, \tag{42}$$

when the braking maneuver is activated. Under both the situations,

$$\tau \dot{a}(t) + a(t) = u(t). \tag{43}$$

In addition, the information and sensing time delay of the ACC controller is not considered in this study, which means the controller can access the instantaneous information to yield the desired acceleration or deceleration.

From Eqs. (37), (39), (40) and (43) the ACC system can be expressed as,

$$\dot{\mathbf{x}} = \mathbf{A}\mathbf{x} + \mathbf{E}a_p + \mathbf{B}u, \tag{44}$$

where $\mathbf{x} \in \mathbf{R}^n$ is the state variable and,

$$\mathbf{x} = \left[h, h_{des}, v, v_p, a\right]^{\mathbf{T}}, \tag{45}$$

$$\mathbf{A} = \begin{bmatrix} 0 & 0 & -1 & 1 & 0 \\ 0 & 0 & 0 & 0 & t_{pth} \\ 0 & 0 & 0 & 0 & 1 \\ 0 & 0 & 0 & 0 & 0 \\ 0 & 0 & 0 & 0 & -\frac{1}{\tau} \end{bmatrix},$$

$$\mathbf{E} = [0, 0, 0, 1, 0]^{\mathbf{T}},$$

$$\mathbf{B} = \left[0, 0, 0, 0, \frac{1}{\tau}\right]^{\mathbf{T}}.$$

In order to achieve the range policy, the following controller for the acceleration commands of the vehicle is proposed which takes into consideration the difference between the desired gap and current distance between the two vehicles, the speed tracking ability of the follower and the movement tendency of the leader. Therefore, the HVA controller, which denotes the distance, velocity and acceleration control respectively, is proposed as,

$$\begin{aligned} u(t) &= \mathbf{k}\mathbf{x} \\ &= k_h(h - h_{des}) + k_v\left(v_p - v\right) + k_a a_p(t - \eta), \end{aligned} \tag{46}$$

where $\mathbf{k} = [k_h, -k_h, -k_v, k_v, k_a]$ is the control gain and $a_p(t - \eta)$ represents the acceleration of the preceding vehicle with wireless communication delay $\eta$. It should be noted that even though $a_p$ is used, we still prefer to call this controller ACC rather than CACC for the reason that the main part consists of range and range rate information, and CACC usually involves richer wireless communication information.

Furthermore, the absolute value of acceleration of the vehicle is supposed not to exceed a certain value in consideration of human driving comfort [22]. The following equations are used for this acceleration constraint,

$$\begin{aligned} a_{\min} &\le u \le a_{\max}, \\ a_{\min} &\le a \le a_{\max}. \end{aligned} \tag{47}$$

Therefore, the control formulation reads,

$$\begin{aligned} u(t) &= sat\,(\mathbf{kx}) \\ &= \begin{cases} a_{\min}, & \text{if } \mathbf{kx} < a_{\min} \\ k_h(h - h_{des}) + k_v\left(v_p - v\right) + k_a a_p\,(t - \eta)\,, & \text{if } a_{\min} \le \mathbf{kx} \le a_{\max}\,, \\ a_{\max}, & \text{if } \mathbf{kx} > a_{\max} \end{cases} \end{aligned} \tag{48}$$

where $sat\,()$ is a standard saturation function.

Two stability criteria, i.e. plant stability and string stability, are required to be satisfied for the ACC design. As the vehicles in the platoons are assumed to be identical, we focus on two consecutive identical vehicles for stability analysis. Notably, due to the system constraints of the control signal (Eq. (47)), the system also has to be stable on actuator saturation condition.

**Plant stability** Let S be the symmetric polyhedron,

$$\mathrm{S}(\mathbf{k}) = \left\{\mathbf{x} \in \mathbf{R}^n : |\mathbf{kx}| \le 1\right\}. \tag{49}$$

D is defined as a set of diagonal matrices whose diagonal elements are either 0 or 1. $m$ is introduced as the dimension of the control input variable, i.e., $u \in \mathbf{R}^m$, and obviously $m = 1$ in this study. Therefore,

$$\mathrm{D} = \{D_1, D_2\}, \tag{50}$$

where $D_1 = 1$ and $D_2 = 0$ [6]. The complements are $D_1^- = 1 - D_1 = 0$ and $D_2^- = 1 - D_2 = 1$.

**Lemma 1** ([6, 20]) *Given* $\mathbf{k}, \mathbf{H} \in R^{m \times n}$*, the following relation*

$$sat\,(\mathbf{kx}) \in \mathbf{Co}\left\{D_i\mathbf{kx} + D_i^-\mathbf{Hx} : i \in \left[1, 2^m\right]\right\}, \tag{51}$$

*holds for* $\mathbf{x} \in \mathrm{S}(\mathbf{H})$.

The lemma implies that if $|\mathbf{Hx}| \le 1$, the saturation function can be represented by a convex combination as

$$sat(\mathbf{kx}) = \sum_{i=1}^{2^m} \mu_i (D_i \mathbf{k} + D_i^- \mathbf{H})\mathbf{x}, \tag{52}$$

for $0 \leq \mu_i \leq 1$ and $\sum_{i=1}^{2^m} \mu_i = 1$.

It should be noted that the absolutes of $a_{\min}$ and $a_{\max}$ are assumed to be identical for simplicity in this chapter,

$$|a_{\max}| = |a_{\min}| = a_{\lim}, \tag{53}$$

where $a_{\lim}$ is the corresponding acceleration limit. Equation (44) can be rewritten as

$$\dot{\mathbf{y}} = \mathbf{Ay} + \mathbf{E}\tilde{a}_p + \mathbf{B}\tilde{u}, \tag{54}$$

where

$$\begin{aligned} \mathbf{y} &= \frac{\mathbf{x}}{a_{\lim}}, \\ \tilde{a}_p &= \frac{a_p}{a_{\lim}}, \\ \tilde{u} &= \frac{u}{a_{\lim}}, \\ \max|\tilde{u}| &= 1. \end{aligned} \tag{55}$$

Plant stability requires that the velocity perturbation of the follower is supposed to attenuate until the system reaches the equilibrium when the preceding one travels with a constant speed. Consider Eq. (54) and define a Lyapunov function for the system as,

$$V = \mathbf{y}^{\mathbf{T}}\mathbf{Py}, \mathbf{P} > 0. \tag{56}$$

Its time-derivative turns out to be,

$$\dot{V} = \left[\mathbf{Ay} + \mathbf{B}\tilde{u}\right]^{\mathbf{T}} \mathbf{Py} + \mathbf{y}^{\mathbf{T}}\mathbf{P}\left[\mathbf{Ay} + \mathbf{B}\tilde{u}\right]. \tag{57}$$

Given the fact that the saturation function can be represented by a convex combination [9], for $i = 1, 2, ..., 2^m$,

$$\dot{V} = \mathbf{y}^{\mathbf{T}} \left\{ \mathbf{A} + \mathbf{B} \times \sum_{i=1}^{2^m} \mu_i (D_i \mathbf{k} + D_i^- \mathbf{H}) \right\}^{\mathbf{T}} \mathbf{Py} \tag{58a}$$

$$+ \mathbf{y}^{\mathbf{T}}\mathbf{P} \left\{ \mathbf{A} + \mathbf{B} \times \sum_{i=1}^{2^m} \mu_i (D_i \mathbf{k} + D_i^- \mathbf{H}) \right\} \mathbf{y},$$

where $0 \le \mu_i \le 1, \sum_{i=1}^{2^m} \mu_i = 1$.

$$\dot{V} = \sum_{i=1}^{2^m} \mu_i \mathbf{y}^{\mathbf{T}} \left\{ \left[\mathbf{A} + \mathbf{B}(D_i \mathbf{k} + D_i^- \mathbf{H})\right]^{\mathbf{T}} \mathbf{P} + \mathbf{P} \left[\mathbf{A} + \mathbf{B}(D_i \mathbf{k} + D_i^- \mathbf{H})\right] \right\} \mathbf{y}, \quad (59)$$

which indicates that if

$$\left[\mathbf{A} + \mathbf{B}(D_i \mathbf{k} + D_i^- \mathbf{H})\right]^{\mathbf{T}} \mathbf{P} + \mathbf{P} \left[\mathbf{A} + \mathbf{B}(D_i \mathbf{k} + D_i^- \mathbf{H})\right] < 0, \quad (60)$$

for $i = 1, 2, ..., 2^m$, then $\dot{V} < 0, \forall \mathbf{y} \in \varepsilon \backslash \{0\}$ where

$$\varepsilon = \left\{ \mathbf{y} \in \mathbf{R}^n : \mathbf{y}^{\mathbf{T}} \mathbf{P} \mathbf{y} \le \gamma \right\}, \gamma > 0. \quad (61)$$

The resulting LMI (linear matrix inequality) condition (Eq. 60)) can be easily solved by LMI toolbox in Matlab. Therefore, plant stability criterion (Eq. (60)) has been satisfied for both unsaturated and saturated situations.

**String stability** In uniform flow equilibrium the vehicles in the platoon all share the same velocity $v^*$ and space headway $h^*$.

$$h = h^*, h_{des} = h^*, v = v^*, v_p = v^*, \quad (62)$$

subject to Eq. (38).

$$h^* = t_{pth} v^* + h_0, v^* \in [0, v_{\max}]. \quad (63)$$

Let $\tilde{h}(t) = h(t) - h^*$, $\tilde{h}_{des}(t) = h_{des}(t) - h^*$, $\tilde{v}(t) = v(t) - v^*$ and $\tilde{v}_p(t) = v_p(t) - v^*$ denote the perturbations of the states from the equilibrium respectively. Substitute Eq. (46) into Eq. (44),

$$\frac{d}{dt} \begin{bmatrix} \tilde{h} \\ \tilde{h}_{des} \\ \tilde{v} \\ \tilde{v}_p \\ a \end{bmatrix} = \begin{bmatrix} 0 & 0 & -1 & 1 & 0 \\ 0 & 0 & 0 & 0 & 2 \\ 0 & 0 & 0 & 0 & 1 \\ 0 & 0 & 0 & 0 & 0 \\ \frac{k_h}{\tau} & -\frac{k_h}{\tau} & -\frac{k_v}{\tau} & \frac{k_v}{\tau} & -\frac{1}{\tau} \end{bmatrix} \begin{bmatrix} \tilde{h} \\ \tilde{h}_{des} \\ \tilde{v} \\ \tilde{v}_p \\ a \end{bmatrix} \quad (64)$$

$$+ \begin{bmatrix} 0 \\ 0 \\ 0 \\ 1 \\ 0 \end{bmatrix} a_p(t) + \begin{bmatrix} 0 \\ 0 \\ 0 \\ 0 \\ \frac{k_a}{\tau} \end{bmatrix} a_p(t - \eta),$$

where $\tilde{v}_p$ is considered to be the input of the system while $\tilde{v}$ is the output. The transfer function can be obtained by the Laplace transform of Eq. (64):

$$\Gamma(s) = \frac{k_a e^{-\eta s} s^2 + k_v s + k_h}{\tau s^3 + s^2 + (2k_h + k_v) s + k_h}. \tag{65}$$

String stability can be ensured if the fluctuation of the velocity around the equilibrium becomes smaller as it propagates along the platoons [50]. That means $|\Gamma(j\omega)| < 1, \forall\omega > 0$ must be satisfied which is equivalent to

$$\alpha_1^2 + \alpha_2^2 - \beta_1^2 - \beta_2^2 < 0, \tag{66}$$

where

$$\begin{aligned} \alpha_1 &= k_h - k_a\omega^2 \cos\eta\omega, \\ \alpha_2 &= k_v\omega + k_a\omega^2 \sin\eta\omega, \\ \beta_1 &= -\omega^2 + k_h, \\ \beta_2 &= \left(t_{pth}k_h + k_v\right)\omega - \tau\omega^3. \end{aligned} \tag{67}$$

Additionally, $\omega$ is arbitrarily checked in the range of (0, 10] for string stability here, and the string stability for the system with saturated controller is beyond interest of this study.

## 4.2 Multi-objective Optimization of ACC

Generally, several objectives are considered as the control goals for MOP of ACC, which include driving safety, comfort and fuel economy. Given the range policy, safety objective is supposed to be related with the ability of tracking the desired trajectory and the velocity of the preceding vehicle [39]. In this section, 2-norm of tracking errors are employed to quantify the safety issue as,

$$G_{th} = \int_0^{t_s} (h - h_{des})^2 \, dt, \tag{68}$$

$$G_{tv} = \int_0^{t_s} \left(v - v_p\right)^2 dt, \tag{69}$$

where $G_{th}$ and $G_{tv}$ denote the goals for tracking errors of space headway and velocity respectively, $h_{des}$ is the desired space headway according to the current velocity based on CTH range policy (Eq. (38)) and $t_s$ is the simulation duration.

Comfort objective of driving is usually associated with the number, size and frequency of vibrations or oscillations in the longitudinal acceleration of the vehicle. At the same time, the maximum deceleration is often related to comfort. Additionally, jerk values are considered to have influence on human driving comfort [36]. Not only focusing on the driver experience, fuel consumption is suggested to be dominated by acceleration of vehicle regardless of the engine operational area [54, 64], which is also linked with the comfort goal. The vehicle with larger absolute of acceleration tends to consume more fuel in comparison with others with the same travel distance and average speed. We can see that the acceleration commands and the actual acceleration values of the vehicle are similar with each other in Eq. (43). Therefore, the quantification of the objectives with respect to driving comfort and fuel economy can be defines as two $L^2$ norm functions of the control command and its derivative,

$$G_u = \int_0^{t_s} u^2 dt, \tag{70}$$

$$G_{du} = \int_0^{t_s} \dot{u}^2 dt. \tag{71}$$

The control MOP design of ACC can be formulated as,

$$\min_{\mathbf{k}_Q \in Q} \{G_{th}, G_{tv}, G_u, G_{du}\}, \tag{72}$$

where $\mathbf{k}_Q = [k_h, k_v, k_a]$ and $Q$ is the gain space with boundaries. Notably, the units for $v$, $v_p$ are m/s when calculating $G_{tv}$.

Additionally, proper value constraints of the objective functions are needed to keep the control performance within a certain range.

$$\begin{aligned} \max G_{th} &\le G_{th,\lim}, \\ \max G_{tv} &\le G_{tv,\lim}, \\ \max G_u &\le G_{u,\lim}, \\ \max G_{du} &\le G_{du,\lim}, \end{aligned} \tag{73}$$

where $G_{th,\lim}$, $G_{tv,\lim}$, $G_{u,\lim}$, $G_{du,\lim}$ are the corresponding limits. The cell mapping method is also used in this study.

## 4.3 Model Predictive Control

The brief description of MPC algorithm for ACC is presented in the following [28]. Note that the MPC algorithm is often designed in the discrete-time domain, the model can be then formulated as,

$$\hat{\mathbf{x}}(k+1) = \hat{\mathbf{A}}\hat{\mathbf{x}}(k) + \hat{\mathbf{B}}\hat{\mathbf{U}}(k), \tag{74}$$

where $\hat{\mathbf{x}}(k) = \left[\tilde{h}(k), \tilde{v}(k), a(k)\right]^T$, $\hat{\mathbf{U}}(k) = \left[u(k), \dot{v}_p(k-m), \tilde{v}_p(k)\right]^T$,

$$\hat{\mathbf{A}} = \begin{bmatrix} 1 & -dt & -\frac{dt^2}{2} \\ 0 & 1 & dt \\ 0 & 0 & 1-\frac{dt}{\tau} \end{bmatrix}, \hat{\mathbf{B}} = \begin{bmatrix} 0 & \frac{dt^2}{2} & dt \\ 0 & 0 & 0 \\ \frac{dt}{\tau} & 0 & 0 \end{bmatrix}, \tag{75}$$

and $k$ denotes the $k$th sampling point and $dt$ represents the sampling time. $m = \frac{\eta}{dt}$ is the number of delayed time steps due to wireless communication lag with the preceding vehicle. $\hat{\mathbf{x}}(k)$ is the state of the system at the $k$th iteration and $\hat{\mathbf{U}}(k)$ denotes the corresponding input where $u(k)$ is the acceleration control command and $\dot{v}_p(k-m)$, $\tilde{v}_p(k)$ are regarded as the measurable disturbances. With the same control objectives as the HVA control shown in Eqs. (68)–(71) and the constant time headway range policy (Eq. (38)), the output equation can be defined as,

$$\hat{\mathbf{y}}(k) = \hat{\mathbf{E}}\hat{\mathbf{x}}(k) + \hat{\mathbf{F}}\hat{\mathbf{U}}(k), \tag{76}$$

where

$$\hat{\mathbf{E}} = \begin{bmatrix} 1 & -t_{pth} & 0 \\ 0 & 1 & 0 \\ 0 & 0 & 1 \end{bmatrix}, \hat{\mathbf{F}} = \begin{bmatrix} 0 & 0 & 0 \\ 0 & 0 & -1 \\ 0 & 0 & 0 \end{bmatrix}, \tag{77}$$

and $\hat{\mathbf{y}}(k)$ is the measurement of the system at the $k$th step. Then the cost function can be obtained as [28],

$$G\left(\hat{\mathbf{y}}, u, \Delta u\right) = \sum_{i=0}^{p-1} \|\Phi y(k+i+1|k)\|_{\omega_y}^2 \tag{78}$$
$$+ \sum_{i=0} \|u(k+i|k)\|_{\omega_u}^2 + \sum_{i=0}^{p-1} \|\Delta u(k+i|k)\|_{\omega_{\Delta u}}^2 + \rho\xi^2,$$

where $\Delta u(k) = u(k) - u(k-1)$, $p$ denotes the length of the predictive horizon of the algorithm and $(k+i|k)$ represents the predicted value according to condition of the system at the $k$th time step. $\xi$ is introduced as slack variable and $\omega_y$, $\omega_u$, $\omega_{\Delta u}$ are the corresponding weighting coefficients. Therefore, the optimization problem can be expressed as,

$$\min G\left(\hat{\mathbf{y}}, u, \Delta u\right), \tag{79}$$

subject to the constraints in Eq. (47).

The optimization problem Eq. (79) can be formulated as a quadratic program (QP) and solved by a Dantzig–Wolfe active set algorithm at each time step. Once an optimal solution sequence $[\Delta u^*(k+i|k)]_{i=1:p}$ is obtained, only the first element of the sequence will be used as the increment of the acceleration command and the actual control signal is thus computed as,

$$u(k+1) = u(k) + \Delta u^*(k+1|k). \tag{80}$$

**Table 3** Parameters for HVA control

| Parameter | Value | Parameter | Value |
|---|---|---|---|
| $v_{\max}$ | 100 kph | $\tau$ | 0.1 s |
| $G_{th,\lim}$ | 2000 | $\eta$ | 0.2 s |
| $G_{tv,\lim}$ | 1500 | $t_{pth}$ | 2 s |
| $G_{u,\lim}$ | 1000 | $a_{\lim}$ | 5 m/s$^2$ |
| $G_{du,\lim}$ | 20 | $h_0$ | 10 m |

In the next iteration, the controller will calculate the new command based on the updated information.

## 4.4 ACC Design

We focus on two consecutive vehicles in the platoon configuration as the vehicles are considered to be identical. Control design with the proposed method is carried out in a scenario where velocity fluctuation of the preceding vehicle is significant with respect to the stability criteria (Eqs. (60) and (66)). The system is subject to a sinusoidal input $v_p(t) = v^* + v_{amp}\sin(\omega t)$ where $v^*$ is assumed to be 50 kph and $v_{amp} = 10$ kph, $\omega = 1$. The initial condition satisfies the equilibrium at $v^*$ with the prefered time headway $t_{pth} = 2$ s. The simulated time span is 50 s with sampling time of 0.1 s. The other relative parameters are provided in Table 3.

The SCM hybrid algorithm [58] is carried out in the gain space with lower and upper bounds as $[-3, -3, -3]$ and $[10, 10, 10]$. Hence,

$$Q = \left\{ \mathbf{k}_Q \in \mathbf{R}^3 | \begin{array}{l} [-3, -3, -3] \le \mathbf{k}_Q \le [10, 10, 10] \\ \text{subject to Eqs. (60) and (66)} \end{array} \right\}. \tag{81}$$

The coarse partition computation of the cell space is by $10 \times 10 \times 10$ and the refinement is by $3 \times 3 \times 3$. Figure 14 shows the Pareto set obtained in this situation. The Pareto front is presented in Figs. 15 and 16, which indicate the conflicting nature of the multi-objective optimization problem, e.g., smaller value of $G_{th}$ always leads to larger $G_{tv}$ etc.

The initial velocities of the two vehicles are set $v^*$ for all the simulations in this study. Figure 17 shows the simulation results of the system with all the gains in the Pareto set, where the responses of the ordinary gains are marked red and those of three gains in extreme cases, when the three control objectives are in the lowest level respectively, are marked as blue solid lines. We can see that the three special gains almost set the boundaries of the responses under all the Pareto optimal controls.

## 4.5 Scenario Simulations

Different traffic scenarios are simulated in this section with both the HVA control and MPC to evaluate the control performance of the proposed method. For the parameters of MPC, the sampling time step is 0.1 s, the prediction and control horizons are both 10 s, the coefficient for slack variable $\rho = 3$ and diag([0.5, 1, 0]) is considered as the coefficient matrix $\omega_y$, $\omega_u = 0.5$ and $\omega_{\Delta u} = 0.05$ [28]. The results of MPC reported in this chapter are obtained by using the MPC toolbox in Matlab. It should be noted that the parameters used in MPC can be tuned in order to pay more attention on different objectives, but lower value of one certain goal leads to increase of another inevitably according to Eq. (78). The comparison of the results is only used to show that the performance achieved by our proposed method is as good as that with MPC, rather than to prove that one totally outperforms the other.

**Steady following** In the first traffic scenario, we consider the input of the system as in the control design where $v_p(t) = v^* + v_{amp}\sin(\omega t)$, $v^* = 50$ kph, $v_{amp} = 10$ kph, $\omega = 1$ and the simulation period is 50 s. Two gains in Pareto set for HVA control are selected as $\mathbf{k}_1 = [8.4833, 4.5833, -1.0500]$ and $\mathbf{k}_2 = [2.8500, 6.3167, -0.6167]$ whose corresponding Pareto fronts and the objective function values are both [203.4, 1061.1, 406.6, 7.9] and [1614.2, 447.0, 674.9, 9.3]. The simulation results under HVA control and MPC are shown in Fig. 18 and the objective function values for MPC is $[G_{th}, G_{tv}, G_u, G_{du}]_{mpc} = [607.1, 838.6, 634.1, 19.6]$. It is observed that compared with MPC, the gain $\mathbf{k}_1$ has some advantage in $G_{th}$, $G_u$ and $G_{du}$, but is weak in $G_{tv}$. $\mathbf{k}_2$ highly outperforms MPC in $G_{tv}$ and $G_{du}$ but with worse objective values of $G_{th}$. Generally, the proposed method is able to achieve system performance as good as MPC and even better focusing on some specific objectives with corresponding gains. Additionally, string stability is shown in Fig. 18 with sinusoidal input for the fact that the perturbations do not amplified in the propagation from the leader to the follower.

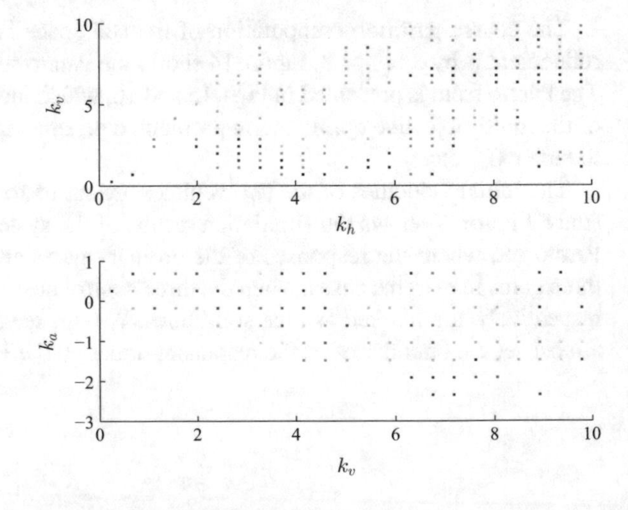


**Fig. 14** The projection of Pareto set in ACC subject to a sinusoidal input. The color in the *upper* and *lower* sub-figure represents $k_a$ and $k_h$ respectively. *Red* indicates the highest level while *blue* indicates the lowest level

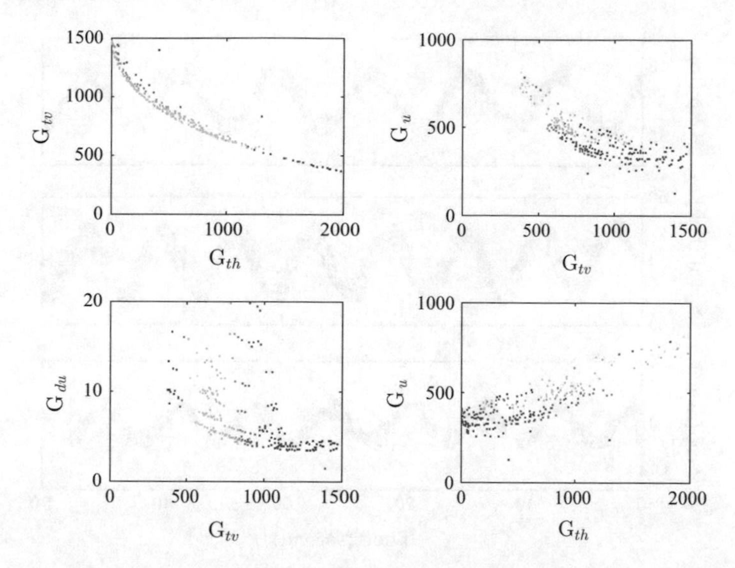


**Fig. 15** The 2D projections of Pareto front of ACC. The *color code* indicates the level of the third goal. *Red* indicates the highest level while *blue* indicates the lowest level. Conflicting nature of the control objectives is shown

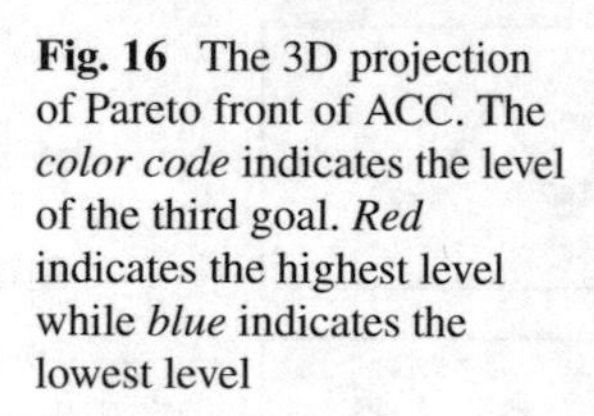

**Fig. 16** The 3D projection of Pareto front of ACC. The *color code* indicates the level of the third goal. *Red* indicates the highest level while *blue* indicates the lowest level

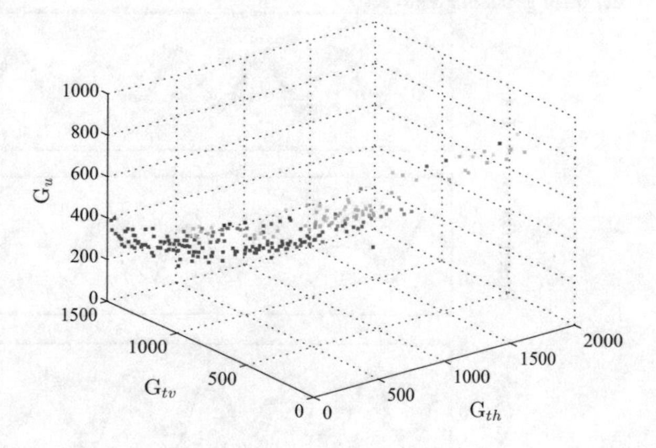


Platoon simulation is carried out in this scenario. One leader with sinusoidal input and ten followers at equilibrium initially are considered. $v_{amp}$ is 25 kph. The control performance of vehicle speed with $\mathbf{k}_1$ is shown in Fig. 19. It can be observed clearly that velocity fluctuations decay as they propagate from the leader to the 10th follower which reflects the string stability and the effectiveness of the proposed controller.

**Approaching** It is assumed that the preceding vehicle in this scenario keeps a constant velocity $v^*$ during the simulation which indicates its acceleration remains zero, and $k_a$ that is part of the gain will be helpless for control in this situation. Both the two vehicles are running at $v^*$ first, and the follower is assumed to be in the

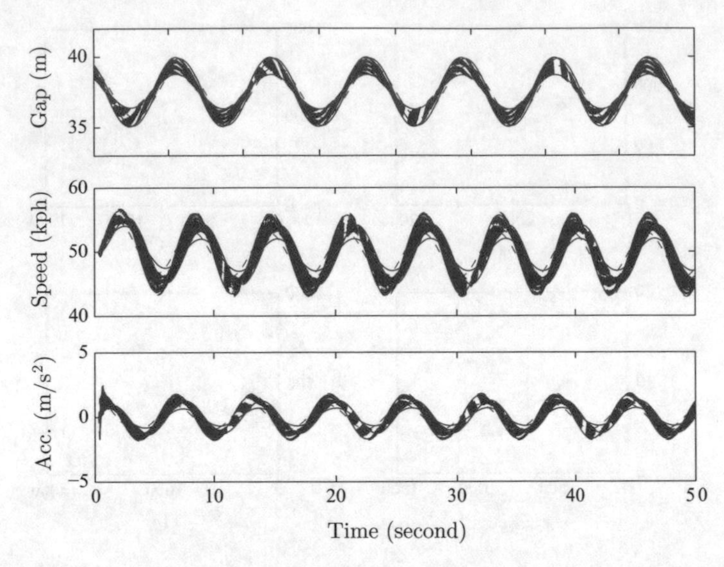


**Fig. 17** The simulation results in time domain responses of the gap, speed and instant acceleration of ACC system with HVA control method. *Solid blue lines* with gains in extremal cases in Pareto set when the three control objectives are in the lowest level respectively. *Dashed red line* with all the other gains in Pareto set

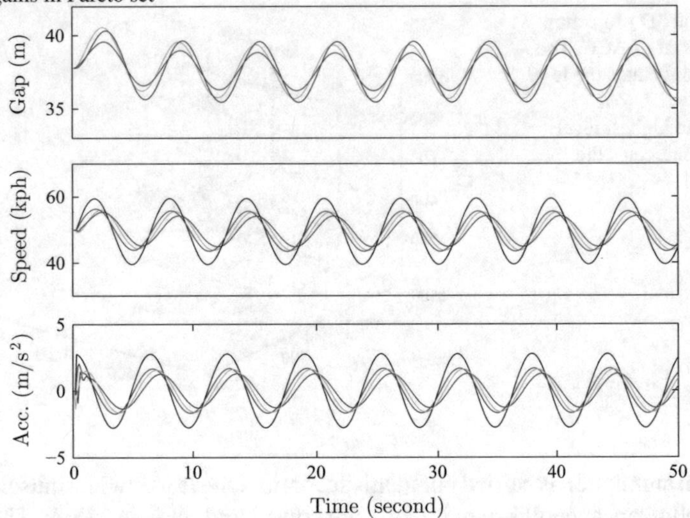


**Fig. 18** The simulation results of the gap, speed and instant acceleration of ACC system with sinusoidal input. The *red* and *blue lines* represent the performance with $\mathbf{k}_1$ and $\mathbf{k}_2$ respectively, *green lines* show that of MPC and the preceding vehicle is shown by *black line*

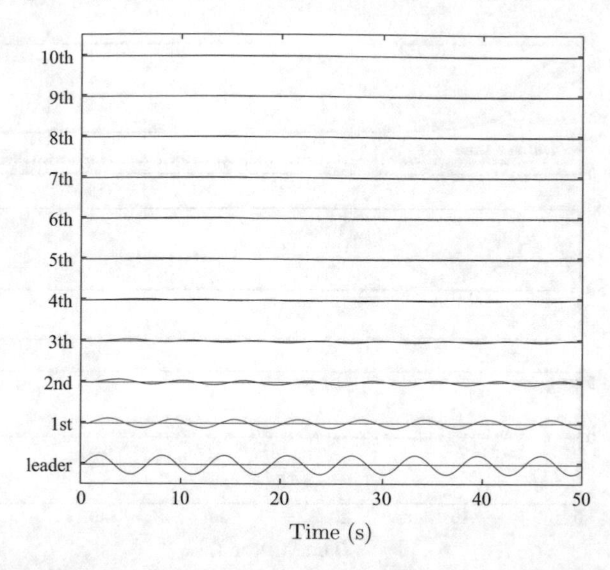


**Fig. 19** The platoon simulation results of vehicle speed with sinusoidal input and gain $\mathbf{k}_1$. The vertical axis shows the corresponding velocities. The *dotted black lines* are the velocity equilibrium while the *red* and *blue lines* represent the fluctuations of velocities of the leader and 10 followers

position that is $h = 100\,\text{m}$ behind the front vehicle at the beginning. $h$ is obviously longer than the equilibrium for distance. $\mathbf{k}_3 = [5.0167, 8.4833, -1.4833]$ and $\mathbf{k}_4 = [7.1833, 8.9167, -1.4833]$, whose corresponding Pareto fronts are [1173.6, 563.5, 607.3, 14.3] and [808.1, 694.3, 507.3, 12.4], are selected for this scenario. Figure 20 shows the performance during the approach of the ACC vehicle to the target one under the control of $\mathbf{k}_3$, $\mathbf{k}_4$ and MPC. The simulation results are $[G_{th}, G_{tv}, G_u, G_{du}]_3 = [6.6600\times10^4, 4294.0, 855.1, 52.3]$, $[G_{th}, G_{tv}, G_u, G_{du}]_4 = [6.3615\times10^4, 4693.0, 983.1, 53.4]$ for HVA control and $[G_{th}, G_{tv}, G_u, G_{du}]_{mpc} = [1.5980\times10^5, 4826.7, 886.8, 430.5]$. The proposed controller is able to help the follower to catch up with the leader more quickly and safely with nice objective function values compared with that of MPC in this scenario.

**Stop and go** The third scenario describes an emergency situation where the leader is assumed to brake heavily at $-5\,\text{m/s}^2$ till it stops from $v^*$ and the initial distance between the two vehicles is 15 m which is much shorter than the equilibrium distance. After idling for a while, the leader starts to accelerate at $2\,\text{m/s}^2$ for 10 s. Figure 21 shows the control performance with two gains in the Pareto set $\mathbf{k}_5 = [2.8500, 3.7167, -0.6167]$ and $\mathbf{k}_6 = [9.7833, 1.9833, 0.2500]$ for HVA control whose Pareto fronts are [749.7, 717.4, 538.9, 7.4] and [37.0, 1335.9, 335.5, 3.6], the simulation results are $[G_{th}, G_{tv}, G_u, G_{du}]_5$=[5851.8, 1167.5, 850.8, 24.2], $[G_{th}, G_{tv}, G_u, G_{du}]_6$=[4670.1, 1343.5, 876.7, 35.5] while for MPC, it turns out to be $[G_{th}, G_{tv}, G_u, G_{du}]_{mpc}$=[10398.0, 1369.6, 918.5, 242.9].

As we can see in Fig. 21, the control performance with MPC exceeds below the danger line (gap $h_0$) that is possible for the vehicles to collide and unacceptable for

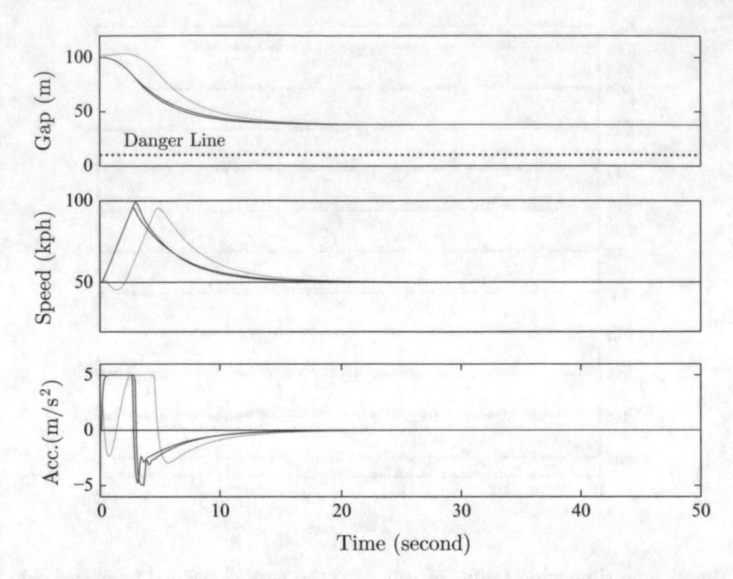


**Fig. 20** The simulation results of the gap, speed and instant acceleration of ACC system in approach scenario. The *red* and *blue lines* represent the performance with $\mathbf{k}_3$ and $\mathbf{k}_4$ respectively, *green lines* show that of MPC and the preceding vehicle is shown by *black line*. The mauve *dotted line* in the *upper* subfigure is considered as the *danger line* below which the two vehicles may collide

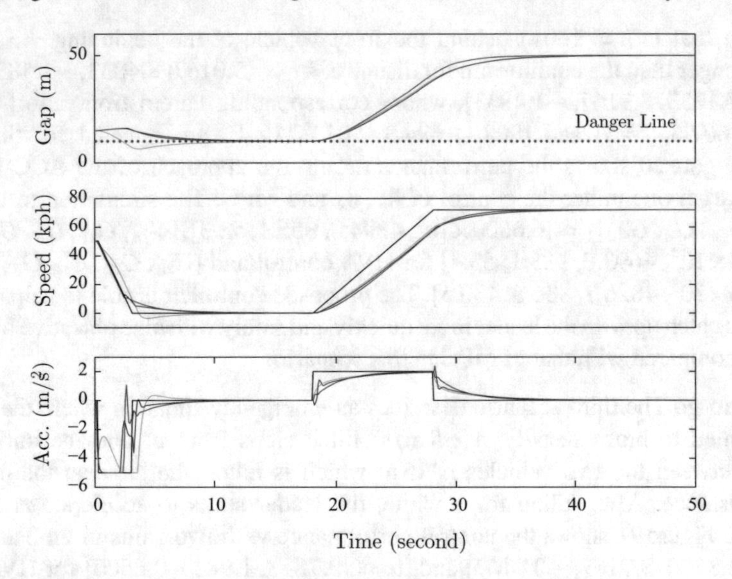


**Fig. 21** The simulation results of the gap, speed and instant acceleration of ACC system in the stop and go scenario. The *red* and *blue lines* represent the performance with $\mathbf{k}_5$ and $\mathbf{k}_6$ respectively, *green lines* show that of MPC and the preceding vehicle is represented by *black lines*. The mauve *dotted line* in the *upper* subfigure is considered as the *danger line* below which the two vehicles may collide

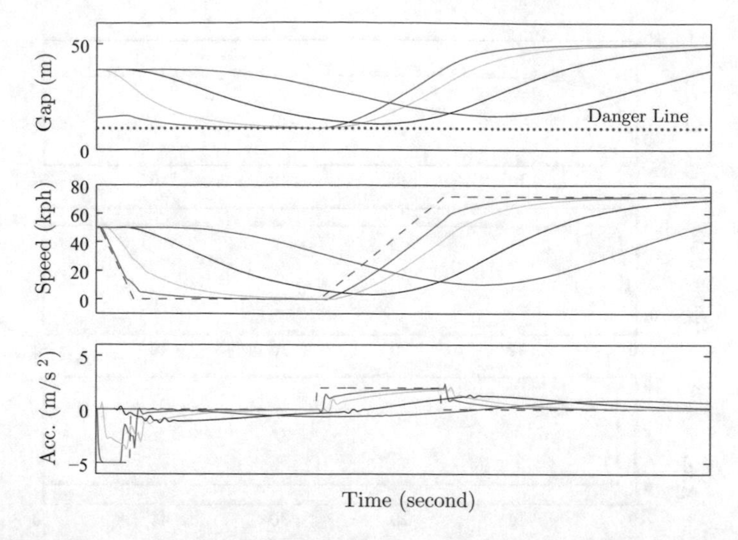


**Fig. 22** The platoon simulation results of the gap, speed and instant acceleration in the stop and go scenario with the gain $\mathbf{k}_5$. The *dotted red lines* are the leader of the platoon while the *dark blue lines* represent the first follower and *light blue lines*, *black lines*, *red lines* mean the 2nd, 5th, 10th follower respectively

safety consideration while those of the two HVA controllers stay beyond the line and thus are more effective in this emergency situation.

The platoon simulation in this emergency situation is also studied as shown in Fig. 22. The leader performs hard decelerating and accelerating while the followers are at equilibrium initially. The decelerating demand declines as the emergency situation propagates backward. The followers are able to perform steady decelerating and accelerating movement with small acceleration.

**Accelerating** In the last scenario, the two consecutive vehicles are assumed to be on a road with the speed limit of 100 kph. The leader accelerates fast at 10 m/s$^2$ to the speed limit first and then keeps cruising. Two cases are considered here with different initial gap while the controller is fixed as $\mathbf{k}_7 = [4.5833, 5.8833, -0.1833]$. Figure 23 shows the control performance with the initial gap of 50 m and Fig. 24 reflects that of 60 m. It can be observed that in the first case the follower is able to catch up with the leader in the simulation to maintain the desired space headway in steady state. However, due to the constraints of the system (Eq. (47)), the follower is not able to follow the preceding vehicle instantaneously given a longer initial gap in the second case and at the mean time, it can not exceed the speed limit to keep the desired distance with the leader, which leads to an extra gap between them from the equilibrium when they have reached the steady state that will not be eliminated if the leader keeps the speed. The gap remains larger than the desired space headway of the follower when they are both at maximum speed.

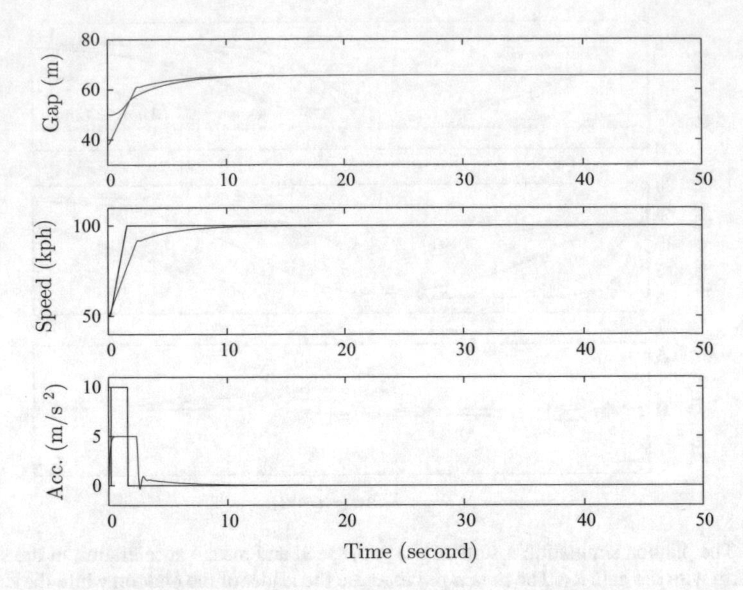


**Fig. 23** The simulation results of the gap, speed and instant acceleration of ACC system in the first case of the accelerating to speed limit scenario. The *red lines* represent the performance with $\mathbf{k}_7$, the preceding vehicle is shown by *black line* while the *blue line* in the *upper* subfigure shows the desired gap of the follower at every moment

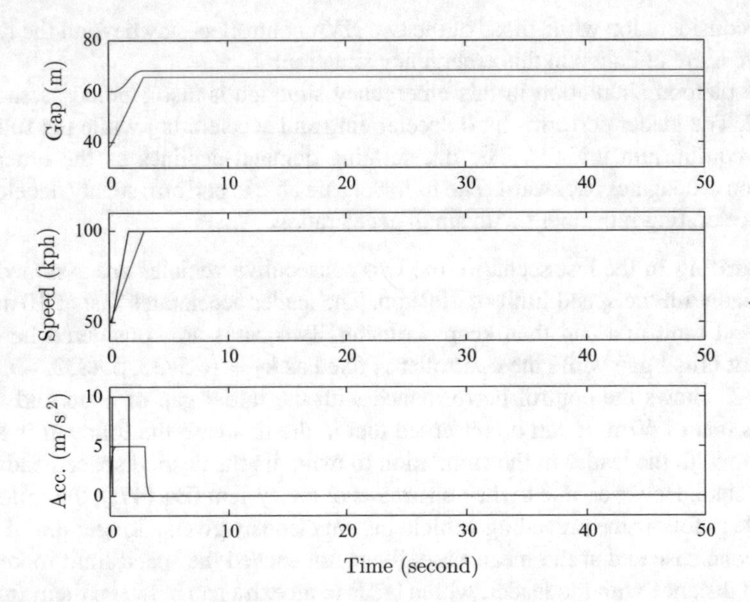


**Fig. 24** The simulation results of the gap, speed and instant acceleration of ACC system in the second case of the accelerating to speed limit scenario. The *red lines* represent the performance with $\mathbf{k}_7$, the preceding vehicle is shown by *black line* while the *blue line* in the *upper* subfigure shows the desired gap of the follower at every moment

## 5 Conclusions

In this chapter, we have presented a multi-objective optimal local trajectory planning algorithm for defensive driving of autonomous ground vehicles and a novel PID-form low-level adaptive cruise control. We have proposed to use the longitudinal terminal velocity and acceleration as the design parameters for the trajectory planning. Different driving performance objectives are optimized simultaneously including traffic safety, transportation efficiency, driving comfort, path consistency. The optimal Pareto solutions are obtained with the cell mapping method. An algorithm is proposed to assist the user to select and implement the optimal designs. In the low-level ACC, the advantages of the proposed HVA control lie in that it is off-line designed that avoids considerable computational burden of on-line control algorithm like MPC and thus is more fast-responsive. The control objectives of HVA are proved to be as good as and even better than those of MPC in many aspects. Extensive numerical simulations show that the proposed algorithms are promising and may provide a new tool for designing the intelligent navigation system that helps AGVs to drive safely in the mixed traffic with erratic human drivers.

**Acknowledgements** The material in this chapter is based on work supported by grants (11172197, 11332008 and 11572215) from the National Natural Science Foundation of China, and a grant from the University of California Institute for Mexico and the United States (UC MEXUS) and the Consejo Nacional de Ciencia y Tecnología de México (CONACYT) through the project "Hybridizing Set Oriented Methods and Evolutionary Strategies to Obtain Fast and Reliable Multi-objective Optimization Algorithms".

## References

1. Bageshwar, V.L., Garrard, W.L., Rajamani, R.: Model predictive control of transitional maneuvers for adaptive cruise control vehicles. IEEE Trans. Vehicul. Technol. **53**(5), 1573–85 (2004)
2. Bando, M., Hasebe, K., Nakayama, A., Shibata, A., Sugiyama, Y.: Dynamical model of traffic congestion and numerical simulation. Phys. Rev. E **51**(2), 1035–1042 (1995)
3. Bose, A., Ioannou, P.: Mixed manual/semi-automated traffic: a macroscopic analysis. Transp. Res. Part C Emerg. Technol. **11C**(6), 439–62 (2003)
4. Brackstone, M., McDonald, M.: Car-following: A historical review. Transp. Res. Part F Traffic Psychol. Behav. **2**(4), 181–196 (1999)
5. Burns, L.D.: Sustainable mobility: a vision of our transport future. Nature **497**(7448), 181–182 (2013)
6. Cao, Y.Y., Lin, Z.L., Shamash, Y.: Set invariance analysis and gain-scheduling control for LPV systems subject to actuator saturation. Syst. Control Lett. **46**(2), 137–51 (2002)
7. Chandler, R.E., Herman, R., Montroll, E.W.: Traffic dynamics: studies in car following. Oper. Res. **6**(2), 165–184 (1958)
8. Chiha, I., Liouane, N., Borne, P.: Tuning PID controller using multiobjective ant colony optimization. Appl. Comput. Intell. Soft Comput. **536326**, 7 (2012)
9. Choi, H.C., Jang, S., Chwa, D., Hong, S.K.: Guaranteed cost control of uncertain systems subject to actuator saturation. In: Proceedings of SICE-ICASE International Joint Conference. p. 6. Piscataway, NJ, USA (2007)

10. Cook, P.A.: Conditions for string stability. In: Proceedings of IEEE Transactions on Automatic Control, vol. 54, pp. 991–998. Netherlands (2005)
11. Das, I.: On characterizing the knee of the Pareto curve based on normal-boundary intersection. Struct. Optim. **18**(2), 107–115 (1999)
12. Ge, J.I., Avedisov, S.S., Orosz, G.: Stability of connected vehicle platoons with delayed acceleration feedback. In: Proceedings of the ASME Dynamic Systems and Control Conference. Palo Alto, California, USA (2013)
13. Gipps, P.G.: A model for the structure of lane-changing decisions. Transp. Res. Part B Methodol. **20B**(5), 403–414 (1986)
14. González, D., Pérez, J., Lattarulo, R., Milanés, V., Nashashibi, F.: Continuous curvature planning with obstacle avoidance capabilities in urban scenarios. In: Proceedings of the 17th International IEEE Conference on Intelligent Transportation Systems, pp. 1430–1435 (2014)
15. Hardy, J., Campbell, M.: Contingency planning over probabilistic obstacle predictions for autonomous road vehicles. IEEE Trans. Robot. **29**(4), 913–929 (2013)
16. Hidas, P.: Modelling lane changing and merging in microscopic traffic simulation. Transp. Res. Part C Emerg. Technol. **10C**(5–6), 351–371 (2002)
17. Hidas, P.: Modelling vehicle interactions in microscopic simulation of merging and weaving. Transp. Res. Part C Emerg. Technol. **13**(1), 37–62 (2005)
18. Howard, T.M., Green, C.J., Kelly, A., Ferguson, D.: State space sampling of feasible motions for high-performance mobile robot navigation in complex environments. J. Field Robot. **25**(6–7), 325–345 (2008)
19. Hsu, C.S.: Cell-to-cell Mapping: A Method of Global Analysis for Nonlinear Systems. Springer, New York (1987)
20. Hu, T.S., Lin, Z.L., Chen, B.M.: An analysis and design method for linear systems subject to actuator saturation and disturbance. Automatica **38**(2), 351–9 (2002)
21. Huang, S., Ren, W.: Autonomous intelligent cruise control with actuator delays. J. Intell. Robot. Syst. Theory Appl. **23**(1), 27–43 (1998)
22. Ioannou, P., Xu, Z.: Throttle and brake control systems for automatic vehicle following. J. Intell. Transp. Syst. **1**(4), 345–77 (1994)
23. Jonsson, J.: Fuel optimized predictive following in low speed conditions. In: Proceedings of Modeling and Control of Economic Systems, p. 119. Klagenfurt, Austria (2003)
24. Katrakazas, C., Quddus, M., Chen, W.H., Deka, L.: Real-time motion planning methods for autonomous on-road driving: state-of-the-art and future research directions. Transp. Res Part C Emerg. Technol. **60**, 416–442 (2015)
25. Kelly, A., Nagy, B.: Reactive nonholonomic trajectory generation via parametric optimal control. Int. J. Robot. Res. **22**(7–8), 583–601 (2003)
26. Kesting, A., Treiber, M., Helbing, D.: General lane-changing model mobil for car-following models. Transp. Res. Record **1999**, 86–94 (2007)
27. Khoie, M., Salahshoor, K., Nouri, E., Sedigh, A.K.: PID controller tuning using multi-objective optimization based on fused genetic-immune algorithm and immune feedback mechanism. In: Proceedings of Advanced Intelligent Computing Theories and Applications, pp. 267–76. Berlin, Germany (2011)
28. Li, S., Li, K., Rajamani, R., Wang, J.: Model predictive multi-objective vehicular adaptive cruise control. IEEE Trans. Control Syst. Technol. **19**(3), 556–566 (2011)
29. Li, X., Sun, J.Q.: Effect of interactions between vehicles and pedestrians on fuel consumption and emissions. Phys. A Stat. Mech. Appl. **416**, 661–675 (2014)
30. Li, X., Sun, J.Q.: Studies of vehicle lane-changing to avoid pedestrians with cellular automata. Phys. A Stat. Mech. Appl. **438**, 251–271 (2015)
31. Li, X., Sun, J.Q.: Effects of turning and through lane sharing on traffic performance at intersections. Phys. A Stat. Mech. Appl. **444**, 622–640 (2016)
32. Li, X., Sun, J.Q.: Effects of vehicle-pedestrian interaction and speed limit on traffic performance of intersections. Phys. A Stat. Mech. Appl. **460**, 335–347 (2016)
33. Li, X., Sun, J.Q.: Studies of vehicle lane-changing dynamics and its effect on traffic efficiency, safety and environmental impact. Phys. A Stat. Mech. Appl. **467**, 41–58 (2017)

34. Li, X., Sun, Z., Cao, D., Liu, D., He, H.: Development of a new integrated local trajectory planning and tracking control framework for autonomous ground vehicles. Mechanical Systems and Signal Processing (2015). doi:10.1016/j.ymssp.2015.10.021
35. Liebner, M., Baumann, M., Klanner, F., Stiller, C.: Driver intent inference at urban intersections using the intelligent driver model. In: Proceedings of IEEE Intelligent Vehicles Symposium, pp. 1162–1167 (2012)
36. Martinez, J.J., de Wit, C.C.: A safe longitudinal control for adaptive cruise control and stop-and-go scenarios. IEEE Trans. Control Syst. Technol. **15**(2), 246–58 (2007)
37. McNaughton, M., Urmson, C., Dolan, J.M., Lee, J.W.: Motion planning for autonomous driving with a conformal spatiotemporal lattice. In: Proceedings of IEEE International Conference on Robotics and Automation, pp. 4889–4895 (2011)
38. Miller, I., Campbell, M., Huttenlocher, D.: Efficient unbiased tracking of multiple dynamic obstacles under large viewpoint changes. IEEE Trans. Robot. **27**(1), 29–46 (2011)
39. Naus, G., van den Bleek, R., Ploeg, J., Scheepers, B., van de Molengraft, R., Steinbuch, M.: Explicit mpc design and performance evaluation of an acc stop-&-go. In: Proceedings of American Control Conference, pp. 224–9. Piscataway, NJ, USA (2008)
40. Naus, G.J.L., Ploeg, J., de Molengraft, M.J.G.V., Heemels, W.P.M.H., Steinbuch, M.: Design and implementation of parameterized adaptive cruise control: an explicit model predictive control approach. Control Eng. Pract. **18**(8), 882–892 (2010)
41. Orosz, G., Moehlis, J., Bullo, F.: Delayed car-following dynamics for human and robotic drivers. In: Proceedings of the ASME International Design Engineering Technical Conferences and Computers and Information in Engineering Conference. Washington, DC, USA (2011)
42. Orosz, G., Shah, S.P.: A nonlinear modeling framework for autonomous cruise control. In: Proceedings of the ASME 5th Annual Dynamic Systems and Control Conference Joint with the JSME 11th Motion and Vibration Conference, vol. 2, pp. 467–471. Fort Lauderdale, FL, United States (2012)
43. Orosz, G., Wilson, R.E., Stépán, G.: Traffic jams: dynamics and control. Philos. Trans. R. Soc. A Math. Phys. Eng. Sci. **368**(1928), 4455–4479 (2010)
44. Pareto, V.: Manual of Political Economy. The MacMillan Press, London (1971)
45. Qin, W.B., Orosz, G.: Digital effects and delays in connected vehicles: Linear stability and simulations. In: Proceedings of the ASME Dynamic Systems and Control Conference. Palo Alto, California, USA (2013)
46. Schütze, O., Laumanns, M., Coello, C.A.C.: Approximating the knee of an MOP with stochastic search algorithms. In: Parallel Problem Solving from Nature, pp. 795–804. Springer, Berlin (2008)
47. Schwesinger, U., Rufli, M., Furgale, P., Siegwart, R.: A sampling-based partial motion planning framework for system-compliant navigation along a reference path. In: Proceedings of IEEE Intelligent Vehicles Symposium (IV), pp. 391–396 (2013)
48. Shim, T., Adireddy, G., Yuan, H.: Autonomous vehicle collision avoidance system using path planning and model-predictive-control-based active front steering and wheel torque control. Inst. Mech. Eng. Part D J. Autom. Eng. **226**(6), 767–778 (2012)
49. Shinar, D., Compton, R.: Aggressive driving: an observational study of driver, vehicle, and situational variables. Accident Anal. Prevention **36**(3), 429–437 (2004)
50. Swaroop, D., Hedrick, J.K.: String stability of interconnected systems. IEEE Trans. Autom. Control **41**(3), 349–57 (1996)
51. Toledo, T., Zohar, D.: Modeling duration of lane changes. Transp. Res. Record **1999**, 71–78 (2007)
52. Treiber, M., Hennecke, A., Helbing, D.: Congested traffic states in empirical observations and microscopic simulations. Phys. Rev. E **62**(2), 1805–1824 (2000)
53. Treiber, M., Kanagaraj, V.: Comparing numerical integration schemes for time-continuous car-following models. Phys. A Stat. Mech. Appl. **419**, 183–195 (2015)
54. Tsugawa, S.: An overview on energy conservation in automobile traffic and transportation with its. In: Proceedings of the IEEE International Vehicle Electronics Conference, pp. 137–42. Piscataway, NJ, USA (2001)

55. Varaiya, P.: Smart cars on smart roads: problems of control. IEEE Trans. Autom. Control **38**(2), 195–207 (1993)
56. Wei, J., Snider, J.M., Gu, T., Dolan, J.M., Litkouhi, B.: A behavioral planning framework for autonomous driving. In: Proceedings of IEEE Intelligent Vehicles Symposium Proceedings, pp. 458–464 (2014)
57. Xiao, L., Gao, F.: Practical string stability of platoon of adaptive cruise control vehicles. In: Proceedings of IEEE Transactions on Intelligent Transportation Systems, vol. 12, pp. 1184–94. USA (2011)
58. Xiong, F., Qin, Z., Xue, Y., Schütze, O., Ding, Q., Sun, J.: Multi-objective optimal design of feedback controls for dynamical systems with hybrid simple cell mapping algorithm. Commun. Nonlinear Sci. Numer. Simul. **19**(5), 1465–1473 (2014)
59. Xiong, F.R., Qin, Z.C., Ding, Q., Hernández, C., Fernández, J., Schütze, O., Sun, J.Q.: Parallel cell mapping method for global analysis of high-dimensional nonlinear dynamical systems. J. Appl. Mech. **82**(11), (2015). doi:10.1115/1.4031149
60. Xu, W., Wei, J., Dolan, J.M., Zhao, H., Zha, H.: A real-time motion planner with trajectory optimization for autonomous vehicles. In: Proceedings of IEEE International Conference on Robotics and Automation, pp. 2061–2067 (2012)
61. Yadlapalli, S.K., Darbha, S., Rajagopal, K.R.: Information flow and its relation to stability of the motion of vehicles in a rigid formation. In: Proceedings of IEEE Transactions on Automatic Control, vol. 51, pp. 1315–19. USA (2006)
62. Yoon, Y., Shin, J., Kim, H.J., Park, Y., Sastry, S.: Model-predictive active steering and obstacle avoidance for autonomous ground vehicles. Control Eng. Pract. **17**(7), 741–750 (2009)
63. You, F., Zhang, R., Lie, G., Wang, H., Wen, H., Xu, J.: Trajectory planning and tracking control for autonomous lane change maneuver based on the cooperative vehicle infrastructure system. Expert Syst. Appl. **42**(14), 5932–5946 (2015)
64. Zhang, J., Ioannou, P.A.: Longitudinal control of heavy trucks in mixed traffic: Environmental and fuel economy considerations. IEEE Trans. Intell. Transp. Syst. **7**(1), 92–104 (2006)
65. Zhang, L., Orosz, G.: Designing network motifs in connected vehicle systems: delay effects and stability. In: Proceedings of the ASME Dynamic Systems and Control Conference. Palo Alto, California, USA (2013)
66. Zhang, Y., Kosmatopoulos, E.B., Ioannou, P.A., Chien, C.C.: Autonomous intelligent cruise control using front and back information for tight vehicle following maneuvers. IEEE Trans. Vehicul. Technol. **48**(1), 319–328 (1999)
67. Ziegler, J., Bender, P., Dang, T., Stiller, C.: Trajectory planning for Bertha - A local, continuous method. In: Proceedings of IEEE Intelligent Vehicles Symposium, pp. 450–457 (2014)

# Augmenting the LSA Technique to Evaluate Ubicomp Environments

**Víctor R. López-López, Lizbeth Escobedo, Leonardo Trujillo and Victor H. Díaz-Ramírez**

**Abstract** LSA is a useful user study technique, it is well known and used to design and evaluate Ubicomp systems. The LSA technique enables researchers to collect data, analyze it, and obtain quantitative and statistical results. A key advantage of using LSA is that it is performed in the user's environment. However, analyzing large amounts of data is considered by some researchers to be a burden and time consuming, prone to human error. In this paper we explore the use of computer vision techniques to automate the data analysis and coding when using LSA. We present a system that uses facial tracking, object recognition and composite correlation filters to detect the *Attention* behavior of a subject. Our results indicate that computer vision can automate the LSA technique and reduce the burden of coding data manually by the researcher. The findings from this study reveal emergent practices of the use of our proposed system to automate the evaluation of Ubicomp environments.

## 1 Introduction

Evaluation of Ubicomp environments and applications is a key component to meet usability, user experience, and to measure the impact of use. A technique commonly

V.R. López-López
Posgrado en Ciencias de la Ingeniería,
Departamento de Ingeniería Eléctrica y Electrónica, Blvd. Alberto Limón Padilla y Av. ITR Tijuana S/N, C.P. 22430 Tijuana, Baja California, Mexico
e-mail: vlopez@tectijuana.edu.mx

L. Escobedo
UCSD, 9452 Medical Center Dr, La Jolla, CA 92037, USA
e-mail: loescobedobravo@ucsd.edu

L. Trujillo (✉)
Instituto Tecnológico de Tijuana,Calzada Del Tecnológico S/N Fraccionamiento Tomas Aquino, 22414 Tijuana, Baja California, Mexico
e-mail: leonardo.trujillo@tectijuana.edu.mx

V.H. Díaz-Ramírez
CITEDI-IPN,Tijuana, Baja California, Mexico
e-mail: vdiazr@ipn.mx

© Springer International Publishing AG 2018

Y. Maldonado et al. (eds.), *NEO 2016*, Studies in Computational Intelligence 731,
https://doi.org/10.1007/978-3-319-64063-1_2

used to evaluate Ubicomp environments is direct observations. This is a user centered technique that allows researchers to collect data of the activities of the users in their normal environment [15]. Due to this context of use, Ubicomp applications are difficult to evaluate. For instance, Weiser suggested that to evaluate Ubicomp, we need to work with people in disciplines such as psychology and anthropology [23]. Lag Sequential Analysis (LSA) is a well-known technique borrowed from psychology methods [8, 13] and widely used to evaluate Ubicomp environments [2–5]. In general, LSA enables a researcher to analyze the data for sequential correlations between observed events.

LSA is a technique for gathering quantitative data by observing users in their real and natural work environment, as they perform their normal activities. It is traditionally used in the field of developmental psychology to study the behavior of persons by measuring the number of times certain behaviors occur, precede or follow a selected behavior. The behaviors are defined by the researchers using a scheme code. Data can be captured live with paper and pencil or coded from previously recorded videos. An advantage of using videos to capture data is that it can be re-coded for different studies or behaviors of researchers interests. Videos can also be used for qualitative observational purposes. The main and significant disadvantages of using LSA is the associated cost, because it can be extremely time consuming, especially for a single researcher. Also, if there is more than one researcher coding videotapes there is the need to verify the reliability of different coders (e.g., using IOA or Cohens Kappa [21]).

In this paper we explore the use of computer vision (CV) techniques to allow the researcher to automate the data collection and coding in LSA. Our aim is to augment the LSA technique, reducing its burden and human errors. The remainder of this paper proceeds as follows. First, we discuss related work. Then, we describe the LSA technique, present our method and our proposed CV system used to augment the LSA technique. Later, we present the results of using our proposed CV system to code videos, using a previous deployment of a Ubicomp system [16]. Finally, we close discussing conclusions and directions of future work.

## 2 Related Work

Ubicomp researchers have given importance to the use of the LSA technique to evaluate the user experience and impact of use of their systems [2–5]. There are technologies to help analyze qualitative and quantitative data of user centered studies.

Research [8, 18] and commercial [7] efforts have emerged to create new technologies and tools to help in data analysis in usability studies. Existing technologies (e.g., MacSHAPA [18], NVivo,[1] MAXQDA,[2] Atlas.ti,[3] QDAMiner,[4] and HyperRe-

[1] http://www.qsrinternational.com/products_nvivo.aspx.

[2] http://www.maxqda.com.

[3] http://atlasti.com.

[4] http://provalisresearch.com/products/qualitative-data-analysis-software.

search[5]) for analysis of data help to put together the data (e.g., interviews, videos, notes, photos) of studies and to apply different techniques to analyze it (e.g., coding, statistics, cites, ethnography, etc.). Those technologies contain interfaces that help organize and analyze data from qualitative research. For example, they have an interface to view a video and encode without having to switch between screens. In Atlas.ti researchers can manage multiple documents and code all data types. For video coding, researchers can select a video segment and assign a code for it. This is, researchers need to do the coding manually. The major advantage is that Atlas.ti allows managing codes among different types of collected data. Using MacSHAPA [18] researchers can play, pause, stop, rewind, fast forward, jog, and shuttle videos at different speeds. MacSHAPA allows to capture timecodes from videos, insert them into the data, and find a video frame that corresponds to a time-stamp selected in the data. However, MacSHAPA does not encode data automatically; instead, it lets researchers develop and change coding categories, store them, and use them to encode data manually.

This body of work presented in related literature helps to analyze the data already captured, for example encode videos or interviews. However, the data collection and coding methods are still the same, manual and prone to human errors. Despite the interest of researchers and professionals in computer and social sciences to use the LSA technique for data analysis, there is little attention on developing intelligent methods to automate data collection and reduce the cost involved in coding. These works inspire us to explore the use of CV techniques to automate the coding of data from videos of user centered studies.

## 3 Lag Sequential Analysis

In LSA the periods of observation are broken into a sequence of intervals called *lags*. For time-sampling methods, each lag represents a fixed period of time; for event-sampling methods, each lag represents the duration of an event.

To start coding with the LSA technique, researchers must choose the events or behaviors of interest (BoI) that will be described using a systematic coding scheme for the video coding. It is common to choose events that would apply equally well before and after the introduction of new technologies into the environment, to compare interactions following a reference. After this step, researchers can start to code the video. Each time a BoI occurs, the researcher makes a time-stamp indicating that a certain BoI occurred. On each time-stamped the researchers annotate the video and participants identification (participant id, time, etc.) and a code for each of the BoI (Fig. 1).

In Fig. 1 each row represents a time-stamp of a BoI that occurred. The researchers codes the time, and then writes a code for each BoI. Each code in the row indicates the state of the behavior in the time-stamp. This process is made for each participant in all the videos of the study. On average, coding a video for just one participant

[5]http://www.researchware.com/products/hyperresearch.html.

| 1 | 2 | 3 | 4 | 5 | 6 | 8 | 9 | 10 | 11 | 12 | 13 | 14 | 15 | 16 | | 17 | |
|---|---|---|---|---|---|---|---|---|---|---|---|---|---|---|---|---|---|
| Informant | Role | DayObsv | Therapy | Date | Time | | INT | AUX | Atention | DIS | EMO | Help | Reward | Mannerisms | Object | INT | Transi |
| EMM | E | 1 | 1 | 13/02/2012 | 00:00:05 | | ALE | NONE | ONT | DIS | CON | NONE | NONE | MOJ | TAR | NONE | TT+SA |
| EMM | E | 1 | 1 | 13/02/2012 | 00:00:54 | | ALE | NONE | ONT | DIS | CON | FISS | NONE | NONE | TAR | NONE | IT |
| EMM | E | 1 | 1 | 13/02/2012 | 00:01:55 | | ALE | NONE | ONT | DIS | CON | FISS+VER | RVER+RVIS | NONE | TAR | NONE | FE |
| EMM | E | 1 | 1 | 13/02/2012 | 00:02:10 | | ALE | NONE | ONT | DIS | CON | NONE | NONE | NONE | TAR | NONE | TE+SA |
| EMM | E | 1 | 1 | 13/02/2012 | 00:02:38 | | ALE | NONE | ONT | DIS | CON | VER | NONE | NONE | TAR | NONE | IE |
| EMM | E | 1 | 1 | 13/02/2012 | 00:02:53 | | ALE | NONE | ONT | DIS | CON | VER+FISS | NONE | NONE | TAR | NONE | IE |
| EMM | E | 1 | 1 | 13/02/2012 | 00:03:37 | | ALE | NONE | ONT | DIS | CON | NONE | NONE | NONE | TAR | NONE | IE |
| EMM | E | 1 | 1 | 13/02/2012 | 00:03:53 | | ALE | NONE | ONT | DIS | CON | NONE | NONE | NONE | TAR | NONE | FE |
| EMM | E | 1 | 1 | 13/02/2012 | 00:03:55 | | ALE | NONE | ONT | DIS | CON | NONE | RVIS | HR | TAR | NONE | TE+SA |
| EMM | E | 1 | 1 | 13/02/2012 | 00:04:05 | | ALE | NONE | ONT | DIS | CON | NONE | NONE | NONE | TAR | NONE | IE |
| EMM | E | 1 | 1 | 13/02/2012 | 00:04:11 | | ALE | NONE | ONT | DIS | CON | VER | NONE | NONE | TAR | NONE | IE |
| EMM | E | 1 | 1 | 13/02/2012 | 00:04:44 | | ALE | NONE | ONT | DIS | CON | VER | NVIS | NONE | TAR | NONE | FE |
| EMM | E | 1 | 1 | 13/02/2012 | 00:04:49 | | ALE | NONE | ONT | DIS | CON | NONE | NONE | HR | TAR | NONE | TE+SA |
| EMM | E | 1 | 1 | 13/02/2012 | 00:05:05 | | ALE | NONE | ONT | DIS | CON | NONE | NONE | NONE | TAR | NONE | IE |
| EMM | E | 1 | 1 | 13/02/2012 | 00:05:08 | | ALE | NONE | ONT | DIS | CON | VER | NONE | NONE | TAR | CM+CV | IE |
| EMM | E | 1 | 1 | 13/02/2012 | 00:05:19 | | ALE | NONE | ONT | DIS | CON | NONE | RVER+RVIS | NONE | TAR | NONE | FE |
| EMM | E | 1 | 1 | 13/02/2012 | 00:05:28 | | ALE | NONE | ONT | DIS | CON | NONE | NONE | NONE | TAR | NONE | TE+SA |
| EMM | E | 1 | 1 | 13/02/2012 | 00:05:58 | | ALE | NONE | ONT | DIS | CON | VER | NONE | NONE | TAR | NONE | IE |
| EMM | E | 1 | 1 | 13/02/2012 | 00:06:00 | | ALE | NONE | ONT | DIS | CON | VER+FIS | NONE | NONE | TAR | CM+CV | IE |
| EMM | E | 1 | 1 | 13/02/2012 | 00:06:03 | | ALE | NONE | ONT | DIS | CON | VER+FISS | NONE | NONE | TAR | NONE | IE |
| EMM | E | 1 | 1 | 13/02/2012 | 00:06:19 | | ALE | NONE | ONT | DIS | CON | NONE | RVIS | NONE | TAR | NONE | FE |
| EMM | E | 1 | 1 | 13/02/2012 | 00:06:31 | | ALE | NONE | ONT | DIS | CON | NONE | RVER+RFIS+RVIS | NONE | TAR | CM+CV | TE+SA |
| EMM | E | 1 | 1 | 13/02/2012 | 00:06:57 | | ALE | NONE | ONT | DIS | CON | NONE | NONE | NONE | TAR | NONE | FT |
| # | # | # | # | # | 00:07:13 | | # | # | # | # | # | # | # | # | # | # | # |
| # | # | # | # | # | # | | # | # | # | # | # | # | # | # | # | # | # |
| URI | E | 1 | 1 | 13/02/2012 | 00:00:01 | | ALE | NONE | OFFT | DIS | CON | NONE | NONE | NONE | TAR | NONE | TT+SA |
| URI | E | 1 | 1 | 13/02/2012 | 00:01:40 | | ALE | NONE | ONT | DIS | CON | VER+FISS | NONE | GRI | TAR | CM+CV | IT |
| URI | E | 1 | 1 | 13/02/2012 | 00:02:03 | | ALE | NONE | ONT | DIS | CON | NONE | NONE | NONE | TAR | CM+CV+V | IT |
| URI | E | 1 | 1 | 13/02/2012 | 00:02:19 | | ALE | NONE | ONT | DIS | CON+ANS | VER+FISS+FIS | NONE | NONE | TAR | NONE | IE |

**Fig. 1** Example of coding in LSA, where the first block of data (e.g., before the #marks) corresponds to the observed behaviors coding of interest in a 17 min 12 s video. The *left side* from the *black line* identifies the video, activity, and the person of interest observed. Each column in the *right side* describes the codification of the observed behaviors from the person of interest identified on the *left side*

takes twice the time of the total duration of the video. Despite the advantages of using LSA, the disadvantages of using it are evident. These are, time consuming and very susceptible to human errors.

## 3.1 LSA to Evaluate Ubicomp Systems

For more than five years we have been creating a LivingLab [20] where we deploy Ubicomp systems to help to improve the quality of life of children with autism, and their teachers and therapists. In this LivingLab we have been recording many hours of video of use of several Ubicomp systems. In this LivingLab we've been designing, developing, and pilot testing Ubicomp technology to augment several dimensions of the therapy cycle of children with autism.

We have been using LSA to quantitatively analyze our data, but at this time we still have many hours of observation that has not been analyzed due to the time required to apply the LSA technique. Consequently we have several unknown results.

In some studies we have been interested in observing *Attention* in children with autism a BoI during therapies that has two states *using* and *not using* our Ubicomp technology. To do this, in our scheme code we described how we recognize *Attention* in the child, observing the activities of a child.

For example, for SensoryPaint [16, 17, 24] we have defined in our scheme code the BoI of *Attention* as *"the child is on task (ONT) when he/she is looking to the interactive surface while performing the activity of paint using a ball, and off task (OFT) when the child is not doing at least one of the actions described in ONT"*. In Fig. 1, *Attention* is showed in the yellow column.

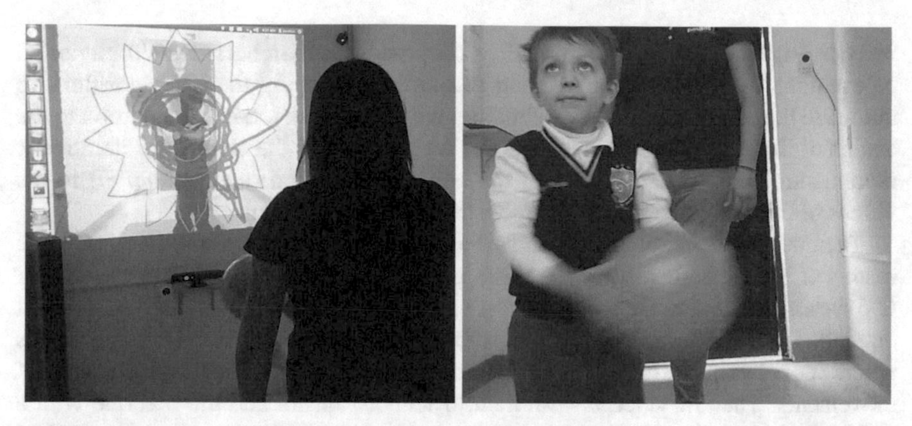

**Fig. 2** A child ONT using SensoryPaint coloring a star

In SensoryPaint children with autism use rubber balls detected by the system that act as paintbrushes of various sizes, textures, and colors, allowing for the painting of lines. Users can either draw in a free form mode or use a template (Fig. 2). Finally, sounds are played in connection with ball movement.

In a first study of using SensoryPaint, 10 participants used it for 9 weeks during daily 15-min sensory therapies twice a week. This is 270 min of use per each child. A total duration in videos to analyze of 2, 700 min or 45 h. This is almost 90 h to code using the LSA technique per researcher.

With all videos analyzed using LSA the main goal is to quantify the BoIs and get statistics. For example, how much time the children were ONT and OFT, how many segments, frequency, mean times, per child, per day, per study, etc. Then we can also get the statistical significance of our results and provide conclusions.

Using LSA in our Ubicomp studies enable us to get useful data to evaluate the impact of use of our Ubicomp systems in a person's natural environment. However, the cost of using LSA could hinder and/or delay its use and the generation of conclusive results.

# 4 Method

Based on our own experience of use of the LSA technique and insights provided in relevant literature, we conducted a qualitative study to understand the challenges researchers have when using the LSA technique. We conducted 3 semi-structured interviews with 3 expert researchers in using LSA in areas such as Ubicomp and Human-Computer Interaction. Then following an interactive user-centered design methodology we used the results of the study to iteratively design 2 low-fidelity prototypes that were presented to the researchers in a mock-up form. We conducted 3 participatory design sessions to discuss our prototypes and uncover new design insights. We used the results of our participatory sessions to re-design our prototype by choosing from the design alternatives proposed to researchers. Then for a period of 6 months we first concentrated our efforts in developing the CV algorithms to automate the coding in the LSA technique, without putting too much attention to the design of the graphical user interface. We were designing the algorithms iteratively, incrementing the difficulty level of the behavior detection (e.g., *Attention*) in the videos of the study to enhance the robustness of the algorithms.

Then we tested the robustness and reliability of the CV algorithms comparing the results of coding manually, serving as a ground truth, and automatically using our CV system. We first make and use a test video, then, when the algorithm was robust enough, a real video from a previous study with SensoryPaint is used. Both videos where recorded in the same equipped room with a child making the same type of movements. The test video is with a child without autism and the real one with a child with autism. Both videos were recorded in a closed room with constant light and the camera located in front of the child in a fixed position. The videos show a child positioned in front of the camera at a 1 m distance. In the test video the child was making controlled movements, performing different gaze directions and head poses. The test video has a duration of 55 s and the real video 19 s with a frame rate of 20 Hz and a resolution of $640 \times 480$ pixels and $1280 \times 720$ pixels respectively; Fig. 3 shows different snapshot of each video.

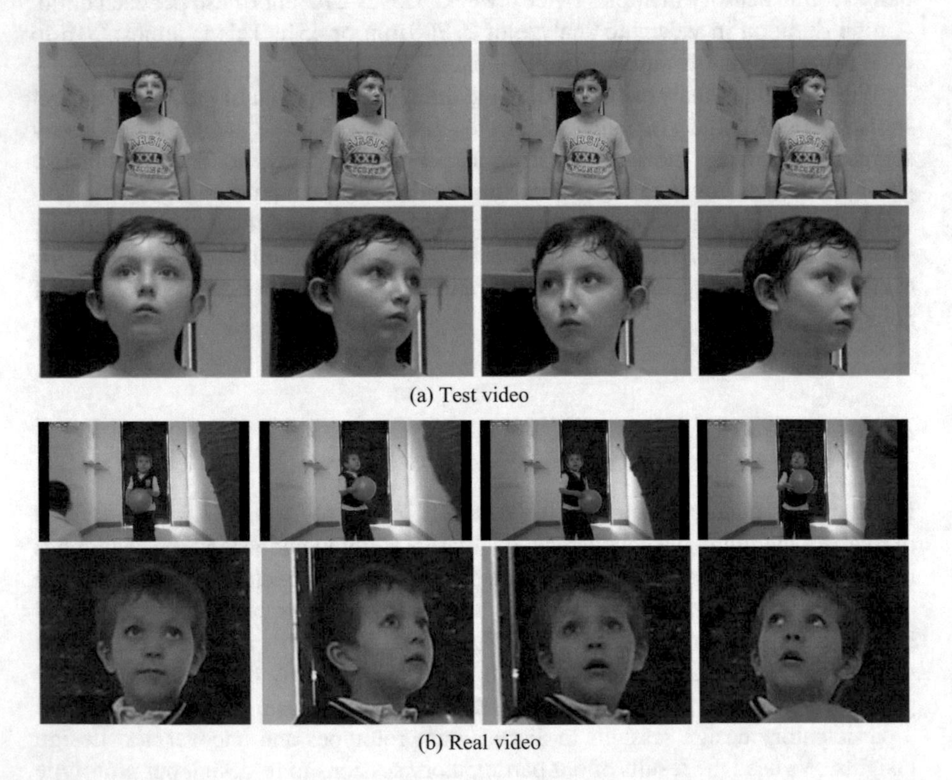

(a) Test video

(b) Real video

**Fig. 3** Illustration of snapshots of: **a** the test video; and **b** the real video. The *top images* shows the whole snapshot and the *bottom* shows a zoom of the face of the child in the video

Finally, we then compare the reliability to detect *Attention*, and the time researchers take to code the videos manually and automatically using developed CV system. Results are reported using the total accuracy and the $\tilde{\chi}^2$ test.

# 5 Computer Vision System to Augment LSA

Considering the hyper/hypo sensibility of the participants of our LivingLab we were interested in a non-invasive method to automatically perform the LSA technique. Our proposed system will code previously recorded videos or could be used online. CV is a robust field in computer science [14], developing non-invasive methods that are particularly applicable in this case.

We first select *Attention* as our BoI and define how the CV system will interpret that BoI. In this sense, *Attention* is when the child is looking at the screen and also is using the ball to paint with SensoryPaint, as we can see in Fig. 2. In this work, we concentrate our efforts to detect the first condition, looking at the screen. In our approach, we pose this problem as a combination of both gaze detection and head pose estimation, two important tasks to CV research.

The proposed approach is to first detect, track and recognize the face of a person of interest (POI), hereafter we will refer to this as face recognition (FR); and then estimate the gaze direction of that POI, hereafter we will refer to this as gaze direction estimation (GDE). The process is divided into three main stages that occur in the following order: Pre-processing, FR and GDE. The entire process is depicted as a block diagram in Fig. 4.

## 5.1 *Pre-processing*

The pre-processing process is semi-automatic, the system first asks the user to manually initialize the system and define the inputs for the FR process. This pre-processing stage is divided into four steps that are performed in the following order:

(a) Load and play a video that will be coded.
(b) Select the frame of the video in which the face of the person of interest is shown.
(c) Crop the face in the image of the person of interest $I^*$ from the selected frame.
(d) Create the filter $f_{I^*}$ of $I^*$ using the synthetic discriminant function (SDF) [6].

The SDF filter is a type of composite correlation filter that is widely used for distortion tolerant object recognition and in tracking applications [1, 12]. These filters represent the impulse response of a linear system designed in such a manner that the coordinates of the system's maximum output are estimates of the target's or in this case the location of $I^*$ within an observed scene (e.g., frame of the video sequence). The systems output can be obtained by a basic correlation, hence the name of composite correlation filter.

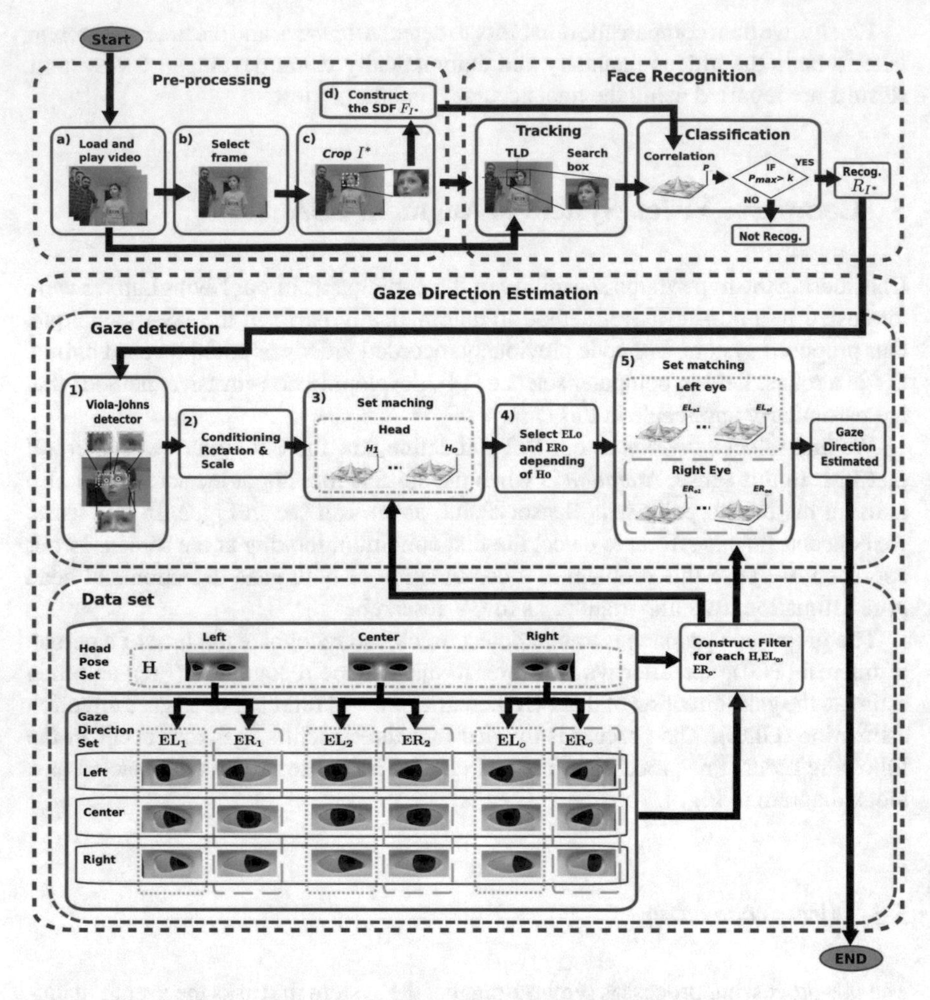


**Fig. 4** Block diagram of the proposed CV system to augment the LSA technique

### 5.1.1 Correlation Filter

Correlation is a mathematical function that indicates the similarity between two signals, or in this case images, defined as

$$c(m,n) = \sum_i \sum_j Im_{fr}(i,j) h_{Im^*}(i-m, j-n), \tag{1}$$

where $c(m,n)$ is called a correlation plane and the maximum value $c_{max}$ indicates the best estimate of the location of $f_{I^*}$ within the frame $I_{frame}$ of the video sequence.

Basically the composite correlation filter may be divided into two phases: the design phase and the use phase. The design phase of a correlation filter is often computationally expensive; however, each filter need only be designed once, and the filter thereafter can be used efficiently in the frequency domain.

### 5.1.2 The Design Phase of a Filter

In the present work, we are interested in the design of a filter capable of matching an object when it is embedded in a disjoint background and the observed scene is corrupted with additive noise. Additionally, the filter must be able to recognize geometrically distorted versions of the target, such as rotated and scaled versions. For example in $I^*$ everything that is not the face of a person can be thought as noise.

Let $T = \{I_i^*(x, y); i = 1, \ldots, N\}$ be a set of available image templates representing different views of the image $I^*$ to match. We assume that the observed scene, in this case $I_{frame}$, contains an arbitrary view of the target embedded into a disjoint background $b(x, y)$ at unknown coordinates $(\tau_x, \tau_y)$, and the image is corrupted with zero-mean additive white noise $n(x, y)$, as follows:

$$I^*(x, y) = I^*(x - \tau_x, y - \tau_y) + b(x, y)\bar{w}(x - \tau_x, y - \tau_y) + n(x, y), \tag{2}$$

where $\bar{w}(x, y)$ is the region of support of the target defined as zero within the area occupied by the target, and unity elsewhere. It is known that the best filter for detecting a single view of the target in Eq. (2), in terms of the signal to noise ratio and the minimum variance of measurements of location errors, is the matched filter whose frequency response is given by [10]

$$H^*(u, v) = \frac{T(u, v) + \mu_b \bar{W}(u, v)}{(T(u, v) + \mu_b \bar{W}(u, v))^2 + P_n(u, v)}. \tag{3}$$

In Eq. (3), $T(u, v)$ and $\bar{W}(u, v)$ are the Fourier transforms of $t(x, y)$ and $\bar{w}(x, y)$, respectively, $\mu_b$ is the mean value of $b(x, y)$, and $P_b(u, v)$ and $P_n(u, v)$ are the power spectral densities of $b_0(x, y) = b(x, y) - \mu_b$ and $n(x, y)$, respectively.

Now, let $h_i(x, y)$ be the impulse response of a MF constructed to match the $i$th available view of the target $t_i(x, y)$ in $T$. Let $H = \{h_i(x, y); i = 1, \ldots, N\}$ be the set of all MF impulse responses constructed for all training images $t_i(x, y)$. We want to synthesize a filter capable to recognize all target views in $T$, by combining the optimal filter templates contained in $H$, and by using only a single correlation operation. The required filter $p(x, y)$ is designed according to the SDF filter model, whose impulse response is given by [12]

$$p(x, y) = \sum_{i=1}^{N} \alpha_i h_i(x, y), \tag{4}$$

where $\{\alpha_i; i = 1, \ldots, N\}$ are weighting coefficients that are chosen to satisfy the following conditions: $\langle p(x, y), t_i(x, y)\rangle = u_i$; where "<, >" denotes inner-product, and $u_i$ are prespecified output correlation values at the origin, produced by the filter $p(x, y)$ in response to the training patterns $t_i(x, y)$.

Let us denote a matrix $\mathbf{R}$ with $N$ columns and $d$ rows ($d$ is the number of pixels), where its $i$th column contains the elements of the $i$th member of $H$ in lexicographical order. Let $\mathbf{a} = [\alpha_i; i = 1, \ldots, N]^T$ be a vector of weighting coefficients. Thus, Eq. 4 can be rewritten as

$$\mathbf{p} = \mathbf{Ra}. \tag{5}$$

Furthermore, let $\mathbf{u} = [u_i = 1; i = 1 \ldots, N]^T$ be a vector of correlation constraints imposed to the filter's output in response to the training patterns $t_i(x, y)$, and let $\mathbf{Q}$ be a $d \times N$ matrix whose $i$th column is given by the vector version of the $i$th view of the target in $T$. Note that the filter's constraints can be expressed as

$$\mathbf{u} = \mathbf{Q}^{+}\mathbf{p}, \tag{6}$$

where superscript "+" denotes conjugate transpose. By substituting Eq. 5 into 6, we obtain $\mathbf{u} = \mathbf{Q}^{+}\mathbf{Ra}$. Thus, if matrix $\mathbf{Q}^{+}\mathbf{R}$ is nonsingular the solution for $\mathbf{a}$, is

$$\mathbf{a} = \left[\mathbf{Q}^{+}\mathbf{R}\right]^{-1}\mathbf{u}. \tag{7}$$

Finally, by substitution of Eq. 7 into 5, the solution for the SDF filter is given by

$$\mathbf{p} = \mathbf{R}\left[\mathbf{Q}^{+}\mathbf{R}\right]^{-1}\mathbf{u}. \tag{8}$$

We propose the following process based on the SDF filter design, depicted in Fig. 5. First, given $t(x, y)$ we create two synthetically rotated versions of $t(x, y)$, one rotated 15 degrees clockwise and another rotated 15 counterclockwise, and construct a three element set $T$. Next, set $H$ is created by synthesizing the MF impulse responses of all patterns in $T$. Finally, an SDF filter template is synthesized using Eq. (8).

#### 5.1.3 The Use Phase of a Filter

As mentioned before, the filter is used by correlating the $\mathbf{p}$ within $I_{frame}$. Then, the maximum value of the correlation plane output $C_{max}$ indicates the best estimate of the location of $\mathbf{p}$ within $I_{frame}$.

### 5.2 *Face Recognition*

The FR stage is divided into two steps which are: tracking and classification. The tracking process allows us to automatically locate $I^*$ in each frame $I_{frame}$ of the

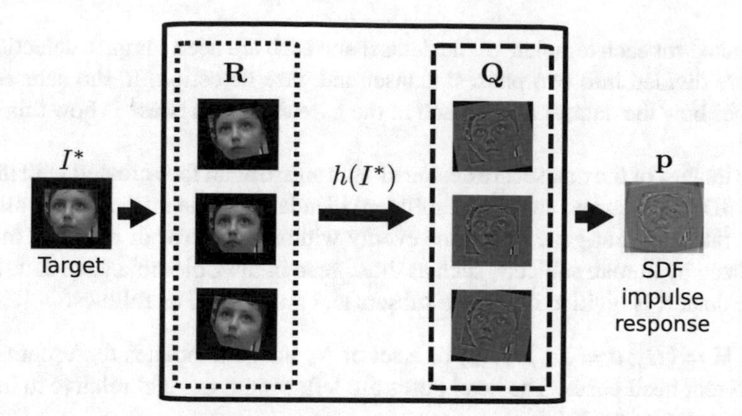


**Fig. 5** Procedure for constructing the SDF filter. *Left* target object; *Center* training templates used to construct the matrices **R** and **Q**; *Right* the SDF image template (The images of **Q** and the SDF image where normalize for illustration)

video sequence. This is accomplished by using the Tracking Learning Detection (TLD) [11] algorithm, considered one of the state-of-the art methods for this task. TLD works by explicitly decomposing the long-term tracking task into tracking, learning and detection. The tracker follows the object (e.g., $I^*$) from frame to frame. The detector localizes all appearances that have been observed so far and corrects the tracker if necessary. The learning estimates detectors errors and updates it to avoid these errors in the future.

In this work, after successfully tracking $I^*$ in each frame, a region of size $n \times n$ is automatically cropped around the tracked location. This region is called the search box $I_{SB}$ and it serves as an input for the classification step.

The classification process is for determining if $I^*$ is present or not in each frame of the video. This is done by first correlating $I^*_f$ within $I_{SB}$ and then applying a threshold to the correlation output.

After applying the correlation filter, a threshold $k \in [0, 1]$ is used to determine if $I^*_f$ is recognized within $I_{SB}$ in each frame of the video. This is achieved by comparing $P_{max}$ with $k$; for example, if $P_{max} > k$ the $I^*$ is recognized $R_{I^*}$ in its respective frame of the video sequence, if $P_{max} \leq k$ the $I^*$ is not recognized.

### 5.3 Gaze Direction Estimation

After the FR stage, the GDE process is applied. The proposed methodology for GDE is based on template matching by finding first the head pose and then the gaze direction considering three possibilities: left, center, or right. This is done by creating a dataset with templates of different head poses **H** and for each of them there is a pair of eye templates $\mathbf{ER_o}$ and $\mathbf{EL_o}$ of different gaze directions. Then, a SDF filter is constructed for each template of the dataset and each are used for gaze detection. The GDE are divided into two phases: dataset and gaze detection. In the dataset phase describes how the dataset is build and in the gaze detection phase is how this dataset is used.

The images of the dataset are designed with an artificial face created with the open source 3D graphics and animation software blender [9]. The artificial face allows us to generate different gaze directions evenly without the error or bias that might be introduced by human subjects, such as those seen in the Colombia gaze dataset [19].

The dataset is divided into three subsets and are defined as follows:

1. Let $\mathbf{H} = \{H_o; o = 1, \ldots, N_h\}$ be a set of $N_h$ image templates representing three different head poses. The head poses are left, center, or right relative to the horizontal axis of the head.
2. Let $\mathbf{EL_o} = \left\{EL_{o,i}; o = 1, \ldots, N_h; i = 1, \ldots, N_h\right\}$ be a set of $N_h \times N_i$ image templates representing three different gaze directions of the left eye for each of the head poses $H_o$. The gaze directions of the left eye are left, center, or right relative to the horizontal axis of the eye.
3. Let $\mathbf{ER_o} = \left\{ER_{o,e}; o = 1, \ldots, N_h; e = 1, \ldots, N_h\right\}$ be a set of $N_h$ image templates representing three different gaze directions of the right eye for each of the head poses $H_i$. The gaze directions of the right eye are left, center, or right relative to the horizontal axis of the eye.

In the Gaze detection phase the dataset is used to find the head pose and gaze direction of a person. This is done by the following steps:

1. Starting from the recognized image $R_{I^*}$, the eyes of the person are detected to provide information of the location of each eye, this is done using the Viola-Johns detector [22] with the cascade classifier trained with eye images[6]; with this information a region of $n_{eyes} \times n_{eyes}$ and $n_{face} \times n_{face}$ are cropped automatically for the eyes and the face respectively.
2. The cropped region of the face and eyes of the recognized image $R_{I^*}$ are conditioned to obtain the same scale and rotation as the dataset, this is to ensure an optimal correlation between the two images or otherwise there will be an error in the values used for the set matching process. The conditioning of the scale is done by matching the distance between each eye from the recognized image and the distance from the eyes of the known dataset, and the conditioning of the rotation is done by matching the angle formed between the eyes and the horizontal axis of the image.
3. Find the pose of the head by correlating each image of the head pose set in the cropped region of the face of the recognized image. The image that produces the highest $P_{max}$ indicates the head pose.
4. Depending of the head pose $H_o$, select the corresponding left $EL_{o,i}$ and right $ER_{o,e}$ eye set.

[6]http://docs.opencv.org/2.4/modules/objdetect.

5. Find the gaze direction by correlating each image of the left $EL_{o,i}$ and right $ER_{o,e}$ eye set in the cropped region of the eyes of the recognized image $R_{I*}$. The images of each set $EL_{o,i}$ and $ER_{o,e}$ produces $C_{EL_{o,i}}$ and $C_{ER_{o,e}}$ correlation value respectively. The pair with the highest sum indicates the estimated gaze direction $Csum_{i,e} = C_{EL_{o,i}} + C_{ER_{o,e}}$ For example, for the pair ($C_{EL_{1,1}} = 0.1, C_{ER_{1,1}} = 0.5$) the $Csum_{1,1} = 0.6$ and ($C_{EL_{1,2}} = 0.2$, $C_{ER_{1,2}} = 0.5$) the $Csum_{2,2} = 0.7$, then the pair $i = 2$ and $e = 2$ achieved the highest $Csum$ value, assuming that 2 represents the code center, the estimated gaze direction is center.

After the final step, the estimated gaze direction will indicate if the person is looking at the left, center, or right for each frame of the video sequence. Then, these results will be stored as a code in a list with the time-stamp; the time is calculated by dividing the number of the frame when the code occur by the frame rate of the video sequence, e.g., for code left at the 50th frame with a 25 frame rate of the video, the time of the code is 2 s. Finally, the list is ordered in seconds by sampling the mode of what occurred in a second.

# 6 Experiments and Results

The CV system is implemented in C++ using the OpenCV v2.4.2 library in Linux Ubuntu 12.04. The TLD algorithm implementation used is from the OpenTLD[7] by Zdenek Kalal. The algorithms were executed on a PC with AMD Turion(tm) II Dual-Core Mobile M520 × 2.

Our results indicate that the use of CV is reliable to detect and code the behavior of *Attention*. The results are divided into two groups for the test and real video that are the GDE and the coding, and each are presented graphically in a plot (e.g., Fig. 6). Figure 6a, c are the GDE plot and they shows the manual coding as the dotted line and the CV algorithm coding as the solid line. The vertical axis shows the value NAN for not recognized, left, center and right direction. The horizontal axis represents time. Figure 6b, d are the coding plot and shows the manual and CV algorithm coding similarly to the GDE plot, but the vertical axis shows the values of *No Attention* and *Attention*. In the best case scenario for both plots, the CV system coding should be close to the manual coding.

In Fig. 6, it seems that all plots show a close result between the manual and the CV algorithm coding. This can be confirmed with numerical results, calculating the error and the $\tilde{\chi}^2$ test for each plot, these results are shown in Table 1. The test video results achieved the lowest error, in particularly the coding of *Attention* and *No Attention* with about of 15% of error, whereas the real video result achieved a slightly larger error of 26%. However, on the $\tilde{\chi}^2$ test results, both videos achieved a $\tilde{\chi}^2$ value lower than the critical value at the significance level of $\alpha = 0.05$, thus failing to reject the

[7] https://github.com/zk00006/OpenTLD.

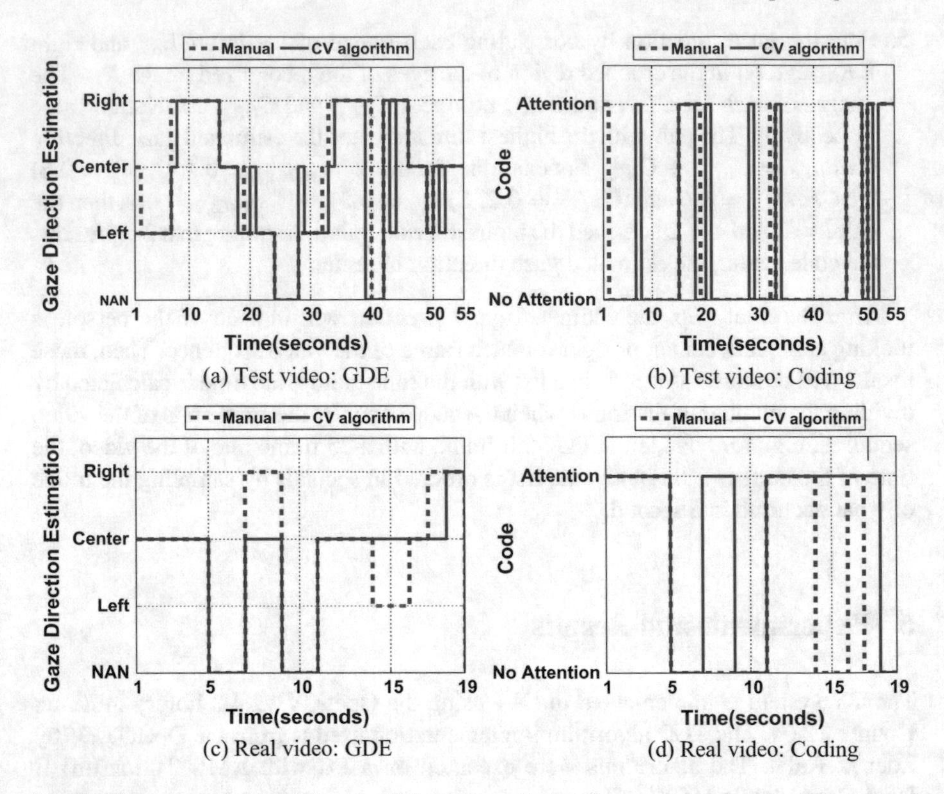


(a) Test video: GDE

(b) Test video: Coding

(c) Real video: GDE

(d) Real video: Coding

**Fig. 6** Illustration of the GDE and codification results for the test and real video. The *top* and *bottom rows* are the test and real video respectively, and the *right* and *left column* are the GDE and the codification, respectively

null hypothesis, meaning that the manual coding (expected value) is equal to the CV coding (observed).

In addition, we can examine the performance of the CV classifier by counting the number of actual and predicted classifications using the confusion matrix, as shown in Table 2, where each subtable presents a confusion matrix of the GDE and codification results for the test and real video. Ordered same as Fig. 6, the top and bottom rows are the test and real video respectively, and the right and left column are the GDE and the codification, respectively. In the best case scenario for each subtable, there should be zero every were except in the diagonal of the table (cells that match actual and predict class).

In Table 2(a), for the NAN class, from 6 cases the CV classifier predicted all 6 correctly; for the Left class, from 15 cases the CV classifier predicted 2 for the Center class; for the Center class, from 15 cases the CV classifier predicted 4 for the Left class; for the Right class, from 19 cases the CV classifier predicted 1 for the NAN class, 1 for the Left class and 2 for the Center class. From this matrix we can observe

**Table 1** Numerical result of the $\tilde{\chi}^2$ test and total error for the codification of the test and real videos

| Video | Test | | Real | |
|---|---|---|---|---|
| Groups | GDE | Coding | GDE | Coding |
| Error(%) | 18.1818 | 14.5454 | 26.3157 | 26.3157 |
| $\tilde{\chi}^2$ test | 1.6087 | 0 | 7.375 | 5.3977 |

that the CV classifier has more difficulty distinguishing the Left and Right classes with the Center class.

In Table 2(b), for the *No Attention* class, from 40 cases the CV classifier predicted 4 for the *Attention* class; for the *Attention* class, from 15 cases the CV classifier predicted 4 for the *No Attention* class. From this matrix we can observe that the CV classifier has difficulty distinguishing between *No Attention* and *Attention* classes.

In Table 2(c), for the NAN class, from 5 cases the CV classifier predicted all 5 correctly; for the Left class, from 2 cases the CV classifier predicted all for the Center class; for the Center class, from 8 cases the CV classifier predicted all correctly; for the Right class, from 4 cases the CV classifier predicted 3 for the Center class. From this matrix we can observe the same problems as in Table 2(a).

In Table 2(d), for the *No Attention* class, from 11 cases the CV classifier predicted 5 for the *Attention* class; for the *Attention* class, from 8 cases the CV classifier predicted all correctly. From this matrix we can observe that the CV classifier has a slight problem distinguishing the *No Attention* class.

In general from Table 2, the Center class leads to misclassifications in the GDE and codification results for the test and real video (let us not forget that the *Attention* class is in a sense the Center class), this can be due to multiple reasons that will be discussed below.

Let us now take a more detailed view of the general results summarized in Fig. 6 and Table 1. Figure 7 shows different frames of the GDE result for the test video (a) and real video (b). Each figure shows 5 frames, each coded manually, with the corresponding time-stamp of each frame given numerically at the top. Below each video frame we show the corresponding result given by the CV system; the false × and true ✓ check marks indicate if there is a match between the manual coding and the result of the CV system.

In Fig. 7a there are 4 correct matches and 1 false. The true matches (at seconds 2, 7 and 29) are visually correct, but the frame at second 41 has no tag assigned by the CV system, that is because the face recognition step returned a *not recognized* due to the threshold that was chosen to eliminate head poses not considered in the database. The false match at second 1 is a case were the change of head pose and/or gaze direction appeared very fast. For example, Fig. 8 shows some of the frames (20 per second) that occur in during the first second of the video. This figure clearly shows that a lot of changes occurred during this short 1 s period, this makes the coding of such time periods much more complex and error prone, even for a human.

**Table 2** Subtables show the confusion matrix for each, the GDE and codification results for both, the test and real video. Ordered the same as Fig. 6, the top and bottom rows are the test and real video respectively, and the right and left column are the GDE and the codification, respectively

| Actual Class | Predicted Class: NAN | Left | Center | Right |
|---|---|---|---|---|
| NAN | 6 | 0 | 0 | 0 |
| Left | 0 | 13 | 2 | 0 |
| Center | 0 | 4 | 11 | 0 |
| Right | 1 | 1 | 2 | 15 |

(a) Test video: GDE

| Actual Class | Predicted Class: No-Attention | Attention |
|---|---|---|
| No-Attention | 36 | 4 |
| Attention | 4 | 11 |

(b) Test video: Coding

| Actual Class | Predicted Class: NAN | Left | Center | Right |
|---|---|---|---|---|
| NAN | 5 | 0 | 0 | 0 |
| Left | 0 | 0 | 2 | 0 |
| Center | 0 | 0 | 8 | 0 |
| Right | 0 | 0 | 3 | 1 |

(c) Real video: GDE

| Actual Class | Predicted Class: No-Attention | Attention |
|---|---|---|
| No-Attention | 6 | 5 |
| Attention | 0 | 8 |

(d) Real video: Coding

In Fig. 7b there are 3 correct matches and 2 errors. The correct matches occur at time-stamps at second 3, 11 and 18, we can visually see they are correct even with slight changes in head poses compared to the images in Fig. 7a. This shows part of the robustness of the algorithm to changes in head poses. The false matches at second 7 and 15 are cases where the conditioning step fails to correct the rotation due to error produced by the Viola-Johns detector, and thus leads to errors in coding.

Also of interest is the time researchers take to code the videos manually, compared to the proposed CV system, this comparison is shown in Table 3. The CV system is substantially faster in both test videos. The difference is larger on the test video because the resolution is smaller compared to the real video and therefore can be processed faster by the CV system. However, it is important to point out that our current implementation should not be considered to be optimal in any way, the CV system could be made faster and more efficient by optimizing our implementation, using parallel hardware, or simply by using a faster computer. On the other hand, the manual time is basically fixed.

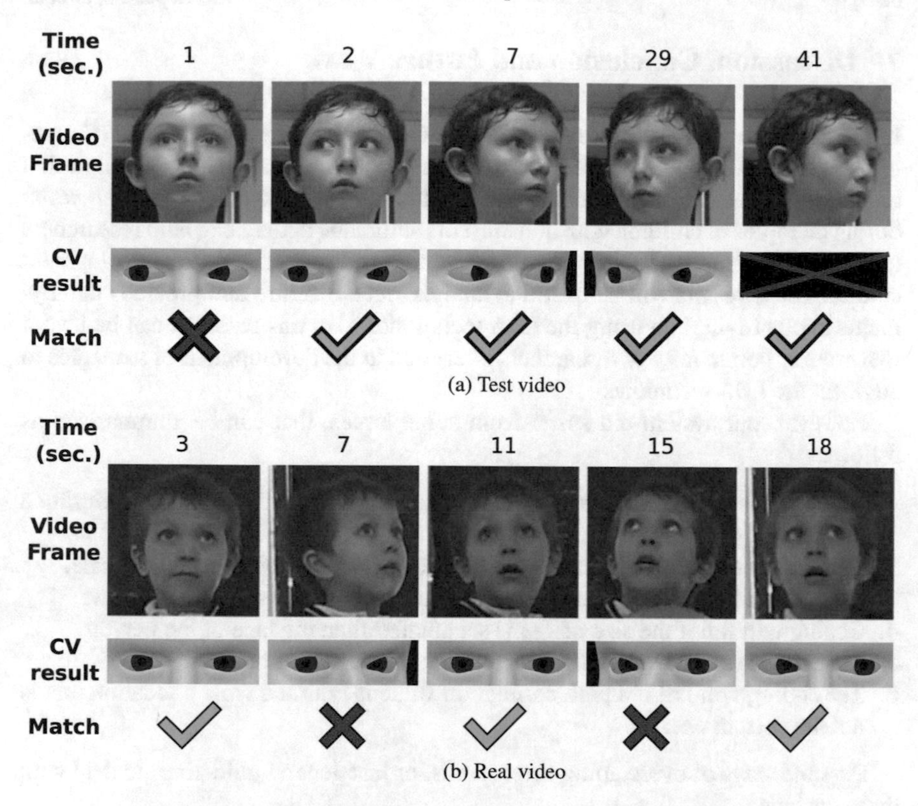


**Fig. 7** Illustration of the frame GDE result for the test video (**a**) and real video (**b**)

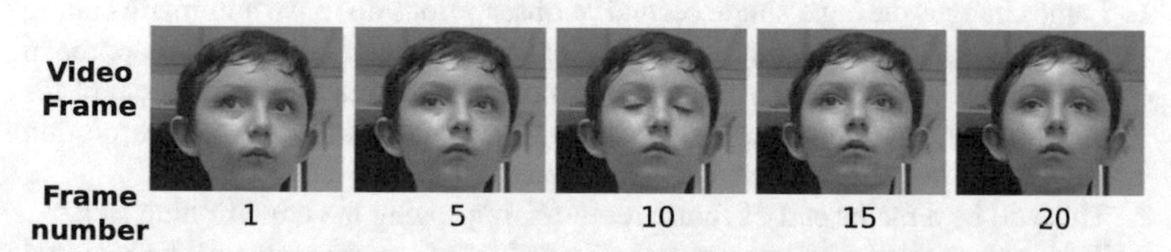


**Fig. 8** Illustration that shows the frame 1, 5, 10, 15 and 20 that correspond to the second 1 of the test video

**Table 3** Numerical result of the time that takes to code the Test and Real video by a person (manually) and by the proposed system (CV algorithm)

| | Coding Time (min.) | |
|---|---|---|
| Video | Manual | CV algorithm |
| Test | 25 | 4.3 |
| Real | 10 | 6.2 |

## 7 Discussion, Conclusion and Future Work

In this research we have presented a study where we explore the use of CV techniques to augment the LSA technique to automatically code behaviors. We show that the use of CV is sufficiently robust and reliable to detect the behavior of *Attention* during therapies of children with autism. This automatic coding can help researchers to reduce significantly the time invested in coding and analyze videos of deployment studies. These results will be useful to analyze data of future and previous deployments in our LivingLab using the LSA technique. Also this research can be useful to a broader community of researchers interested in the development of strategies to augment the LSA technique.

However, our system did suffer from some issues, that can be summarized as follows:

1. Coding fails when the change of head pose and/or gaze direction is fast during a one second period.
2. In some cases the conditioning step fails to correct the rotation and/or scale.
3. Coding can fail if the $k$ threshold of the face recognition step is set to low.
4. Coding will fail if the size of $I_{SB}$ is set smaller than the face of the person.
5. Sometimes the TLD tracking fails to track the person of interest.
6. The coding will fail if a high correlation value is returned from a location that is not the correct one.

Possible ways of overcoming these issues, or just general guidelines to deal with them, include:

1. Large changes during a single second of observations do make automatic coding a bit less accurate, but this is also the case for manual coding. One approach would be to consider each frame independently, but this is too much detail for the goal of LSA. This issue just seems to be a problem of the desired resolution of analysis, not of the automatic CV system.
2. This will be a major part of future research, improving the conditioning step.
3. This is an empirical parameter, and a larger set of experiments will be required to tune it.
4. This is due to the face detection algorithm, other algorithms will be tested to achieve an improvement.
5. We are also considering the use of more robust tracking algorithms, but not many exist in current literature.
6. We can enhance the composite filters by including examples of the negative class during filter construction [6].

**Acknowledgements** We thank participants in this work and the availability to use the data for this work. Also the grants SEP-TecNM (México) 5620.15-P and 5621.15-P, CONACYT (México) Basic Science Research Project No. 178323, and the FP7-Marie Curie-IRSES 2013 European Commission program through project ACoBSEC with contract No. 612689. First author was supported by CONACYT doctoral scholarship No. 302532.

## References


1. Bolme, D.S., Beveridge, J.R., Draper, B.A., Lui, Y.M.: Visual object tracking using adaptive correlation filters. In: 2010 IEEE Conference on Computer Vision and Pattern Recognition (CVPR), pp. 2544–2550 (2010)
2. Caro, K.: Exergames for children with motor skills problems. SIGACCESS Access. Comput. **108**, 20–26 (2014)
3. Consolvo, S., Arnstein, L., Franza, B.R.: User study techniques in the design and evaluation of a ubicomp environment. In: Proceedings of the 4th International Conference on Ubiquitous Computing. UbiComp '02, pp. 73–90. Springer, London (2002)
4. Escobedo, L., Ibarra, C., Hernandez, J., Alvelais, M., Tentori, M.: Smart objects to support the discrimination training of children with autism. Pers. Ubiquit. Comput. **18**(6), 1485–1497 (2013)
5. Escobedo, L., Tentori, M., Quintana, E., Favela, J., Garcia-Rosas, D.: Using augmented reality to help children with autism stay focused. IEEE Pervasive Comput. **13**(1), 38–46 (2014)
6. Gaxiola, L., Díaz-Ramírez, V.H., Tapia, J., Diaz-Ramirez, A., Kober, V.: Robust face tracking with locally-adaptive correlation filtering. In: Bayro-Corrochano, E., Hancock, E. (eds.) Progress in Pattern Recognition, Image Analysis, Computer Vision, and Applications. Lecture Notes in Computer Science, vol. 8827, pp. 925–932. Springer International Publishing (2014)
7. Gibbs, G.R.: Computer assisted qualitative data analysis: Nvivo, maxqda, atlas.ti, qdaminer, hyperresearch. In: IfM?s 21st Annual Research Methodology Workshop (2014)
8. Gunter, P.L., Jack, S.L., Shores, R.E., Carrell, D.E., Flowers, J.: Lag sequential analysis as a tool for functional analysis of student disruptive behavior in classrooms **1**(3), 138–148 (1993)
9. Hess, R.: Blender Foundations: The Essential Guide to Learning Blender 2.6. Focal Press (2010)
10. Javidi, B., Wang, J.: Design of filters to detect a noisy target in nonoverlapping background noise. J. Opt. Soc. Am. A **11**(10), 2604–2612 (1994)
11. Kalal, Z., Mikolajczyk, K., Matas, J.: Tracking-learning-detection. IEEE Trans. Pattern Anal. Mach. Intell. **34**(7), 1409–1422 (2012)
12. Kerekes, R.A., Vijaya Kumar, B.V.K.: Selecting a composite correlation filter design: a survey and comparative study. Opt. Eng. **47**(6), 067,202–067,202–18 (2008)
13. Klonek, F.E., Lehmann-Willenbrock, N., Kauffeld, S.: Dynamics of resistance to change: a sequential analysis of change agents in action. J. Change Manag. **14**(3), 334–360 (2014)
14. Nowozin, S., Lampert, C.H.: Structured learning and prediction in computer vision. Found. Trends. Comput. Graph. Vis. **6**(3–4), 185–365 (2011)
15. Preece, J., Rogers, Y., Sharp, H.: Beyond Interaction Design: Beyond Human-Computer Interaction. Wiley, New York (2001)
16. Ringland, K.E., Zalapa, R., Neal, M., Escobedo, L., Tentori, M., Hayes, G.R.: Sensorypaint: a multimodal sensory intervention for children with neurodevelopmental disorders. In: Proceedings of the 2014 ACM International Joint Conference on Pervasive and Ubiquitous Computing. UbiComp '14, pp. 873–884. ACM, New York, NY, USA (2014)
17. Ringland, K.E., Zalapa, R., Neal, M., Escobedo, L., Tentori, M., Hayes, G.R.: Sensorypaint: a natural user interface supporting sensory integration in children with neurodevelopmental disorders. In: CHI '14 Extended Abstracts on Human Factors in Computing Systems. CHI EA '14, pp. 1681–1686. ACM, New York, NY, USA (2014)
18. Sanderson, P., Scott, J., Johnston, T., Mainzer, J., Watanabe, L., James, J.: Macshapa and the enterprise of exploratory sequential data analysis (esda). Int. J. Hum.-Comput. Stud. **41**(5), 633–681 (1994). doi:10.1006/ijhc.1994.1077
19. Smith, B., Yin, Q., Feiner, S., Nayar, S.: Gaze Locking: Passive Eye Contact Detection for HumanObject Interaction. In: ACM Symposium on User Interface Software and Technology (UIST), pp. 271–280 (2013)
20. Tentori, M., Escobedo, L., Balderas, G.: A smart environment for children with autism. IEEE Pervasive Comput. **14**(2), 42–50 (2015)

21. Viera, A., Garrett, J.: Understanding interobserver agreement: the kappa statistic. Fam. Med. **37**(5), 360–363 (2005)
22. Viola, P., Jones, M.: Rapid object detection using a boosted cascade of simple features. In: Proceedings of the 2001 IEEE Computer Society Conference on Computer Vision and Pattern Recognition, 2001. CVPR 2001, vol. 1, pp. I–511–I–518 (2001)
23. Weiser, M.: Some computer science issues in ubiquitous computing. Commun. ACM **36**(7), 75–84 (1993)
24. Zalapa, R., Tentori, M.: Movement-based and tangible interactions to offer body awareness to children with autism. Ubiquitous Computing and Ambient Intelligence. Context-Awareness and Context-Driven Interaction: 7th International Conference, UCAmI 2013, Carrillo, Costa Rica, December 2-6, 2013, Proceedings, pp. 127–134. Springer International Publishing (2013)

# Mixed Integer Programming Formulation for the Energy-Efficient Train Timetables Problem

**Rodrigo Alexander Castro Campos, Sergio Luis Pérez Pérez, Gualberto Vazquez Casas and Francisco Javier Zaragoza Martínez**

**Abstract** Railway traffic is the biggest individual electricity consumer in Germany, amounting to 2% of the country's total electricity usage. However, up to 20% of the annual electricity cost depends on the highest power value drawn within the billing period. In this paper, we optimize the timetables of railway traffic in order to avoid high peaks in power consumption, while preserving at the same time some usability and safety considerations. We propose an exact mixed integer programming model together with a systematic way of simplifying the model in order to obtain feasible solutions that are not far from the optimum. We also discuss two possible dynamic programming approaches that may be used for solving small instances with a specific structure. Our approach became *Team Optimixtli*'s winning entry in the *Discrete Optimization Challenge: Energy-Efficient Train Timetables*. This competition was part of the Open Research Challenge 2015 organized by the Friedrich-Alexander-Universität Erlangen-Nürnberg (FAU) in Germany.

## 1 Introduction

German railway traffic operates 98% of its long-haul traffic with electric trains. This amounts to 2% of the Germany's total electricity usage and is also the biggest individual electricity consumer in the country. However, up to 20% of the annual

R.A. Castro Campos (✉) · S.L. Pérez Pérez · G. Vazquez Casas
Graduate Program in Optimization, Universidad Autónoma Metropolitana
Azcapotzalco, Mexico City, Mexico
e-mail: racc@correo.azc.uam.mx

S.L. Pérez Pérez
e-mail: slpp@correo.azc.uam.mx

G. Vazquez Casas
e-mail: gvc@correo.azc.uam.mx

F.J. Zaragoza Martínez
Departamento de Sistemas, Area de Optimización Combinatoria, Universidad Autónoma Metropolitana Azcapotzalco, Mexico City, Mexico
e-mail: franz@correo.azc.uam.mx

© Springer International Publishing AG 2018

Y. Maldonado et al. (eds.), *NEO 2016*, Studies in Computational Intelligence 731,
https://doi.org/10.1007/978-3-319-64063-1_3

electricity cost depends on the highest power value drawn within the billing period. This is done in order to promote a balanced distribution of power consumption [5].

The electricity consumed by the railway traffic comes from two main sources: self-generated energy and public electricity producers [6]. Self-generation of energy occurs during the braking of the trains in the railway. While this energy is not stored due to security reasons, it can be safely transferred to other trains in the system that are accelerating (or otherwise consuming energy) at the same time when the energy is produced. In order to use most of the generated energy and also flatten out the highest peak of energy usage within a given period, the accelerating trains must be synchronized with the braking trains. To achieve this synchronization, we need to compute an efficient timetable that also respects safety constraints and preserves the quality of the railway traffic service to the average user.

In this paper, we consider the Energy-Efficient Train Timetables Problem, which consists in modifying a given timetable in order to minimize the highest peak of power consumption. This problem was presented by the Friedrich-Alexander-Universität Erlangen-Nürnberg (FAU), in the Discrete Optimization challenge of the Open Research Challenge 2015 [2]. In this problem, we are able to alter the departure times of the trains, but the travel times and the order in which trains arrive to the train station are fixed. Similar problems have been considered in different contexts, like the Montréal metro system or when travel times can be modified [1, 7]. A technical report from one of the participating teams is also available [4].

In this paper we present our winning approach to the Discrete Optimization challenge of the Open Research Challenge 2015. The paper is organized as follows. Section 2 is a very slightly edited version of the challenge as found on FAU Open Research Challenge's website,[1] it is included here only for completeness. Section 3 describes an exact mixed integer programming formulation for the optimization problem, the process of transforming the given data into a suitable format for a MIP solver, the instances, and the results (either optimal or failures) obtained. Section 4 discusses several proposals of how to obtain good solutions in those cases where the MIP solver failed. Section 5 describes our main proposal (which we call the *constant granularity* mixed integer programming formulation), the obtained solutions, and the further improvements. Section 6 lists our best results and Sect. 7 contains our conclusions. Finally, in the Appendix we display graphically our best obtained results.

## 2 The Challenge

Railway traffic is the biggest individual electricity consumer in Germany. It amounts to 2% of the country's total electricity usage, which equals the consumption of the city of Berlin. However, the annual electricity cost (1 billion Euros per year) is not only determined by the total amount of energy used. Electricity contracts of big

[1] https://openresearchchallenge.org/discreteOptimization/ChairofEconomics/The+Challenge.

customers normally also incorporate a price component that is dependent on the highest power value drawn within the billing period, i.e. it depends on the timely distribution of the energy usage. This power component accounts for up to 20% of the energy bill.

In the FAU Open Research Challenge *Energy-Efficient Train Timetables*, the aim was to optimize the timetables of railway traffic in order to avoid high peaks in power consumption as they lead to high electricity costs. This can be done by desynchronizing too many simultaneous departures as well as synchronizing accelerating trains with braking trains to make better use of the recuperated braking energy. Starting with a raw timetable (before its publication), the task is to shift the departures of the trains from the stations within small time intervals (several minutes) to come to the final optimized timetable.

## 2.1 The Task

The goal is to compute optimal timetables with respect to the above goal. To this end, we were given problem instances featuring the following data:

1. A railway network in the form of a directed multi-graph.
2. The planning horizon under consideration.
3. A set of trains to schedule, and for each such train:

   a. the intermediate stations and tracks traveled,
   b. feasible departure intervals for each departure from a station,
   c. minimum stopping times in the stations (for getting on and off),
   d. travel times for the tracks,
   e. power profiles for each track traveled, i.e. a time-power-curve.

4. For each track in the network, the order in which it is passed by which trains.
5. The necessary safety distances.

The task is to ensure a favorable system-wide power load over the given planning horizon to minimize the power component of the electricity bill. The system-wide power load is determined by summing up all the individual power profiles of the trains under consideration. Obviously, this total power profile changes when shifting the trains in time. To minimize the power cost of the railway system, we have to minimize the average power drawn by all trains together over any consecutive 15-minute interval within the planning horizon.

To achieve this, the task is to come up with models and algorithms to optimize the departure times of the trains from the stations, keeping to the following requirements:

1. Each train travels the stations and tracks in the order given in data.
2. Each train departure is within the feasible interval for the corresponding station.
3. The minimum stopping time at each station is respected.

4. The trains pass the tracks in the given order and respect the safety distances.
5. Passenger connections at the stations are respected.

The winning team for a given problem instance would be the one with the lowest 15-minute average of power consumption over the planning horizon. This value is summed up over the 10 instances given in the contest. To win the whole contest, a team would need the lowest overall sum over these 10 instances.

## 2.2 Description of the Optimization Problem

The degrees of freedom in this optimization problem are the departure times of the trains from the stations. As a part of each problem instance, we were given a series of train runs. Each of them consists of a series of legs, i.e. journeys between two consecutive stations at which the train stops. We have to adapt the current departure times to form a new timetable which has to respect several side constraints:

Maximal deviation from the current timetable: Each leg has an earliest and a latest possible departure time. The new departure has to lie in this interval.

Minimal stopping time: The description of each leg contains a minimum stopping time. This is the minimal time (in minutes) which the train has to wait at the destination of the leg before it can depart for the next leg. The corresponding value for the final leg can be ignored.

Safety distances: Each leg possesses a safety distance value (in minutes). This is the minimal time which the subsequent train passing the same track has to wait before it can depart for the corresponding leg. The order of the trains passing a given track has to be retained.

Passenger connections: At each station, we have to ensure that the passenger connections established in the original timetable are retained. If the arrival time of one train and the departure time of another lie in an interval of 5–15 min, their new arrival and departure times have to lie in this interval, too.

Under these timetable constraints, we have to find a new timetable that minimizes the system-wide power cost. The description of each leg contains the time-power curve induced by the corresponding train. We assume that this curve as well as the travel time of the leg are fixed. Summation of the power curves for all trains over the complete planning horizon (starting at minute 0 and ending with the latest possible arrival time) leads to the system-wide power curve. Its power cost is determined by averaging its values over 15-minute intervals. The first such interval starts at minute 0, the next one at minute 15, etc. The power cost is now given by the value of the highest such 15-minute average multiplied by a constant cost factor. That means the objective is to minimize the highest arising 15-minute average of the system-wide power curve.

The solutions have to fulfill the following requirements in addition to the above rules:

1. Trains may only depart at full minutes in the planning horizon (which starts at minute zero, second zero). In the timetable data, we are given an earliest and a latest departure time for each leg (given in full minutes). The departure time for this leg has to be a full minute from this interval.
2. When computing the 15-minute averages, we have to set negative values of the summed, system-wide power curve to zero. If braking trains provide more energy than accelerating trains can use at the same time, this energy is lost. That means: One calculates the system-wide curve by summing over the individual curves of the trains, depending on their departure times. Then one computes the maximum of this curve and zero. And then one averages over 15-minute intervals – the first of them starting at minute zero, second zero, and ending at minute 15, second zero, interval ends including (the next would start at minute 15, second zero and end at minute 30, second zero).
3. The calculation of the 15-minute average power values follows the trapezoidal rule for the approximate calculation of integrals as described in the following. Each 15 min-interval consists of 901 seconds (as the last second is also part of the next interval). Thus, to compute the average of the system-wide power curve, we sum over the 899 *inner* seconds of the interval (with weight 1) and add the value for the first and the last second with a weight of $\frac{1}{2}$. This value is then divided by 900 – the length of the interval. The resulting value is the 15 min average which is to be optimized (Explanation: without the division by 900, this value represents the total energy consumed by the trains over that interval – under the assumption that the system-wide power curve is a piecewise-linear function. Division by the length of the time interval yields the average power drawn from the system).

## 3 Mixed Integer Program Formulation

We describe the instances of the challenge, an exact mixed integer programming formulation for the optimization problem, and the results (either optimal or failures) obtained. A similar formulation was given by Haahr and Kidd [4].

### 3.1 The Instances

The challenge contains ten instances with the statistics shown in Table 1: the number of trains, the number of legs, the number of tracks, and the number of stations.

Each instance consists of two JSON files with the same number:

1. `instance_data_number.json.txt` contains the timetable information and
2. `power_data_number.json.txt` contains the power profiles for each leg.

**Table 1** Main statistics for the ten instances

| Instance | Trains | Legs | Tracks | Stations |
|---|---|---|---|---|
| 1 | 13 | 206 | 58 | 30 |
| 2 | 26 | 241 | 78 | 43 |
| 3 | 37 | 352 | 139 | 76 |
| 4 | 48 | 676 | 150 | 84 |
| 5 | 71 | 863 | 274 | 131 |
| 6 | 73 | 712 | 221 | 109 |
| 7 | 126 | 1524 | 308 | 147 |
| 8 | 182 | 3709 | 100 | 51 |
| 9 | 237 | 1808 | 605 | 265 |
| 10 | 277 | 2641 | 666 | 306 |

The information on file `instance_data_number.json.txt` is organized with the following fields:

1. `int TrainID` uniquely identifies each train.
2. `int TrackID` uniquely identifies each track.
3. `int LegID` uniquely identifies each train run on a given track.
4. `int StartStationID` uniquely identifies the origin of the leg.
5. `int EndStationID` uniquely identifies the destination of the leg.
6. `int EarliestDepartureTime` is the earliest departure time of the leg.
7. `int LatestDepartureTime` is the latest departure time of the leg.
8. `int CurrentDepartureTime` is the current departure time.
9. `int TravelTime` is the travel time to the next station in minutes.
10. `int MinimumStoppingTime` is the minimum time the trains needs to stay in the following station (if one exists) in minutes.
11. `int MinimumHeadwayTime` is the minimum time the next train on the same route has to wait until it can depart.

These data are subject to the following constraints:

1. The trains can only depart every full minute from the `EarliestDepartureTime` until `LatestDepartureTime`.
2. Two consecutive trains using the same track have to satisfy a safety constraint. The departure of the later train must be greater than or equal to the `DepartureTime` of the earlier train plus the minimum headway time specified for the earlier train.
3. This constraint implicitly enforces the same order of the trains on each track as established in the original timetable. That means the order of the trains is fixed and is not part of the optimization.
4. At each station, the new timetable has to respect the passenger connections established in the original timetable. If the arrival of one train at a given station and the departure of another train take place within an interval of 5 to 15 min in the old timetable, this relation has to be preserved in the new timetable.

The information on file `power_data_number.json.txt` is organized with the following fields:

1. `int LegID` correspond to the same `LegID` as in the `instance_data_number.json.txt` and is unique.
2. `float Powerprofile[]` contains exactly `TravelTime*60+1` time steps (seconds), where the power is measured (in MW).

## 3.2 Constants and Decision Variables

The general parameters for a given instance $\mathscr{I}$ are:

1. Let $T$ be the number of *trains* and $\mathscr{T}$ be the set of trains.
2. Let $L$ be the number of *legs* and $\mathscr{L}$ be the set of legs.
3. Let $K$ be the number of *tracks* and $\mathscr{K}$ be the set of tracks.
4. Let $S$ be the number of *stations* and $\mathscr{S}$ be the set of stations.

For a given leg $\ell \in \mathscr{L}$, we define some constants:

1. Let $s_\ell \in \mathscr{S}$ be its *source* (origin) station.
2. Let $t_\ell \in \mathscr{S}$ be its *target* (end, destination) station.
3. Let $e_\ell \geq 0$ be its *earliest departure* time.
4. Let $l_\ell \geq 0$ be its *latest departure* time.
5. Let $c_\ell \geq 0$ be its *current departure* time.
6. Let $r_\ell \geq 0$ be its *running* (travel) time.
7. Let $w_\ell \geq 0$ be its *minimum waiting* (stopping) time at the target destination.
8. Let $h_\ell \geq 0$ be the *minimum headway* time for the next train on the same track.
9. For $0 \leq i \leq 60r_\ell$, let $p_{\ell,i}$ be the power consumption measured at second $i$ from departure.

It is given that $e_\ell \leq c_\ell \leq l_\ell$ for all $\ell \in \mathscr{L}$. Finally, we define the decision variable $x_\ell \in \mathscr{Z}_+$ as the *actual departure* time for leg $\ell \in \mathscr{L}$.

## 3.3 Constraints

### 3.3.1 Departure Constraints

The following *departure constraints* must hold:

$$e_\ell \leq x_\ell \leq l_\ell \text{ for all } \ell \in \mathscr{L}. \tag{1}$$

### 3.3.2 Safety Constraints

For each track $k \in \mathscr{K}$, let $\ell_{k,1}, \ell_{k,2}, \ldots, \ell_{k,p}$ be the $p$ legs using track $k$ sorted by increasing departure time, that is, $c_{\ell_{k,1}} < c_{\ell_{k,2}} < \cdots < c_{\ell_{k,p}}$. Then the following *safety constraints* must hold:

$$x_{\ell_{k,i}} - x_{\ell_{k,i-1}} \geq h_{\ell_{k,i-1}} \text{ for all } 1 < i \leq p. \tag{2}$$

### 3.3.3 Waiting Constraints

For each train $t \in \mathscr{T}$, let $\ell_{t,1}, \ell_{t,2}, \ldots, \ell_{t,q}$ be the $q$ legs using train $t$ sorted by increasing departure time, that is, $c_{\ell_{t,1}} < c_{\ell_{t,2}} < \cdots < c_{\ell_{t,q}}$. Then the following *waiting constraints* must hold:

$$x_{\ell_{t,i}} - x_{\ell_{t,i-1}} \geq r_{\ell_{t,i-1}} + w_{\ell_{t,i-1}} \text{ for all } 1 < i \leq q. \tag{3}$$

### 3.3.4 Connection Constraints

For each station $s \in \mathscr{S}$, let $\ell^-_{s,1}, \ell^-_{s,2}, \ldots, \ell^-_{s,p}$ be the $p$ legs arriving to station $s$ and let $\ell^+_{s,1}, \ell^+_{s,2}, \ldots, \ell^+_{s,q}$ be the $q$ legs departing from station $s$. Then the following *connection constraints* must hold:

$$5 \leq x_{\ell^+_{s,j}} - (x_{\ell^-_{s,i}} + r_{\ell^-_{s,i}}) \leq 15 \text{ for all } 1 \leq i \leq p \text{ and } 1 \leq j \leq q, \tag{4}$$

provided that legs $\ell^-_{s,i}$ and $\ell^+_{s,j}$ correspond with different trains and their departure times satisfy $5 \leq c_{\ell^+_{s,j}} - (c_{\ell^-_{s,i}} + r_{\ell^-_{s,i}}) \leq 15$.

## 3.4 Objective Function

All given instances have a planning horizon of 17 fifteen-minutes intervals. For a given feasible solution $x$, let $f(x, i)$ be its average power consumption on interval $0 \leq i \leq 16$. The value of solution $x$ is then $f(x) = \max\{f(x, i) : 0 \leq i \leq 16\}$. Finally, the objective value is $z^* = \min\{f(x) : x \text{ is feasible}\}$. It is not completely obvious that $z^*$ is a linear function of $x$.

Assume first that we could write $f(x, i)$ as a linear function of $x$, then it would be easy to obtain $z^*$ in the following way:

$$z^* = \min z \tag{5}$$

$$\text{subject to}$$

$$f(x, i) \leq z \text{ for all } 0 \leq i \leq 16. \tag{6}$$

In order to write each $f(x, i)$ as a linear function of $x$, we proceed in three steps. First, for each $0 \leq s \leq 17 \times 15 \times 60$, let $\pi_s \geq 0$ be the total power consumption on second $s$ (multiplied by $\frac{1}{2}$ if $s$ is a multiple of 900). Therefore:

$$f(x, i) = \frac{1}{900} \sum_{s=900i}^{900(i+1)} \pi_s. \tag{7}$$

Second, for each leg $\ell \in \mathscr{L}$, and for each $e_\ell \leq m \leq l_\ell$, let $y_{\ell,m} \in \{0, 1\}$ be a binary variable indicating whether leg $\ell$ departed on minute $m$. This can be achieved as follows:

$$\sum_{m=e_\ell}^{l_\ell} y_{\ell,m} = 1 \tag{8}$$

$$\sum_{m=e_\ell}^{l_\ell} m y_{\ell,m} = x_\ell. \tag{9}$$

Note that the first equation above selects exactly one $y_{\ell,m}$ equal to one and all the others equal to zero, and therefore the second equation above assigns $x_\ell = m$.

Third, we need to relate the variables $\pi_s$ with the variables $y_{\ell,m}$. In particular, $y_{\ell,m} = 1$ contributes to the value of $\pi_s$ if $60m \leq s \leq 60(m + r_\ell)$. If $s$ is not a multiple of 900 we have:

$$\pi_s \geq \sum_{\ell \in \mathscr{L}} \left( \sum_{m=\lceil s/60 - r_\ell \rceil}^{\lfloor s/60 \rfloor} p_{\ell,s-60m} y_{\ell,m} \right). \tag{10}$$

Otherwise, if $s$ is a multiple of 900 we have:

$$\pi_s \geq \frac{1}{2} \sum_{\ell \in \mathscr{L}} \left( \sum_{m=\lceil s/60 - r_\ell \rceil}^{\lfloor s/60 \rfloor} p_{\ell,s-60m} y_{\ell,m} \right). \tag{11}$$

Note that in the inner sum, if $m < e_\ell$ or $m > l_\ell$ we can assume $y_{\ell,m} = 0$. Also note that these inequalities allow their right-hand sides to be negative, but $\pi_s \geq 0$.

### 3.5 Results

The resulting models for the ten instances have the number of constraints given in Table 2 of each type: departure, safety, waiting, and connection constraints.

We attempted to solve the models using the state-of-the-art MIP solver Gurobi [3]. Five instances were solved to optimality, while the other five instances quickly ran out of memory. The results are shown on Table 3: we give first the value of the

**Table 2** Number of constraints by type for each of the ten instances

| Instance | Departure | Safety | Waiting | Connection |
|---|---|---|---|---|
| 1 | 206 | 148 | 193 | 21 |
| 2 | 241 | 163 | 215 | 31 |
| 3 | 352 | 213 | 615 | 68 |
| 4 | 676 | 526 | 628 | 274 |
| 5 | 863 | 589 | 792 | 210 |
| 6 | 712 | 491 | 639 | 184 |
| 7 | 1524 | 1216 | 1398 | 1035 |
| 8 | 3709 | 3609 | 3527 | 17261 |
| 9 | 1808 | 1203 | 1571 | 985 |
| 10 | 2641 | 1975 | 2364 | 1773 |

**Table 3** The optimal results for instances 2–6

| Instance | Current value | Optimal value | Improvement (%) |
|---|---|---|---|
| 2 | 4.620948333 | 2.939638333 | 36.38 |
| 3 | 20.635706111 | 15.179507777 | 26.44 |
| 4 | 19.983620000 | 14.966736111 | 25.10 |
| 5 | 27.551612222 | 19.416015000 | 29.53 |
| 6 | 26.349433888 | 20.431125556 | 22.46 |

current (given) solution, then the optimal value obtained with our model, and finally the improvement as a percentage.

In their report [4], Haahr and Kidd were able to find (almost) optimal solutions in 8 out 10 instances, although they also ran into trouble with instance 1, the smallest.

# 4 Ideas for Finding Good Solutions

We sketch two main ideas of how to obtain good solutions and good lower bounds in those cases where the MIP solver failed. They are simplifications of the objective function and of the data. We also sketch a dynamic programming approach to solve the problem.

## *4.1 Simplification of Objective Function*

Our first simplification is straightforward: instead of computing the objective function with the trapezoidal rule, we scale the first and last power consumptions by half and then simply sum the power consumption over each 900 seconds interval. In other words, do not use Eq. 11 while looking for solutions. This simple operation removes some computational burden.

## 4.2 *Simplification of Power Consumption*

Our second simplification is also simple: instead of using the data for each second, group the data with certain granularity and average it. Hence, instead of having a variable for each second's power consumption, we would only have a variable for each subinterval. Of course, having fewer variables may lead to a faster solution. In all our figures, a red dot means power consumption, a blue dot means power generation. Figure 1 shows the power consumption wave of a typical leg. This particular leg is taken from instance 1.

An interesting property of granularity and averaging is as follows: Consider a fixed fifteen-minutes interval $i$ and a fixed schedule $x^i$ for the legs running during that interval. Partition interval $i$ into a set $A$ of subintervals and further refine this partition into a set $B$ of subintervals. Let $f_A(x^i, i)$ be the power consumption evaluated using power averages over each subinterval in $A$ and let $f_B(x^i, i)$ be the power consumption evaluated using power averages over each subinterval in $B$. Then $f_A(x^i, i) \leq f_B(x^i, i)$. Furthermore, if $x_A^*$ is a minimizer of $f_A$ and $x_B^*$ is a minimizer of $f_B$, then $f_A(x_A^*, i) \leq f_A(x_B^*, i) \leq f_B(x_B^*, i)$. In particular, for an arbitrary partition $A$ we have the lower bound $f_A(x_A^*, i) \leq f(x^*, i)$.

### 4.2.1 Constant Granularity

Since each interval consists of 900 seconds, it is natural to group the power consumption data in $\frac{900}{d}$ intervals of $d$ seconds. We discuss this idea further in the next section. Figure 2 shows the power consumption wave of Fig. 1 averaged every $d = 5$ seconds. Note that if $d_1 | d_2$ and $d_2 | 900$, then $f_{d_2}(x^i, i) \leq f_{d_1}(x^i, i)$. In particular, given a $d | 900$ we have $f_d(x_d^*, i) \leq f_1(x^*, i) = f(x^*, i)$. Hence, we have a lower bound to the original problem.

A very similar idea is also present in [4].

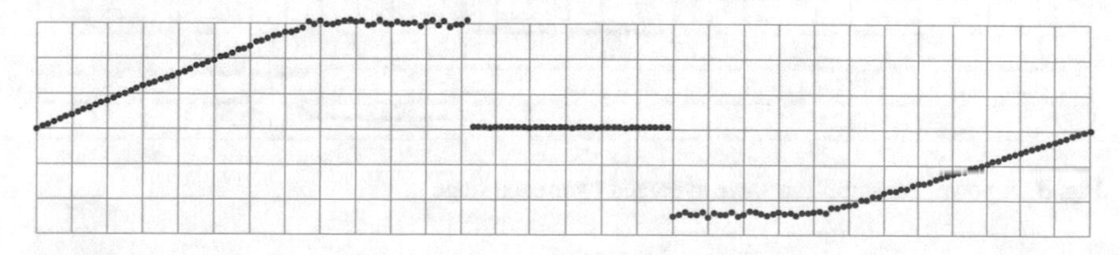

**Fig. 1** A typical power consumption wave

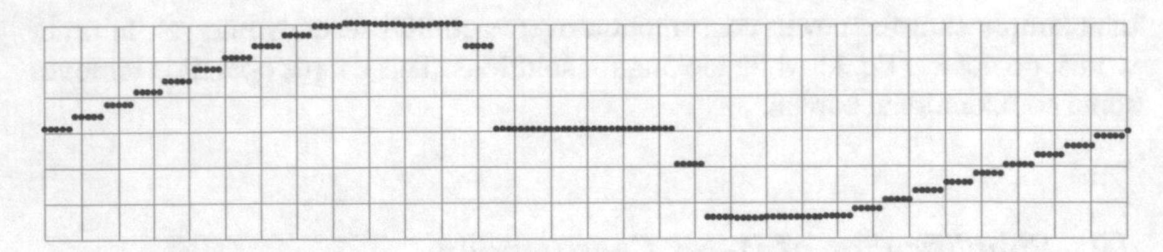

**Fig. 2** A power consumption wave averaged over 5 seconds subintervals

### 4.2.2 Variable Granularity

On further inspection of the power consumption data, we noted that each leg goes through five stages (or less) of power consumption:

1. Almost linearly increasing power consumption.
2. Almost constant maximum power consumption.
3. Almost constant cruise speed power consumption.
4. Almost constant maximum power generation.
5. Almost linearly decreasing power generation.

Therefore, it would also be natural to split the power consumption data over these five stages and average within each of them. We noted that sometimes only the first, third, and fifth stages occur. We also noted that sometimes the third stage generates power. Figure 3 shows the power consumption wave of Fig. 1 averaged over each stage.

### 4.2.3 Piecewise Linear Functions

Given the behavior described above, one could also model the power consumption as a piecewise linear function with five (or three) pieces. Each of these linear functions could be computed using, for instance, linear regression. Figure 4 shows the power consumption wave of Fig. 1 approximated by a linear function over each stage.

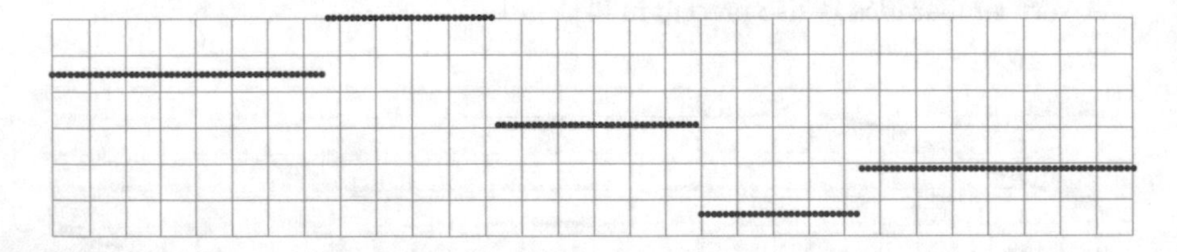

**Fig. 3** A power consumption wave averaged over each stage

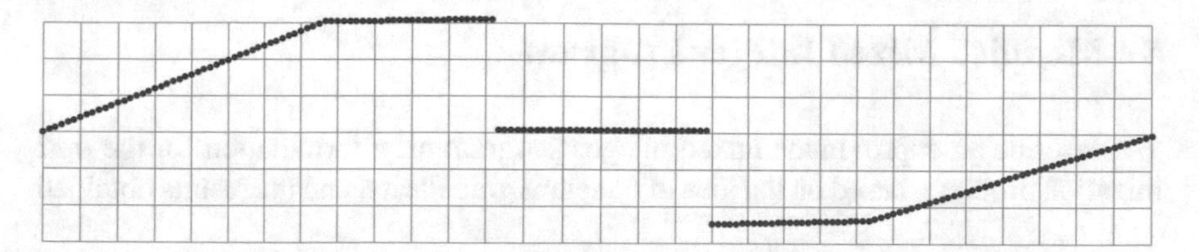

**Fig. 4** A power consumption wave approximated by a piecewise linear function

## 4.3 Dynamic Programming

For each $0 \leq i \leq 16$, let $\mathcal{L}_i$ be the set of legs that may be running during the $i$-th fifteen-minutes interval in the planning horizon and let $\mathcal{X}_i$ be the set of *feasible* assignments to the variables corresponding to legs in $\mathcal{L}_i$. Note that a leg may belong to one or more consecutive intervals. Also note that the power consumption during the $i$-th fifteen-minutes interval only depends on the assignment $x^i \in \mathcal{X}_i$. We say that $x^{i-1} \in \mathcal{X}_{i-1}$ and $x^i \in \mathcal{X}_i$ are *compatible* if they coincide in $\mathcal{L}_{i-1} \cap \mathcal{L}_i$, and satisfy all required constraints.

For $x^i \in \mathcal{X}_i$, let $g(x^i, i)$ be the minimum of the maximum power consumption over the fifteen-minutes intervals $0, \ldots, i$ given that interval $i$ is scheduled with $x^i$. Then

$$g(x^0, 0) = f(x^0, 0) \text{ for all } x^0 \in \mathcal{X}_0 \tag{12}$$

and, for all $1 \leq i \leq 16$ and all $x^i \in \mathcal{X}_i$,

$$g(x^i, i) = \min_{x^{i-1} \in \mathcal{X}_{i-1}} (\max(g(x^{i-1}, i-1), f(x^i, i)) : x^{i-1} \text{ and } x^i \text{ are compatible}). \tag{13}$$

Finally, let $g(i) = \min(g(x^i, i) : x^i \in \mathcal{X}_i)$. It follows that

$$g(0) \leq g(1) \leq \cdots \leq g(16) = z^*. \tag{14}$$

### 4.3.1 Dynamic Programming with Boundaries

It turns out that the sets $\mathcal{X}_i$ are too large and, therefore, the computation of all $g(x^i, i)$ would be too slow. We obtain an improvement if, instead of considering all legs in $\mathcal{L}_i$, we only consider the assignments to *boundary* legs, that is, those legs in $\mathcal{L}_i$ that constrain the assignments of legs in either $\mathcal{L}_{i-1}$ or $\mathcal{L}_{i+1}$. In this way, there would be fewer $g(x^i, i)$ computed, but each of these computations would be more expensive. In either case, the computation could be accelerated by implementing it in parallel for each stage $i$. This approach should be further investigated, at least for instances with few trains running simultaneously.

# 5 Modified Mixed Integer Program

We describe an approximate mixed integer programming formulation for the optimization problem, based on the idea of constant granularity, and the results obtained.

## 5.1 *Constant Granularity Mixed Integer Program*

The mixed integer program that we propose is a modification of our previous exact model. All variables $x_\ell$ and $y_{\ell,m}$ retain their meaning, but variables $\pi_s$ change their meaning into the total power consumption over a subinterval of a few seconds. Therefore, we only explain how do we compute a modified objective function.

Let $d$ be a divisor of 900. For a given feasible solution $x$, let $f_d(x, i)$ be its average power consumption on interval $0 \leq i \leq 16$, computed using the *average* power consumption in subintervals of $d$ seconds. The value of solution $x$ is then $f_d(x) = \max\{f_d(x, i) : 0 \leq i \leq 16\}$. Finally, the objective value is $z_d^* = \min\{f_d(x) : x \text{ is feasible}\}$.

If we write $f_d(x, i)$ as a linear function of $x$, then it is easy to obtain $z_d^*$ as follows:

$$z_d^* = \min z \tag{15}$$

$$\text{subject to}$$

$$f_d(x, i) \leq z \text{ for all } 0 \leq i \leq 16. \tag{16}$$

In order to write each $f_d(x, i)$ as a linear function of $x$, we proceed in three steps. First, for each $0 \leq s \leq \frac{1}{d} 17 \times 15 \times 60$, let $\pi_s \geq 0$ be the total power consumption on the time interval $[ds, d(s + 1)]$ in seconds. Therefore:

$$f_d(x, i) = \frac{1}{900} \sum_{s=900i/d}^{900(i+1)/d-1} \pi_s. \tag{17}$$

Second, for each leg $\ell \in \mathscr{L}$, and for each $e_\ell \leq m \leq l_\ell$, let $y_{\ell,m} \in \{0, 1\}$ be a binary variable indicating whether leg $\ell$ departed on minute $m$. This can be achieved as follows:

$$\sum_{m=e_\ell}^{l_\ell} y_{\ell,m} = 1 \tag{18}$$

$$\sum_{m=e_\ell}^{l_\ell} m y_{\ell,m} = x_\ell. \tag{19}$$

Third, we need to relate the variables $\pi_s$ with the variables $y_{\ell,m}$. In particular, $y_{\ell,m} = 1$ contributes to the value of $\pi_s$ if $60m \leq ds \leq 60(m + r_\ell)$.

$$\pi_s \geq \sum_{\ell \in \mathscr{L}} \left( \sum_{m=\lceil ds/60 - r_\ell \rceil}^{\lfloor ds/60 \rfloor} \left( \sum_{k=0}^{d-1} p_{\ell,ds-60m+k} \right) y_{\ell,m} \right). \tag{20}$$

Note that in the middle sum, if $m < e_\ell$ or $m > l_\ell$ we can assume $y_{\ell,m} = 0$. Also note that these inequalities allow their right-hand sides to be negative, but $\pi_s \geq 0$. Finally, observe that now there is only one variable $\pi_s$ every $d$ seconds. In other words, this modified model has much fewer variables than the exact model.

## 5.2 Solving the Modified Model

An optimal solution found for granularity greater than 1 is not necessarily optimal for granularity equal to 1. In fact, an optimal solution for a coarse granularity is often worse than good but not optimal solution for a fine granularity, as the former is too specialized for the coarse model. However, when a mediocre solution did not improve quickly with the exact model, it often happened that the first improved solutions for coarse granularities were much better than it in the exact model. Furthermore, those better solutions were found in a matter of seconds.

Using a good starting solution affects Gurobi in strange and noticeable ways. The search tree prune and also the Gurobi heuristics become more effective, decreasing the overall runtime and more importantly, the memory consumption. Many were the times when Gurobi ran out of memory while solving the exact model, and a better upper bound found using granularities allowed the exact model to delay such situation in later executions, increasing the chance of finding even better solutions.

## 5.3 Results

By using constant granularity, we were able to obtain good solutions for all instances that were not solved before. Table 4 summarizes these results. For each instance not previously solved to optimality, we show the current (given) value, the value $d$ of the granularity used (often more than one), the value $f_d(x_d)$ obtained from the modified model, the evaluation $f(x_d)$ of the solution of the modified model with the actual objective function, the improvement of this value over the current value (as a percentage), and the gap between the exact model and the modified model (as a percentage). From the last column we can see that we obtained some solutions very close to optimal (about 4% for instance 1 and about 0.1% or less for instances 7, 8, 9, and 10). The results in **bold** were used to continue our work.

**Table 4** Values obtained trough the modified model

| Inst. | Current value | $d$ | $f(x_d)$ | Impr. (%) | $f_d(x_d)$ | Gap (%) |
|---|---|---|---|---|---|---|
| 1 | 1.778052222 | 20 | **1.09065** | 38.66 | 1.04635 | 4.06 |
| 7 | 21.700938333 | 60 | **16.33821** | 24.71 | 16.32999 | 0.05 |
| 7 | 21.700938333 | 50 | 16.67676 | 23.15 | 16.29276 | 2.30 |
| 7 | 21.700938333 | 45 | 16.54990 | 23.74 | 16.29961 | 1.51 |
| 7 | 21.700938333 | 9 | 16.38669 | 24.49 | 16.30985 | 0.47 |
| 8 | 2.772963333 | 60 | 2.56884 | 7.36 | 2.23526 | 12.99 |
| 8 | 2.772963333 | 50 | 2.55792 | 7.76 | 2.19013 | 14.38 |
| 8 | 2.772963333 | 45 | 2.51708 | 9.23 | 2.19154 | 12.93 |
| 8 | 2.772963333 | 9 | 2.44258 | 11.91 | 2.28935 | 6.27 |
| 8 | 2.772963333 | 2 | **2.42464** | 12.56 | 2.42195 | 0.11 |
| 9 | 113.391771111 | 60 | 98.40077 | 13.22 | 98.10779 | 0.30 |
| 9 | 113.391771111 | 45 | 98.17457 | 13.42 | 98.09943 | 0.08 |
| 9 | 113.391771111 | 9 | 98.16771 | 13.43 | 98.09419 | 0.07 |
| 9 | 113.391771111 | 6 | 98.16771 | 13.43 | 98.09419 | 0.07 |
| 9 | 113.391771111 | 2 | **98.13391** | 13.46 | 98.09419 | 0.04 |
| 10 | 79.585383888 | 60 | 68.16482 | 14.35 | 68.10658 | 0.09 |
| 10 | 79.585383888 | 45 | 68.14369 | 14.38 | 68.08756 | 0.08 |
| 10 | 79.585383888 | 9 | **68.14079** | 14.38 | 68.10320 | 0.06 |

**Table 5** Our best results for each of the ten instances

| Inst. | Current value | Our best value | Impr. (%) | Lower bound | Gap (%) |
|---|---|---|---|---|---|
| 1 | 1.778052222 | 1.071813889 | 39.72 | 0.994686978 | 7.20 |
| 2 | 4.620948333 | 2.939638333 | 36.38 | 2.939638333 | 0.00 |
| 3 | 20.635706111 | 15.179507777 | 26.44 | 15.179507777 | 0.00 |
| 4 | 19.983620000 | 14.966736111 | 25.10 | 14.966736111 | 0.00 |
| 5 | 27.551612222 | 19.416015000 | 29.53 | 19.416015000 | 0.00 |
| 6 | 26.349433888 | 20.431125556 | 22.46 | 20.431125556 | 0.00 |
| 7 | 21.700938333 | 16.321493889 | 24.79 | 16.308135556 | 0.08 |
| 8 | 2.772963333 | 2.424517778 | 12.57 | 2.424517778 | 0.00 |
| 9 | 113.391771111 | 98.133795000 | 13.46 | 98.133726889 | 0.00 |
| 10 | 79.585383888 | 68.121432222 | 14.40 | 68.115126000 | 0.01 |
| T | 318.370429443 | 259.006075555 | 18.65 | 258.909215756 | 0.04 |

## 6 Our Best Obtained Results

For each instance, we collect in Table 5 the current (given) objective value, our best objective value, the improvement (as a percentage), our best lower bound, and the optimality gap (as a percentage). The last line is the total over all instances.

Haahr and Kidd's team (Sodor Diesels) obtained an honourable mention in FAU's Optimization Challenge by totalling 259.03, very close to our total. The main difference was that we obtained better solutions for instances 1 and 9.

## 7 Conclusions

We approached FAU's Optimization Challenge from a mathematical programming perspective, that is, building a mixed integer programming model and solving it via an MIP solver. This proved to be easier said than done: being our computational resources very limited, the MIP solver hits rapidly the maximum memory that we had available.

Therefore, we tried several other algorithmic approaches, but again our limited computational resources disallowed nice techniques such as dynamic programming. It was then that we decided for a heuristic approach, but one that was fit for the purpose.

The idea of simplifying the power consumption waves via averaging with a fixed granularity was very successful. First, within our limited resources, it allowed the MIP solver to find near-optimal solutions to this modified problem that were already big improvements over the given schedule. Second, it came as a bonus that the modified problem gave us simultaneously good upper and lower bounds to the optimal solution. Third, it also allowed us to restart the MIP solver using these solutions in order to obtain near-optimal solutions to the original optimization problem.

An unexpected deficiency of the near-optimal solutions obtained is related to peak power consumption. This happens in our solutions to instances 3–7 and 9, where the new peak power consumption is greater than the initial peak power consumption. In a more realistic setting, there should be a constraint related to peak power consumption.

Another deficiency of our near-optimal solutions is that sometimes our schedule wastes more generated power than the original schedule. This usually happens at the end of the planning horizon: the last fifteen-minutes interval (and often the previous one too) does not reach the maximum average power consumption (in fact, its consumption is almost negligible). We could think of a secondary objective: to minimize the total power consumption (or, equivalently, to minimize the total unused power generation).

**Acknowledgements** We thank the Mexican National Council on Science and Technology (CONACyT) for providing scholarships to most students in our graduate program (including two team members) and also for providing a scholarship to another team member through the National System of Researchers (SNI). We thank the Mexiquense Council on Science and Technology (COMECyT) for providing a scholarship to our last team member. We thank Universidad Autónoma Metropolitana Azcapotzalco for funding Research Project SI004-13 (Algorithms and Models for Network Optimization Problems) and also for allowing us the use of some computing facilities through the Systems Department. We thank Gurobi Optimization for giving us free licenses to their software. We thank the Friedrich-Alexander-Universität Erlangen-Nürnberg for organizing the Discrete Optimization Challenge and the participating teams for their great effort. Last, but not least, we thank honorary team member Gabrijela Zaragoza for proposing a nice team name, a portmanteau of *optimization* and the nahuatl word *mixtli* (cloud).

## Appendix

In what follows, we display power consumption in two figures for each instance: the first for the given schedule, the second for our best result (Figs. 5, 6, 7, 8, 9, 10, 11, 12, 13, 14, 15, 16, 17, 18, 19, 20, 21, 22, 23, 24).

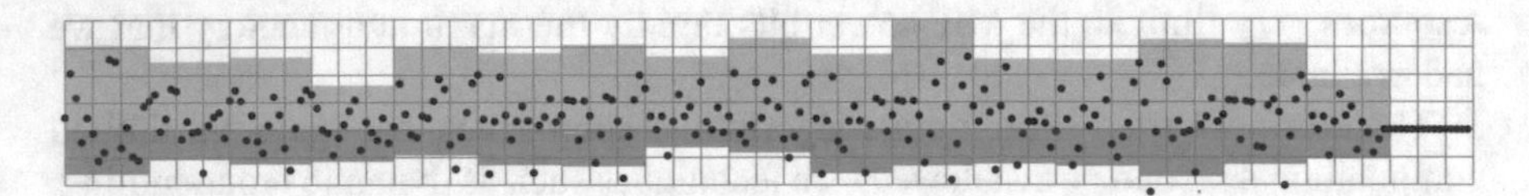

**Fig. 5** The original input for instance 1

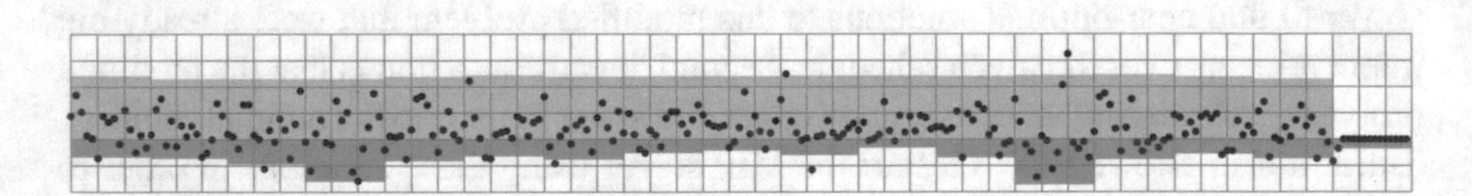

**Fig. 6** The best output for instance 1

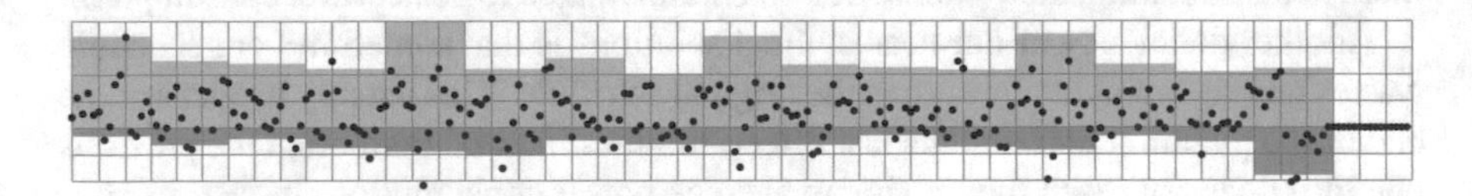

**Fig. 7** The original input for instance 2

As before, a red dot means power consumption, a blue dot means power generation. However, to avoid clutter, each dot represents an average power consumption over one minute. The vertical scale accommodates the maximum instantaneous

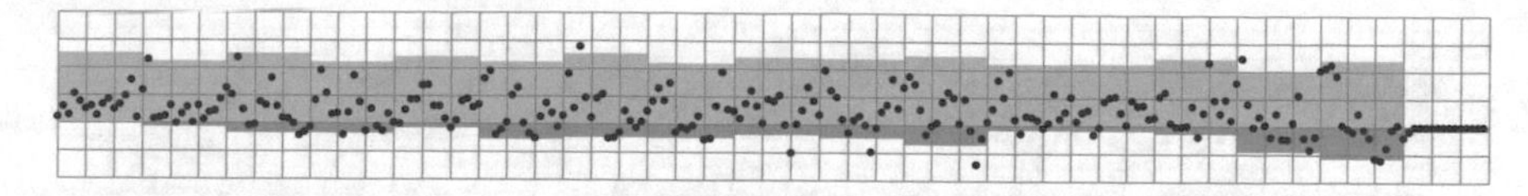

**Fig. 8** The optimal output for instance 2

power consumption as given in the current schedule,[2] while the horizontal scale accommodates the 17 fifteen-minutes intervals. This scale holds for both figures.

We also display in light red the average power consumption (and in light blue the average *unused* power generation) on each fifteen-minutes interval. In this case, the vertical scale accommodates the maximum average power generation on a fifteen-minutes interval as given in the current schedule. This scale also holds for both figures.

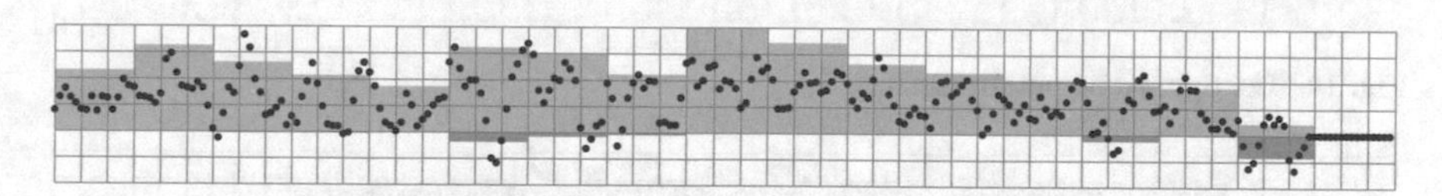

**Fig. 9** The original input for instance 3

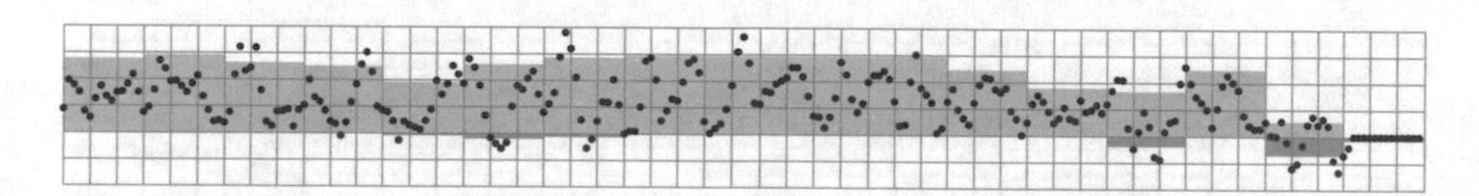

**Fig. 10** The optimal output for instance 3

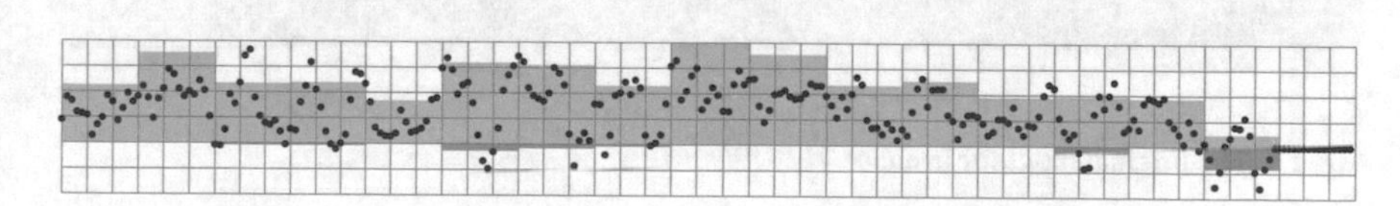

**Fig. 11** The original input for instance 4

[2] A black dot in the second figure means that our solution exceeds this instantaneous maximum.

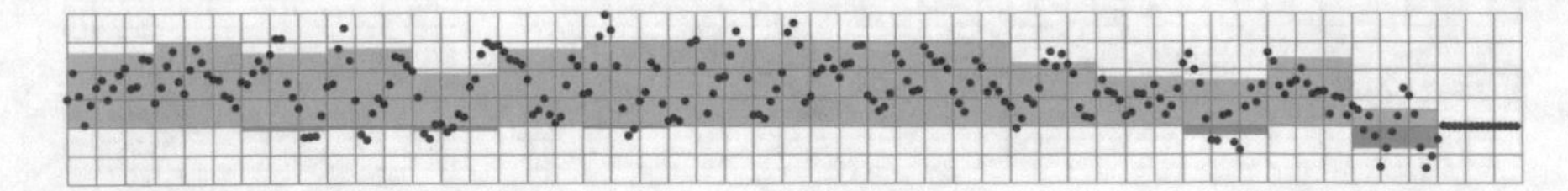

**Fig. 12** The optimal output for instance 4

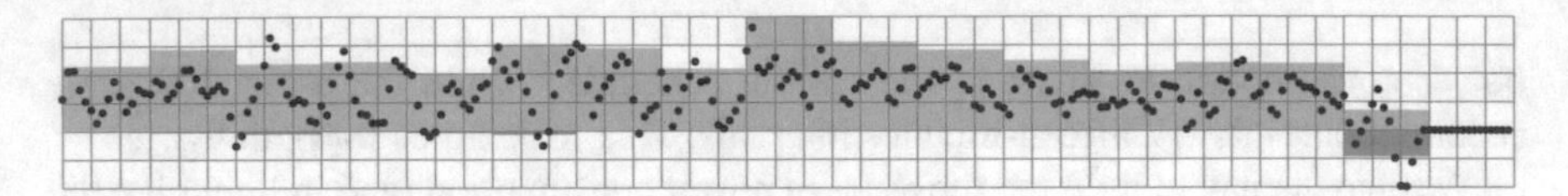

**Fig. 13** The original input for instance 5

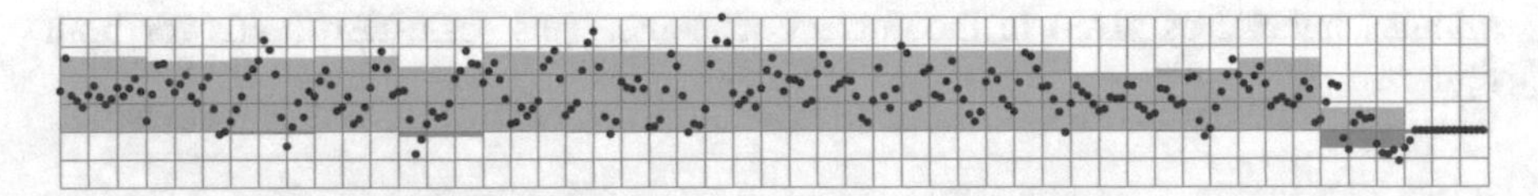

**Fig. 14** The optimal output for instance 5

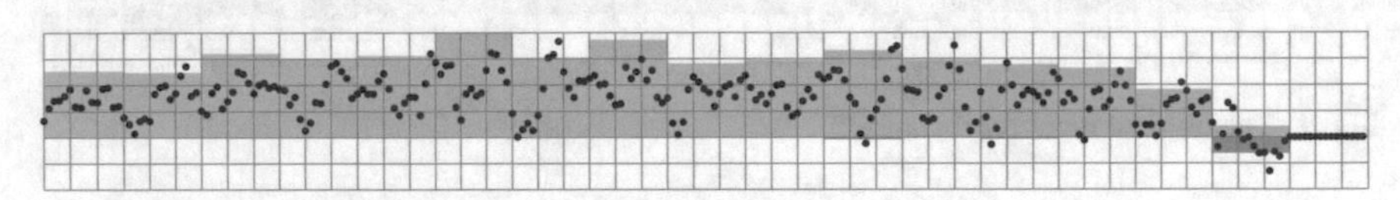

**Fig. 15** The original input for instance 6

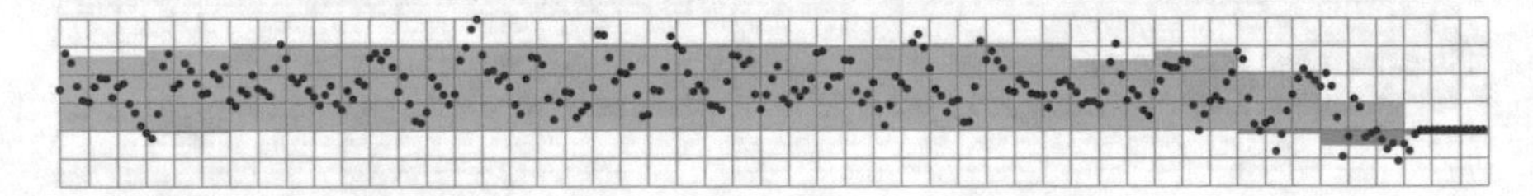

**Fig. 16** The optimal output for instance 6

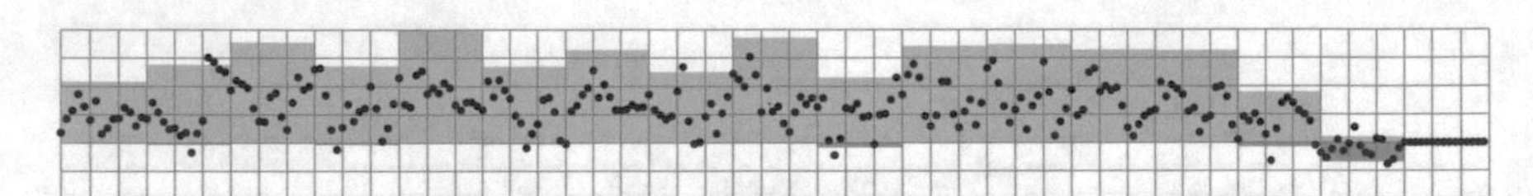

**Fig. 17** The original input for instance 7

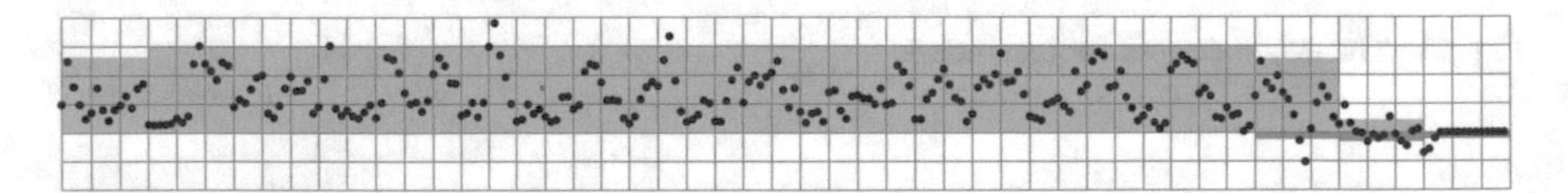

**Fig. 18** The best output for instance 7

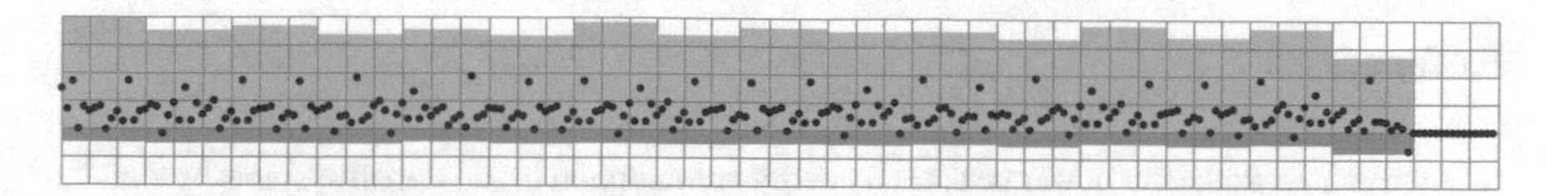

**Fig. 19** The original input for instance 8

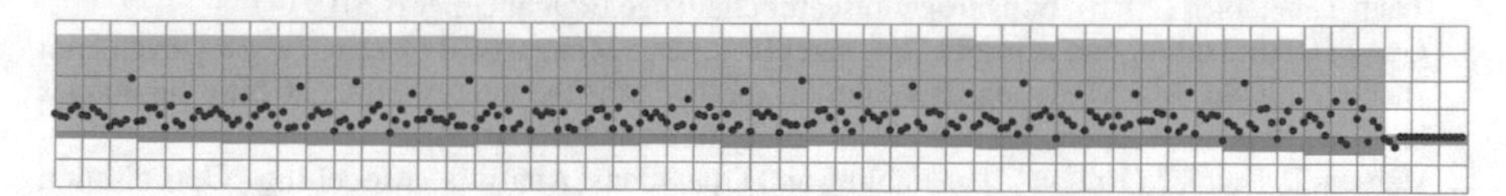

**Fig. 20** The optimal output for instance 8

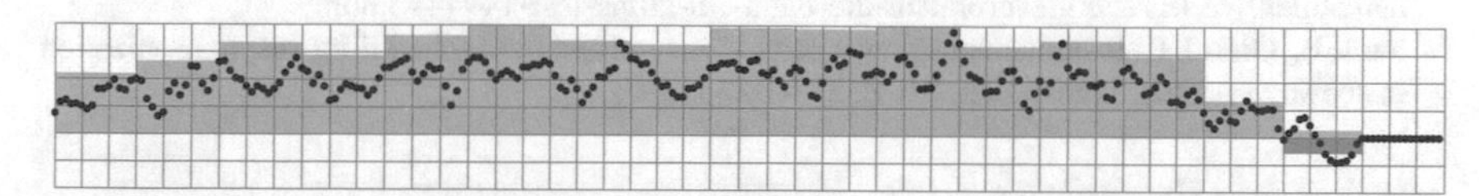

**Fig. 21** The original input for instance 9

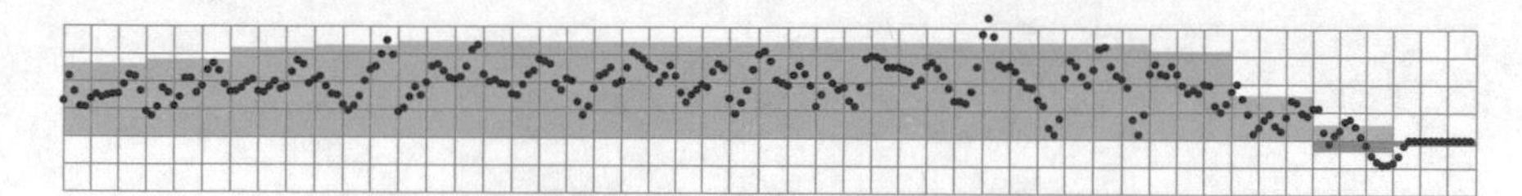

**Fig. 22** The best output for instance 9

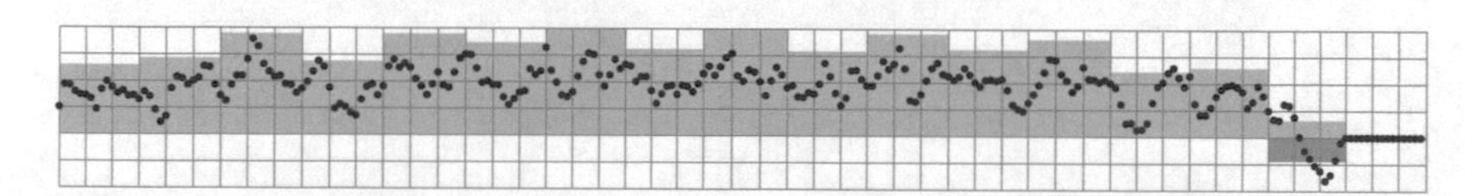

**Fig. 23** The original input for instance 10

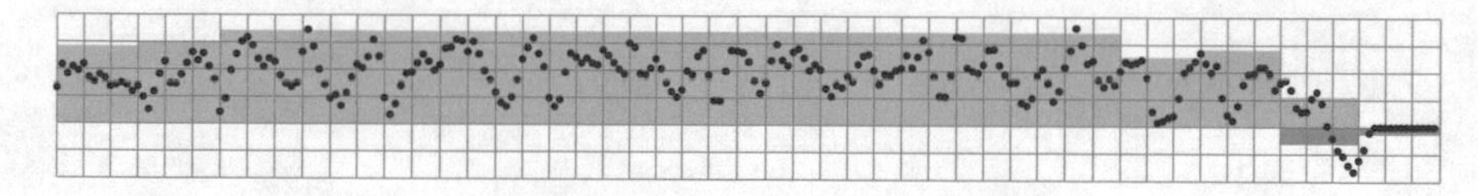

**Fig. 24** The best output for instance 10

## References

1. Albrecht, T.: Reducing power peaks and energy consumption in rail transit systems by simultaneous train running time control. Comput. Railw. IX pp. 885–894 (2010)
2. Friedrich-Alexander-Universität Erlangen-Nürnberg: Discrete Optimization: Energy-efficient Train Timetables (2015). https://openresearchchallenge.org/challengers2015
3. Gurobi Optimization, Inc.: Gurobi Optimizer Reference Manual (2016). http://www.gurobi.com
4. Haahr, J., Kidd, M.: Modeling and solution methods for the energy-efficient train timetables problem, Technical report. Technical University of Denmark (2015)
5. Hansen, I., Pachl, J.: Railway Timetabling and Operations: Analysis - Modelling - Optimisation - Simulation - Performance Evaluation. Eurail press (2014)
6. Ilgmann, G.: Die Bahn im Klima-Test. Frankfurter Allgemeine Zeitung (2007). http://www.faz.net/aktuell/wissen/energieverbrauch-die-bahn-im-klima-test-1494443.html
7. Sans, B., Girard, P.: Instantaneous power peak reduction and train scheduling desynchronization in subway systems. Transp. Sci. **31**(4), 312–323 (1997)

# Distributing Computing in the Internet of Things: Cloud, Fog and Edge Computing Overview

**P.J. Escamilla-Ambrosio, A. Rodríguez-Mota, E. Aguirre-Anaya, R. Acosta-Bermejo and M. Salinas-Rosales**

**Abstract** The main postulate of the Internet of things (IoT) is that everything can be connected to the Internet, at anytime, anywhere. This means a plethora of objects (e.g. smart cameras, wearables, environmental sensors, home appliances, and vehicles) are 'connected' and generating massive amounts of data. The collection, integration, processing and analytics of these data enable the realisation of smart cities, infrastructures and services for enhancing the quality of life of humans. Nowadays, existing IoT architectures are highly centralised and heavily rely on transferring data processing, analytics, and decision-making processes to cloud solutions. This approach of managing and processing data at the cloud may lead to inefficiencies in terms of latency, network traffic management, computational processing, and power consumption. Furthermore, in many applications, such as health monitoring and emergency response services, which require low latency, delay caused by transferring data to the cloud and then back to the application can seriously impact their performances. The idea of allowing data processing closer to where data is generated, with techniques such as data fusion, trending of data, and some decision making, can help reduce the amount of data sent to the cloud, reducing network traffic, bandwidth and energy consumption. Also, a more agile response, closer to real-time, will be achieved, which is necessary in applications such as smart health, security and

P.J. Escamilla-Ambrosio (✉) · E. Aguirre-Anaya · R. Acosta-Bermejo · M. Salinas-Rosales
Instituto Politécnico Nacional Centro de Investigación en Computación, Mexico City, Mexico
e-mail: pescamilla@cic.ipn.mx

E. Aguirre-Anaya
e-mail: eaguirre@cic.ipn.mx

R. Acosta-Bermejo
e-mail: racosta@cic.ipn.mx

M. Salinas-Rosales
e-mail: msalinasr@cic.ipn.mx

A. Rodríguez-Mota
Instituto Politécnico Nacional Escuela Superior de Ingeniería Mecánica y Eléctrica, Unidad Zacatenco, Mexico City, Mexico
e-mail: armesimez@gmail.com

© Springer International Publishing AG 2018

Y. Maldonado et al. (eds.), *NEO 2016*, Studies in Computational Intelligence 731,
https://doi.org/10.1007/978-3-319-64063-1_4

traffic control for smart cities. Therefore, this chapter presents a review of the more developed paradigms aimed to bring computational, storage and control capabilities closer to where data is generated in the IoT: fog and edge computing, contrasted with the cloud computing paradigm. Also an overview of some practical use cases is presented to exemplify each of these paradigms and their main differences.

**Keywords** Internet of things · Cloud computing · Edge computing · Fog computing · Distributed processing

## 1 Introduction

The first idea towards the Internet of things concept, was initially expressed as "computers everywhere", formulated by Ken Sakamura at the University of Tokyo in 1984 [1], and latter referred to as "ubiquitous computing" by Mark Weiser in 1988 (Xerox PARC) [2]. However, in 1999 Kevin Ashton was the first to coin the term "Internet of things" (IoT) in the context of supply chain management [3]. In 2001 the IoT concept was further developed by David Brock in an article of the Auto-ID Center of the Massachusetts Institute of Technology [4]. Since then a large number of researchers have followed and developed this idea, embodied in a wide variety of scientific articles, books and conferences. In all of them, the vision of integrating intelligence in the objects around us persists. This is achieved by providing objects with sensing, acting, storage, and processing capabilities, but overall providing them with interconnectivity via the Internet, all this in order to provide services to different users. It is precisely in the users, where lies the main difference of the IoT with respect to the traditional Internet, since while in the traditional Internet users are people connected to the Internet via a PC, laptop or some mobile device (smart phone or tablet, for example), in the IoT users can be other objects or "smart things" that require some service or interaction. This interaction can take place even with additional sources of information, such as social networks. Moreover, the IoT concept involves some very important activities which are frequently forgotten, the analysis, storage and utilisation of the data collected by different devices/objects. This opens up the possibility of developing a myriad of applications, for example: goods and people tracking, smart houses, smart cities, remote command and control, location based services, remote patient monitoring, and environmental monitoring; to name just a few of them.

In this context, the IoT represents the radical evolution of the traditional Internet into an interconnected network of "smart objects" that not only collect data from the environment (sensing) and interact with the physical world (actuation, command and control), but also use the Internet to provide services for information transfer, analytics, applications, and communications. Furthermore, as Gubbi et al. [5] described in their definition, the IoT is also the *"Interconnection of sensing and actuating devices providing the ability to share information across platforms through a unified framework, developing a common operating picture for enabling innovative applications.*

*This is achieved by seamless ubiquitous sensing, data analytics and information representation with cloud computing as the unifying framework.”*

Generally, interoperability between connected devices is a responsibility of a single vendor that ensures it either through using proprietary interfaces and protocols, through installing add-on software clients on devices, or through the use of a gateway device [6]. However, in the future, devices can be expected to gain sufficient intelligence to interoperate directly, without the need for dedicated gateways [7].

Across the IoT, devices create data that are sent to the main application to be sent on, consumed and used. Depending on the device, the network and power consumption restraints, data can be sent in real time or in batches at any time [8]. Due to the heterogeneity of the data sources, nature of data generated and heterogeneity of data processing servers, an IoT system can also be viewed as a computational grid consisting of a very large number of devices generating data for processing and a large number of resources capable of processing such data [9]. For smart devices and sensors, each event that they perceive or that is registered can and will create data. This data can then be sent over the network back to the central application. At this point, it must be decided which standard the data will be created in and how it will be sent over the network. For delivering this data back, Message Queue Telemetry Transport (MQTT) [10], Hypertext Transfer Protocol (HTTP) [11], and Constrained Application Protocol (CoAP) [12] are the most common standard protocols used [8].

For years, cloud computing was the most used base technology for a lot of enterprise architectures and companies decided to move all their data, computation, processing and so on from “on-premise” infrastructure to the cloud itself. The cloud seems to offer infinite storage space and scaling for computation without any concerns from a company point of view which can configure all the features to change automatically. The result could be less time to spend on handling “on premise” infrastructures and less money to invest [13]. Cloud computing enables convenient, on-demand network access to a shared pool of configurable computing resources, such as networks, servers, storage, applications, and services, that can be rapidly provisioned and released with minimal management effort or service provider interaction [14]. However, for many operations, a cloud-only model is not necessarily the perfect fit. The cost of cloud storage is dropping, but transmitting and storing massive amounts of data in the cloud for analysis quickly becomes prohibitively expensive. Usually, all data are transmitted at their current fidelity, with no ability to sift out what data are of the highest business value and what data can be disregarded. Cloud-based IoT platforms also require a constant network connection, making them less than ideal for companies with remote operations, who cannot afford to cease operations when their connection goes down [15]. As a consequence, a new set of paradigms for IoT data handling have been proposed.

In this scenario, this chapter is developed around two main goals. The first one is to provide a review of the main concepts and paradigms related to data and process management in the IoT, aiming to provide an articulated view of the IoT environment as it is currently plagued by multiple definitions and general concepts. Thus, three different important approaches, cloud, fog and edge computing, are presented, together with a discussion of their main characteristics, advantages, disadvantages and differences between each other. The second goal is to provide an overview of distributed data processing approaches along the cloud, the fog and the edge paradigms concentrating the discussion on developments proposed towards optimizing resources, bandwidth usage and best fit approaches which take advantage of the characteristics that each layer of computing can provide. Finally, the main issues, challenges and opportunities of distributed computing and data processing in the IoT are discussed.

## 1.1 The Need for Distributed Computing

In 2003 Eric Schmidt, Executive Chairman of Google, claimed that up to that year 5 exabytes of data had been generated since the origin of humanity [16]. Besides, nowadays it is claimed that this much data is generated every 2 days and this rate is only increasing [16]. Furthermore, with the advent of the IoT CISCO's estimated that the number of connected objects is going to reach ~50 billion in 2020 [17]. This means that the data deluge problem is going to be worsened. The question is then, if the technology available will evolve fast enough to deal with this forthcoming extraordinary production of data. On the one hand, the evolution of compute and storage technologies is governed by Moore's Law, which stipulates that compute and storage technologies will double in capability/capacity every 18 months. On the other hand, Nielsen's Law projects that the Internet bandwidth doubles every 24 months [18]. Unfortunately, by comparing the two Laws it is shown that bandwidth grows slower than computer power. Hence, this anticipates a future IoT where data will be produced at rates that will far outpace the network's ability to backhaul the information from the network edge, where it is produced by the billions of connected things, to the cloud where it will need to be processed and probably stored. Although Moore's Law contributes partly to the data problem, the same Law can provide the solution. This solution will consists in increasing the functions of the network itself with compute and storage capabilities at the edge of the network. This means to bring cloud computing closer to the data sources allowing the network to perform processing, analysis, and storage of data in lieu of blindly pushing all data up to the cloud. As Bayer and Wetterwald [19] describe this is not just about aggregation or concatenation of sensed physical data (like a gateway will do), but really about distributing intelligence, where effective real time and deterministic processing is needed to implement a functionality, instead of leaving all the processing and analytics to be performed at the cloud side.

From the several reasons that make distributed intelligence a need for IoT, scalability, network resource preservation, close loop control, resilience and clustering are the main aspects [19]. In the case of scalability, the centralised approach is not sufficient to handle an increasing volume of end devices and its geographical specificities. Referring to network resource preservation, distributed processing helps relieving the constraints on the network by sending to the cloud or operation centre only the necessary information and by doing most of the data processing at the remote site much closer to the data's source. In applications where a low latency is critical, such as real-time systems, for close loop control, large delays found in many multi-hop networks and overloaded cloud server farms prove to be unacceptable, and the local, high performance nature of distributed intelligence can minimise latency and timing jitter. Only local processing could satisfy the most stringent requirements, very often combined with advanced networking technologies like deterministic networking. Resilience is of most importance for critical processes that must run even if communication with the operation centre is not effective. In this case, an architecture based on distributing processing is the only valid solution. Finally, moving from individual devices to clusters helps to manage many units as a single one.

Under this context, as new technologies and services become available, it is important to consider that system deployments are in most of the cases application defined. Therefore, in the following paragraphs main IoT's data and process paradigms are described followed by some use cases.

## 2 Cloud, Fog and Edge Computing Definitions

As the IoT paradigm matures, core concepts continue to evolve providing better insights about real world IoT implementations. However, a clear consensus in some areas still does not exist, as many current developments are vendor specific solutions. Thus, in this section cloud, edge and fog computing concepts are defined as a result of a review of different author's proposals.

Before proceeding, it is important to notice that we have observed a loosely use of the "edge" term among definitions, making sometimes difficult to differentiate between concepts. Hence, we propose to use the terms Internet edge and IoT edge to differentiate from the devices at the edge of the Internet network from those at the edge of an IoT system. On the one hand, the Internet edge is the network infrastructure that provides connectivity to the Internet, and that acts as the gateway for the enterprise to the rest of the cyber space. As the gateway to the Internet, the Internet edge infrastructure plays a critical role in supporting the services and activities that are fundamental to the operation of the modern enterprise [20]. Some common Internet edge elements are: edge routers, switches and firewalls. On the other hand, the edge of the IoT includes those system end-points that interact with and communicate real-time data from smart products and services, examples of such devices include a wide array of sensors, actuators, and smart objects [21]. For example, in an IoT patient health monitoring system, the edge of the IoT would be the body area sensor network acquiring patient's data, such as patient's body temperature, respiration rate, heart beat and body movement; while the edge of the Internet would be the Internet-connected gateway which concentrates all the data and sends them to the cloud.

## 2.1 Cloud Computing

A commonly accepted definition for cloud computing was provided by the National Institute for Standards and Technology (NIST) in 2011 [22] as "*a model for enabling ubiquitous, convenient, on-demand network access to a shared pool of configurable computing resources (e.g., networks, servers, storage, applications, and services) that can be rapidly provisioned and released with minimal management effort or service provider interaction.*" In this regard, as described in [23] cloud computing refers to both the applications delivered as services over the Internet and the hardware and systems software in the data centres that provide those services. The services themselves have long been referred to as Software as a Service (SaaS). The data centre hardware and software is what is call a cloud. When a cloud is made available in a pay-as-you-go manner to the public, it is referred to as a public cloud; the service being sold is utility computing. The term private cloud is used to refer to internal datacentres of a business or other organization that are not made available to the public. Thus, cloud computing is the sum of SaaS and utility computing, but does not normally include private clouds. Cloud computing offers the next characteristics [24]: (a) Elastic resources, which scale up or down quickly and easily to meet demand; (b) Metered service so a user only pay for what was used; (c) Self-service, all the IT resources that a user needs with self-service access.

Another useful definition of cloud computing is that one provided by Vaquero et al. [25]: *clouds are a large pool of easily usable and accessible virtualized resources (such as hardware, development platforms and/or services). These resources can be dynamically reconfigured to adjust to a variable load (scale), allowing also for an optimum resource utilization. This pool of resources is typically exploited by a pay-per-use model in which guarantees are offered by the Infrastructure Provider by means of customized Service-Level Agreements (SLAs).*

Therefore, one of the key characteristics of the cloud computing model is the notion of resource pooling, where workloads associated with multiple users (or tenants) are typically collocated on the same set of physical resources. Hence, essential to cloud computing is the use of network and compute virtualisation technologies. Cloud computing provides elastic scalability characteristics, where the amount of resources can be grown or diminished based on user demand [16], with minimal management effort or service provider interaction [22].

Depending on the type of provided capability, there are three cloud service delivery models [24]:

- Infrastructure as a Service (IaaS). Provides companies with computing resources including servers, networking, storage, and data center space on a pay-per-use basis and as an Internet-based service.
- Platform as a Service (PaaS). Provides a cloud-based environment with everything required to support the complete lifecycle of building and delivering web-based (cloud) applications, without the cost and complexity of buying and managing the underlying hardware, software, provisioning, and hosting (users do not install any of these platforms or support tools on their local machines).

**Fig. 1** Cloud service delivery models

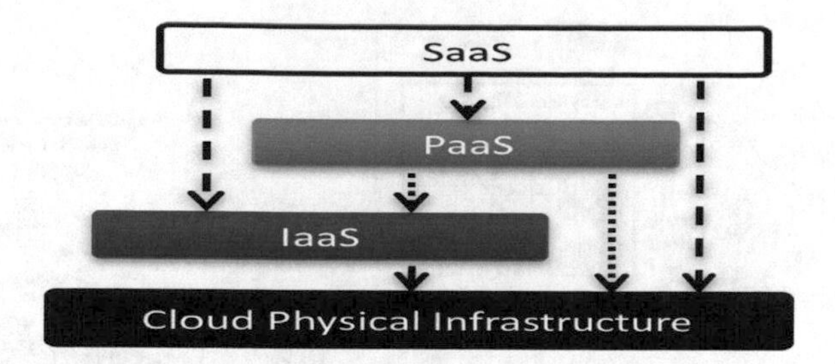


- Software as a Service (SaaS). Cloud-based applications run on distant computers "in the cloud" that are owned and operated by others and that connect to users' computers via the internet and, usually, a web browser.

The PaaS model may be hosted on top of IaaS model or on top of the cloud infrastructures directly, while the SaaS model may be hosted on top of PaaS, IaaS or directly hosted on cloud infrastructure, see Fig. 1.

The cloud computing model is well-suited for small and medium businesses because it helps adopting IT without upfront investments in infrastructure, software licenses and other relevant requirements.

Despite the maturity that cloud computing has reached through the years, and the many potential benefits and revenues that could be gained from its adoption, the cloud computing model still has several open issues. Fog computing and edge computing are new paradigms that have emerged to attend some of the cloud computing issues, these paradigms are defined in the following sections together with some technological approaches for cloud computing that support them.

### 2.1.1 Mobile Cloud Computing

As mobile phones and tablets are getting "smarter," their usage and preference over traditional desktops and laptops has increased dramatically. At the same time, the availability of a huge number of intelligent mobile applications has attracted more people to use smart mobile devices. Some of these applications, such as speech recognition, image processing, video analysis, and augmented reality are computing-intensive and their implementation in portable devices is still impractical due to the mobile device resource limitations. In contrast, the high-rate and highly-reliable air interface allows to run computing services of mobile devices at remote cloud data centres. Hence the combination of mobile computing with cloud computing has resulted in the emergence of what is called mobile cloud computing (MCC) technology. In MCC computing and communications-intensive application workloads, also as storage, are moved from the mobile device to powerful and centralised computing platforms located in clouds [26]. These centralised applications are then accessed over wireless connections based on a thin native client or web browser on the mobile

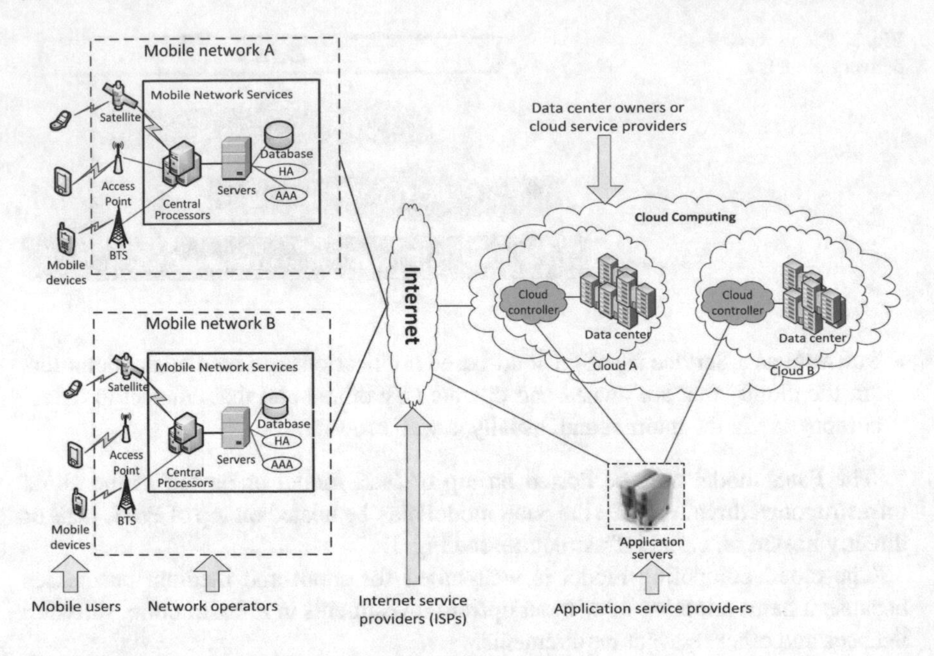


**Fig. 2** Mobile cloud computing architecture [26]

devices. This alleviates the challenge of resource constraint and battery life shortage of mobile devices.

The general architecture of MCC is represented in Fig. 2. As can be seen in the figure, mobile devices are connected to the mobile networks via base stations (e.g., base transceiver station, access point, or satellite) that establish and control the connections and functional interfaces between the networks and mobile devices. Mobile users' requests and information (e.g., ID and location) are transmitted to the central processors that are connected to servers providing mobile network services. Mobile network operators provide services to mobile users as authentication, authorisation, and accounting based on subscribers' data stored in databases. After that, the subscribers' requests are delivered to a cloud through the Internet. In the cloud, cloud controllers process the requests to provide mobile users with the corresponding cloud services [26].

Although MCC has several advantages, it has an inherent limitation, namely, the long propagation distance from the end user to the remote cloud centres, which will result in excessively long latency for mobile applications. MCC is thus not adequate for a wide-range of emerging mobile applications that are latency-critical. Presently, new network architectures are being designed to better integrate the concept of cloud computing into mobile networks.

## 2.2 *Fog Computing*

Recently, as described in [27] to address some of the limitations of cloud computing under the IoT paradigm, the research community has proposed the concept of fog computing, aiming at bringing cloud service features closer to what is referred to as "things," including sensors, embedded systems, mobile phones, cars, etc. These things form part of what in this paper has been labelled as IoT edge devices. It is important to notice that, for this section, when referring to the "edge" or "edge devices" we refer to the IoT edge, unless otherwise stated, trying to keep the reference authors' original essence but differentiating from the Internet edge concept.

The first formal definition of fog computing was stated in 2012 by Bonomi et al. [28] from CISCO as: "*Fog computing is a highly virtualised platform that provides compute, storage and networking services between end devices and traditional cloud computing data centres, typically, but not exclusively located at the edge of the network.*" Since then, different definitions have emerged under distinct scenarios and contexts.

In [29] fog computing is defined as *"a model to complement the cloud for decentralising the concentration of computing resources (for example, servers, storage, applications and services) in data centres towards users for improving the quality of service and their experience."* The cloud and fog synergistically interplay in order to enable new types and classes of IoT applications that otherwise would not have been possible when relying only on stand-alone cloud computing [16]. In this scenario it is expected that a huge number of heterogeneous ubiquitous and decentralised devices will communicate and potentially cooperate among them and with the network to perform storage and processing tasks without the intervention of third-parties. These tasks can be for supporting basic network functions or new services and applications that run in a sandboxed environment [25]. This platform can extend in locality from IoT end devices and gateways all the way to cloud data centres, but is typically located at the network edge. Fog augments cloud computing and brings its functions closer to where data is produced (e.g., sensors) or needs to be consumed (e.g., actuators) [16], enable computing directly at the edge of the network, delivering new applications and services. The computational, networking, storage and acceleration elements of this new model are known as fog nodes. These are not completely fixed to the physical edge, but should be seen as a fluid system of connectivity [30].

Fog-centric architecture serves a specific subset of business problems that cannot be successfully implemented using only traditional cloud based architectures or solely intelligent edge devices. Although fog computing is an extension of the traditional cloud-based computing model where implementations of the architecture can reside in multiple layer of a network's topology, all the benefits of cloud should be preserved with these extension to fog, including containerisation, virtualisation, orchestration, manageability, and efficiency [30].

Furthermore, several surveys have tried specifically to define this new paradigm, its challenges, possible applications as well as scenarios of application, see for example [25, 31, 32]. Although there are several definitions, all of them agree in the sense that the fog computing model moves computation from the cloud closer to the edge of the network (Internet), and potentially right up to the edge of the IoT, this is to the things: sensors and actuators. Moreover this idea is enhanced by considering fog computing as a system-level horizontal architecture that distributes resources and services of computing, storage, control and networking anywhere along the continuum from cloud to things, with the aim of accelerating the velocity of decision making [30].

As indicated by the OpenFog Consortium [30], fog computing is often erroneously called edge computing, but there are key differences. Fog works with the cloud, whereas edge is defined by the exclusion of cloud. Fog is multilayer and hierarchical, where edge tends to be limited to three or four layers. In addition to computation, fog also addresses networking, storage, control and acceleration. Figure 3 shows a subset of the combination of fog and cloud deployments to address various use cases as framed by the layered view of IoT systems [30]. Each fog element may represent a hierarchy of fog clusters fulfilling the same functional responsibilities. Depending on the scenario, multiple fog and cloud elements may collapse into a single physical

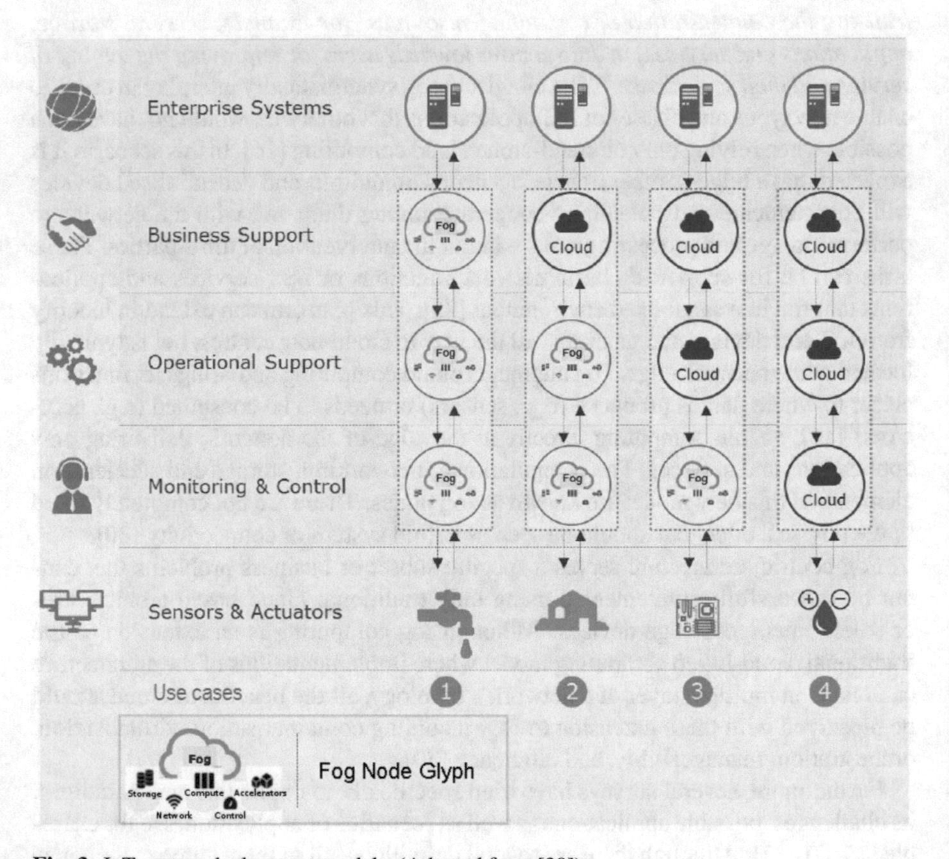


**Fig. 3** IoT system deployment models (Adapted from [30])

deployment. Each fog element may also represent a mesh of peer fog nodes in use cases like connected cars, electrical vehicle charging, and closed loop traffic systems. In these use cases, fog nodes may securely discover and communicate with each other for exchanging context-specific intelligence.

## 2.3 *Edge Computing*

As mentioned in the previous section, it is common in the literature to find that edge and fog computing are defined as the same concept. Nevertheless, these two concepts share several characteristics and both aim to bring cloud resources and services closer to the things which are generating data, in this work edge and fog computing are differentiated in the sense that in the former edge refers to the edge of the Internet, while in the later edge refers to edge of the IoT. Taken this into consideration in the last section a definition of fog computing was provided, while the definition of edge computing is presented in this section.

In [33], it is mentioned that the term edge computing was first coined around 2002 and was mainly associated with the deployment of applications over Content Delivery Networks (CDN). The main objective of this approach was to benefit from the proximity and resources of CDN edge servers to achieve massive scalability. An edge node includes routers, mobile base stations and switches that route network traffic [29]. In this case, "edge" must be understood as the edge of Internet, as defined above. These devices perform sophisticated processing in order to handle the packets coming in over the different subnetworks they aggregate. Examples of edge devices are appliances at the frontend of a data centre that perform functions such as XML acceleration, load balancing, or other content processing, as well as devices at the entry point of an enterprise that perform security-related functions such as firewalls, intrusion detection, and virus checking [34]. This concept places applications, data and processing at the logical extremes of a network rather than centralising them. Placing data and data-intensive applications at the edge reduces the volume and distance that data must be moved [30].

However, notice that devices which have direct connection to the Internet can also be considered as the Internet edge, referred to as edge devices in [35]. For example, a smart phone is the edge between body things and cloud, a gateway in a smart home is the edge between home things and cloud, a micro data center and a cloudlet is the edge between a mobile device and cloud. Figure 4 shows a simplified view of this paradigm.

### 2.3.1 Mobile Edge Computing

The concept of mobile edge computing (MEC) was first proposed by the European Telecommunications Standard Institute (ETSI) in 2014 [36], as a new platform that *"provides IT and cloud-computing capabilities within the Radio Access Network*

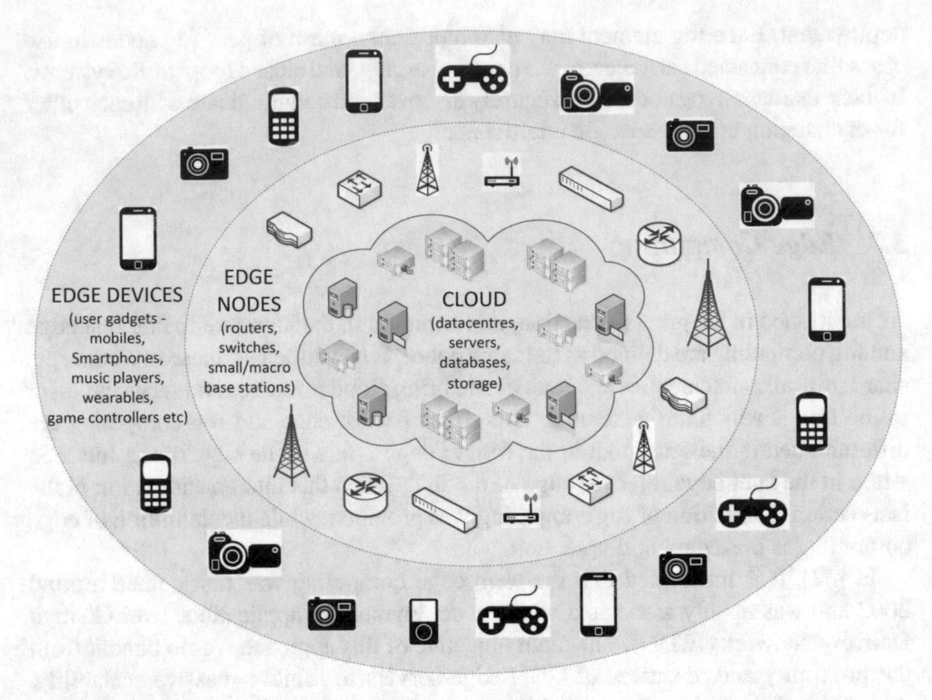


**Fig. 4** Simplified view of the edge computing paradigm [35]

*(RAN) in close proximity to mobile subscribers"*. In recent years, driven by the visions of the Internet of things and 5G communications, MEC has had a renewed interest causing a paradigm shift in mobile computing, from the centralised MCC towards MEC. The main feature of MEC is to co-locate computing, network control and storage resources at the edge of the mobile Radio Access Network [37] aiming at reducing latency. Hence, MEC is not replacing but complimenting the cloud computing model: the delay sensitive part of an application can be executed on MEC servers whereas delay tolerant compute intensive part of applications can be executed on the cloud server. MEC aims to enable connected mobile devices to execute real-time compute-intensive applications directly at the network edge. The characteristics features that distinguish MEC are its closeness to end-users, mobility support, and dense geographical deployment of the MEC servers [15]. In Fig. 5, a representation of the MEC architecture is presented.

Recently, the concept of fog computing was proposed by CISCO as a generalisation of MEC where the definition of edge devices gets broader, ranging from smartphones to set-top boxes. Frequently, these two areas overlap and the terminologies are used interchangeably. However, as mentioned before, fog computing goes beyond the Internet edge, to the continuum from the cloud up to the IoT edge. Besides, these two concepts are interrelated and share several characteristics in common. The taxonomy proposed in [38], shown in Fig. 6, although it was developed for MEC, the parameters taken into account can be considered as valid for cloud and

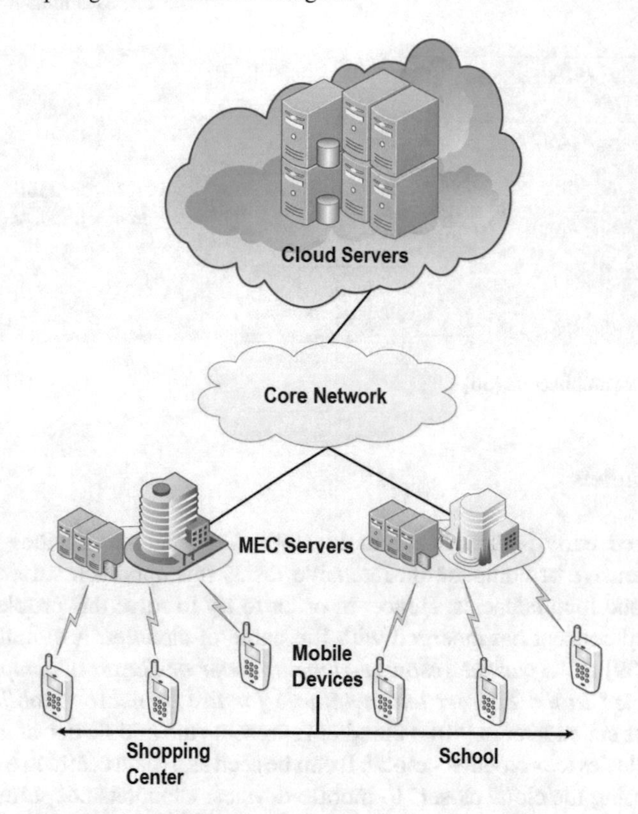


**Fig. 5** Mobile edge computing architecture [38]

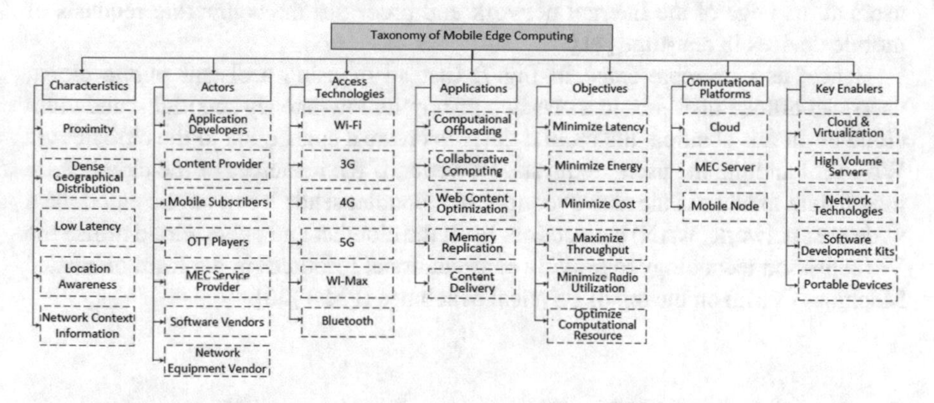


**Fig. 6** Taxonomy of mobile edge computing [38]

fog computing. These parameters include: (a) Characteristics, (b) Actors, (c) Access Technologies, (d) Applications, (e) Objectives, (f) Computation Platforms, and (g) Key Enablers. A description of the concepts included in each parameter can be found in [38].

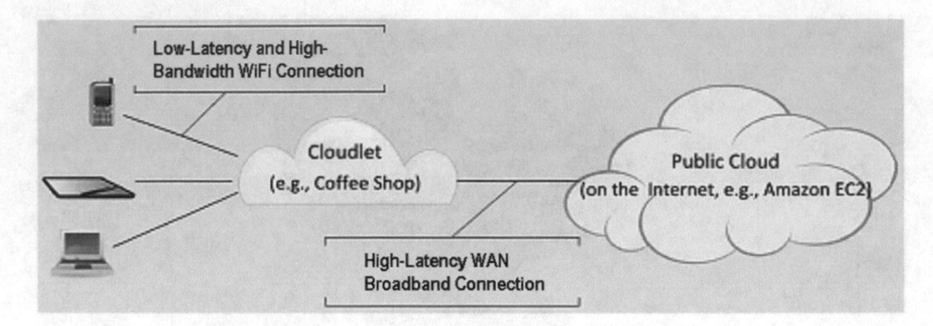


**Fig. 7** Cloudlet architecture [36]

### 2.3.2 Cloudlets

As mentioned early, although a remote cloud helps mobile devices to perform memory-intensive or computation-intensive tasks, this approach suffers from low bandwidth and long latencies. Hence, in order to try to solve this problem, another cloud related concept has emerged with the name of cloudlet. A cloudlet was first defined in [39] as *"a trusted, resource-rich computer or cluster of computers that's well-connected to the Internet and available for use by nearby mobile devices."* Moreover, as stated later in [40] a cloudlet represents the middle tier of a 3-tier hierarchy: mobile device - cloudlet - cloud. It can be seen as a 'data centre in a box' whose goal is to 'bring the cloud closer' to mobile devices. Cloudlets adopt the same idea of edge computing, which is to deploy cloud resources in the proximity of mobile users at the edge of the Internet network and processes the computing requests of mobile devices in real-time [41].

Hence, as it is represented in Fig. 7, instead of using a distant public cloud, users can offload their jobs to a one-hop proximity low-latency and high-bandwidth cloudlet. In this solution, the mobile devices create a connection to the cloudlet via WiFi for handling the users' requests. However, if the resources of the cloudlet are insufficient to address the user's request, the cloudlet sends it to a public cloud via a wide area network (WAN) connection. Both the cloudlet and public cloud utilise the Virtualisation technology to provide computational resources in the form of Virtual Machines (VMs) on the top of Physical Machines (PMs) [36].

## 3 Cloud, Fog and Edge Computing Implementations Overview

Having presented and defined the concepts of cloud, fog and edge computing, in this section an overview of selected implementations of these paradigms is presented.

The idea is to give to the reader a flavour of each one of the technologies when they are put on practice, from the point of view of distributed data processing.

## 3.1 Cloud Computing Implementation Review

The cloud computing model has evolved through time and has been adopted to implement a wide spectrum of applications ranging from high computationally intensive applications down to light weight services. In this section, two application implementations are reviewed, which the authors consider are relevant and demonstrate the usability of the cloud computing model.

### 3.1.1 Cloud-Based Pervasive Patient Health Monitoring

Abawajy and Hassan presented the IoT and cloud-based architecture for remote healthcare monitoring [42], shown in Fig. 8. The approach is referred to as pervasive patient health monitoring (PPHM) system infrastructure. The suitability of the proposed PPHM infrastructure was demonstrated through a case study considering real-time monitoring of a patient suffering from congestive heart failure using ECG. As can be seen in Fig. 8, the proposed architecture is three-tier with the following components: Collection Station, Data Centre, and Observation Station.

The Collection Station consists of an IoT subsystem that is tasked with remote physiological and activity monitoring of patients. The core monitoring infrastructure of the IoT subsystem is a wireless body sensor network (BSN). The personal server provides a link between the IoT subsystem and the cloud infrastructure. The personal server is a dedicated per-patient machine (e.g., a tablet or smartphone) with built in features such as a GPS module, Bluetooth radio module, and SQLite database. It is

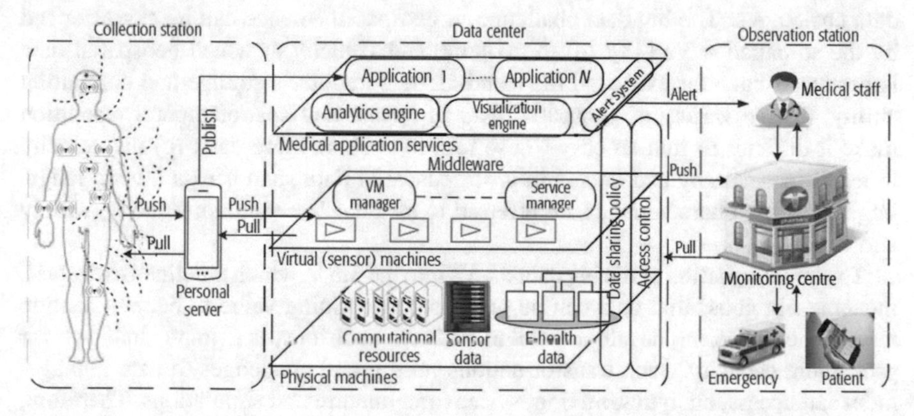


**Fig. 8** Internet of things and cloud-based architecture for remote healthcare monitoring [42]

assumed that the personal server can compatibly interact with various local networks such as WiFi and LTE.

The Data Center Subsystem is a cloud based system where heavy functions that require storing, processing, and analysing the collected patient health data coming from the IoT subsystem is performed. The use of cloud storage offers benefits of scalability and accessibility on demand at any time from any place. The cloud also is used to hosts the middleware system, virtual sensors, and application services that allow medical staff to analyse and visualise patients' data as well as to identify and raise alerts when events requiring urgent intervention are observed.

The Observation Station is where data-driven clinical observation and intervention take place through a monitoring centre. Healthcare actors involved in the monitoring process include doctors, patients, and nursing staff, in clinical observation, patient diagnosis, and intervention processes. The monitoring centre manages all access requests for patient data. Accordingly, if an authorised user issues a data request to the cloud, this is handled by the monitoring centre. If the requested data is available in the sensor data storage, then the data will be returned to the user.

An interesting aspect of the approach proposed by Abawajy and Hassan is that they demonstrated its suitability through a case of study, which consisted in monitoring in real-time a patient suffering from congestive heart failure. Experimental evaluation of the proposed PPHM infrastructure showed that it is a flexible, scalable, and energy-efficient remote patient health monitoring system. Details of the system implementation and results obtained can be found in the paper [42].

### 3.1.2 Cloud Computing Applied to Big Geospatial Data Challenges

The second implementation example presented here is that proposed by Yang et al., [43]. In their work the authors propose a cloud computing based framework to address big data challenges in the area of geospatial science, see Fig. 9. Based on this framework the authors investigate how cloud computing can be utilised to address big data challenges. The big data challenge in geospatial science can be characterised by the so-called 4 Vs [43]: (a) the volume and velocity at which geospatial data is produced have far exceeded the stand-alone computer's storage and computing ability; (b) the variety of geospatial data in format and spatiotemporal resolution make it difficult to find an easy-to-use tool to analyse these data; (c) the veracity in terms of accuracy and uncertainty of geospatial data span across a wide range. Together these characteristics are referred to as the 4 Vs: volume, velocity, variety and veracity.

The transformation of the big data's 4 Vs into the 5th V, which is defined as 'value', meaning big geospatial data can be processed for adding value to better scientific research, engineering development and business decisions, is a grand challenge for processing capacity. Such transformations pose grand challenges to data management and access, analytics, mining, system architecture and simulations. Therefore, as cloud computing provides computing as a utility service with five advantageous characteristics: (a) rapid and elastic provisioning computing power; (b) pooled com-

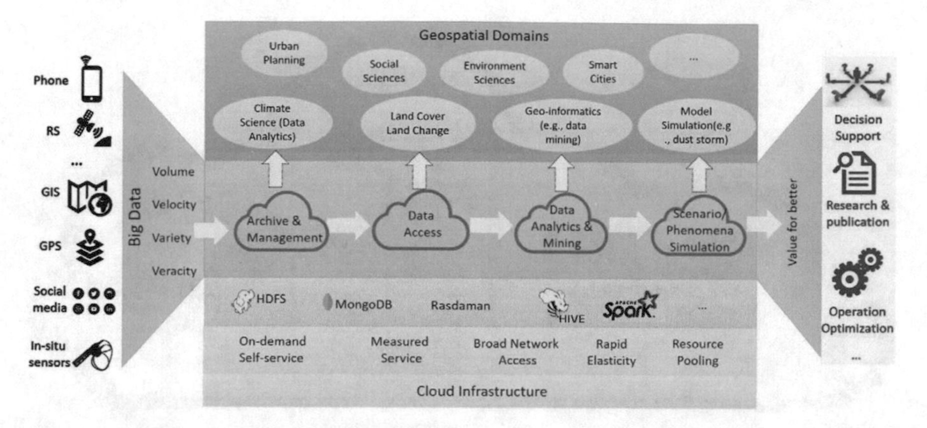


**Fig. 9** Cloud computing based framework for big data processing in geospatial science [43]

puting power to better utilise and share resources; (c) broadband access for fast communication; (d) on demand access for computing as utility services; and (e) pay-as-you-go for the parts used without a significant upfront cost like that of traditional computing resources; it can be utilised to enable solutions for the transformation of the 4 Vs into the 5th V (value). In order to exemplify how the proposed architecture can enable this transformation, four examples of big data processing are reviewed in the paper, these include: climate studies, geospatial knowledge mining, land cover simulation, and dust storm modelling.

For the matter of space, in this overview only the case of climate studies big data processing is reviewed. The combined complexities of volume, velocity, variety, and veracity can be addressed with cloud-based advanced data management strategies and a service-oriented data analytical architecture to help process, analyse and mine climate data. For example, climate simulation poses challenges on obtaining enough computing resources for scientific experiments when analysing big simulation data or running a large number of model simulations according to different model inputs. This problem can be addressed using cloud computing as follows [43]: (a) the climate models can be published as a service (Model as a Service, MaaS) and enough VMs can be provisioned with specific model configurations for each ensemble modelling run on demand; (b) the application is deployed as a service with a web portal to support model operation and monitoring; and (c) the workflow involving different analytics is operated as a service (Workflow as a Service, WaaS) with intuitive graphical user interfaces (GUIs). With this, the big climate data analytics are supported by cloud computing at the computing infrastructure level.

The proposed cloud-based architecture for service-oriented workflow system for climate model study can be seen in Fig. 10. This architecture includes [43]: (a) the model service is responsible for compiling and running models on VMs, which are provisioned based on the snapshot of the system containing the modelling software environment to run a model; (b) the VM monitor service provides the cloud platform with VM status information for resource scheduling; (c) the data analysis service

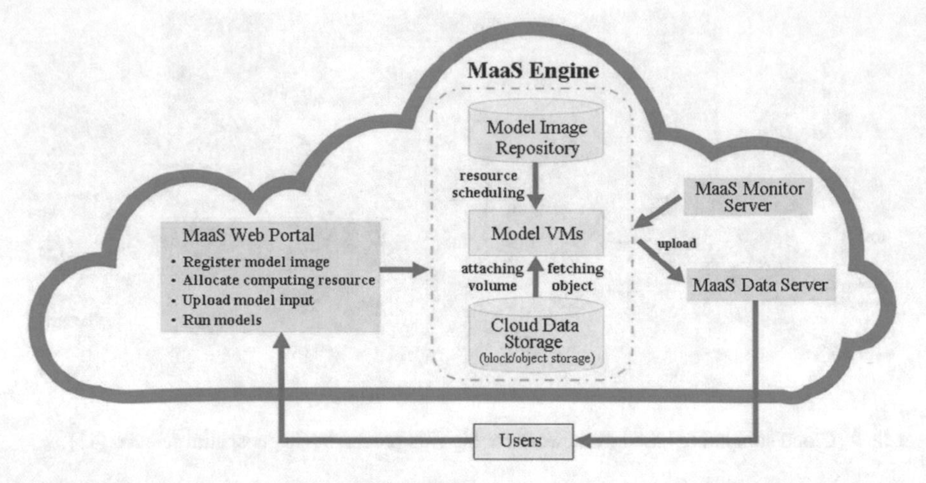


**Fig. 10** Cloud-based service-oriented workflow system for climate model study [43]

feeds the model output as the input for analytics, while analyzing data in parallel to address data intensive issues. Data publishing service enables users to access the analysis results in real time via the Internet. All of these services are controllable through a GUI, which enables users to drag and connect services together to build a complex workflow so the system can automatically transition to the applications specified by the workflow and run on the cloud with automatically provisioned VMs [43].

For further details on the other three big data processing problems addressed using cloud computing, the reader is referred to [43].

## *3.2 Fog Computing Implementation Review*

With the emergence of fog computing, which brings cloud service features closer to the thing generating data, the question now is how processing, storage an analytics tasks are segmented among the fog and the cloud. Deciding what tasks go to fog and what goes to the backend cloud are in general terms application specific. Naturally, certain functions are better fitted to be performed at fog nodes, while other functions are better suited to be performed at the cloud side. Still, customary backend cloud computing will continue to remain an important part of IoT systems as fog computing emerges. This segmentation could be planned, but also can be adjusted dynamically according to the network state, for example, based on changes in processor loads, link bandwidths, storage capacities, fault events, security threats, cost targets, etc., [30]. Hence, several approaches exists which propose the decentralisation of resources and services of computing, storage, control and networking tasks along the continuum from the cloud to the things. In the next sections some proposals for fog computing implementations are reviewed.

### 3.2.1 Reference Architecture for Fog Computing

Perhaps one of the main fog computing proposals is the reference architecture for fog computing from the OpenFog Consortium [30]. This is a system-level horizontal architecture which considers functional boundaries in fog computing as fluid, meaning that multitude of combinations can exists to physically deploy fog-cloud resources based on domain-specific solutions. These deployments will fit in one of four possible domain scenarios. Depending on the scenario, multiple fog and cloud elements may collapse into a single physical deployment. Figure 11 presents the four fog hierarchical domain deployment scenario models. For a complete description of each one of these scenario models the reader is referred to [30].

Real world deployments may involve multi-tenants, fog, and cloud deployments owned by multiple entities. As the OpenFog Consortium argues, many of the fog computing usages will occur as represented in the scenarios represented by 2 and 3 in Fig. 11. The three-layer fog hierarchies shown are for illustrative purposes only. Real world fog deployments may have more or fewer levels. Different vertical application use cases may use a fog hierarchy differently. For example, in a smart city, there may be fog nodes in a region, neighborhood, street corner, and building level. In a smart factory, the hierarchy may be divided by assembly lines, manufacturing cells, and machines [30].

### 3.2.2 Fog Computing Deployment Based on Docker Containerization

An interesting study where practical implementation aspects of fog architectures are reviewed is the one proposed in [44]. This approach considers a fog computing solution based on two primary directions of gateway node improvements: (i) fog-oriented framework for IoT applications based on innovative scalability extensions of the open-source Kura gateway (Kura is an open-source project for IoT-cloud integration and management [45]), and (ii) Docker-based containerization (Docker is an open-source software that automates the deployment of applications inside software containers [46]) over challenging and resource-limited fog nodes, i.e., Raspberry Pi devices [47].

The standard IoT-cloud architecture using Kura is the three-layer shown in Fig. 12a. This architecture has an intermediate fog layer of geographically distributed gateways nodes, positioned at the edge of network localities that are densely populated by IoT sensors and actuators. This architecture is extended under the Kura framework with the introduction of local brokers for scalability purposes, as is shown in Fig. 12b. While in the standard Kura IoT gateways simply aggregate all the data gathered by their sensors and send them to a MQTT broker running on the cloud, delegating all the associated computations to the global cloud, in the proposed extended

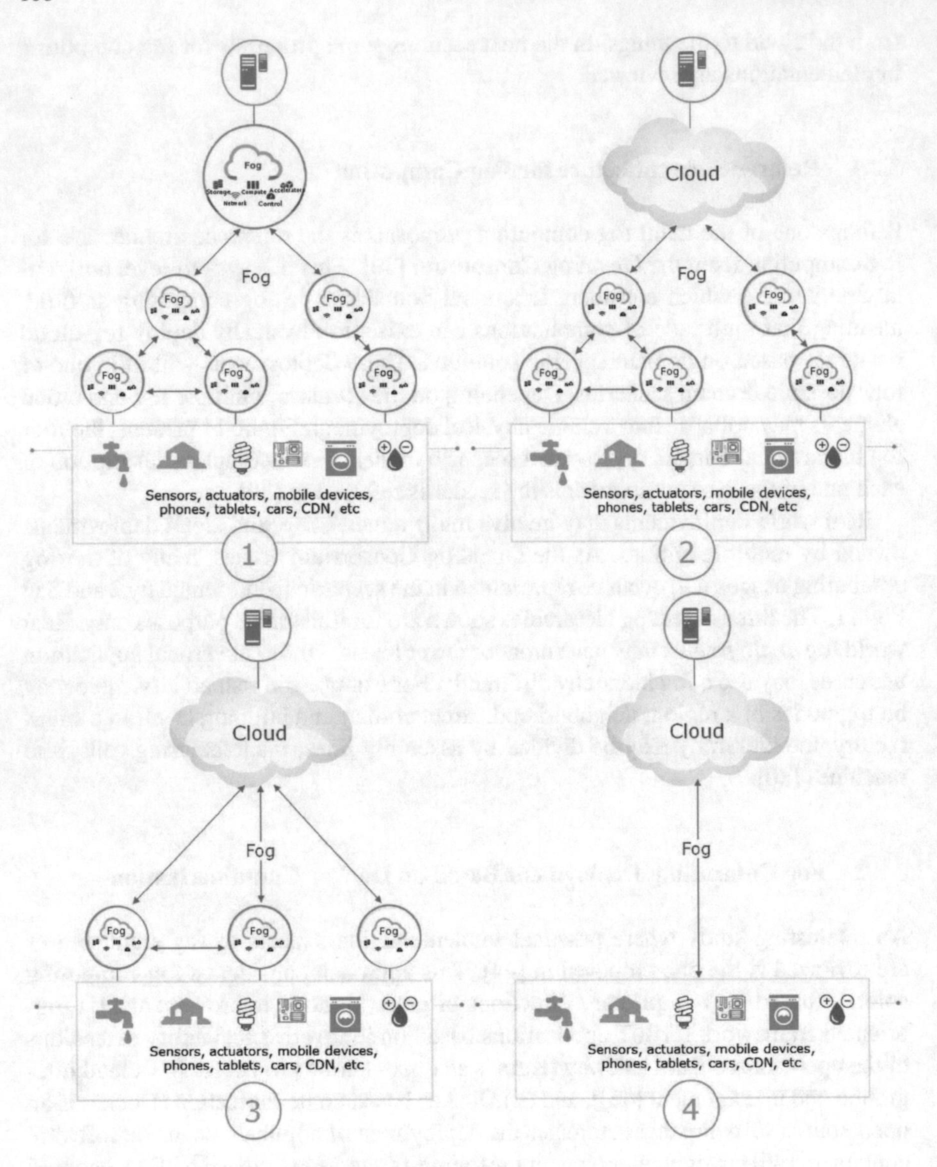


**Fig. 11** Fog hierarchical deployment models [30]

architecture the Kura IoT gateways work as fog nodes which can scalably serve as local MQTT brokers and dynamically coordinate among themselves, without the need of the continuous support of global cloud resources.

Therefore, as depicted in Fig. 12b, MQTT brokers are included on each gateway in order to collect sensed data at the gateway side and, after local processing and inferencing, to send filtered/processed/aggregated data to the cloud. As described by

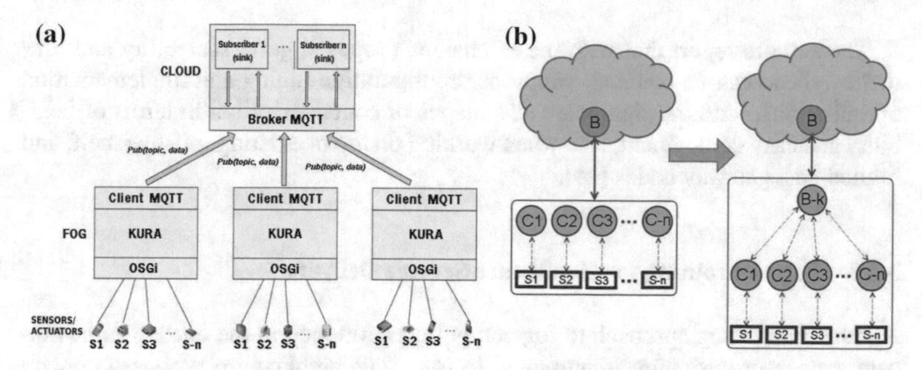


**Fig. 12** **a** IoT-cloud architecture based on Kura; **b** inclusion of an MQTT broker on each gateway [44]. Sx = Sensor x, Cx = Client MQTT x, B-k = Broker at side k, B = Broker at cloud

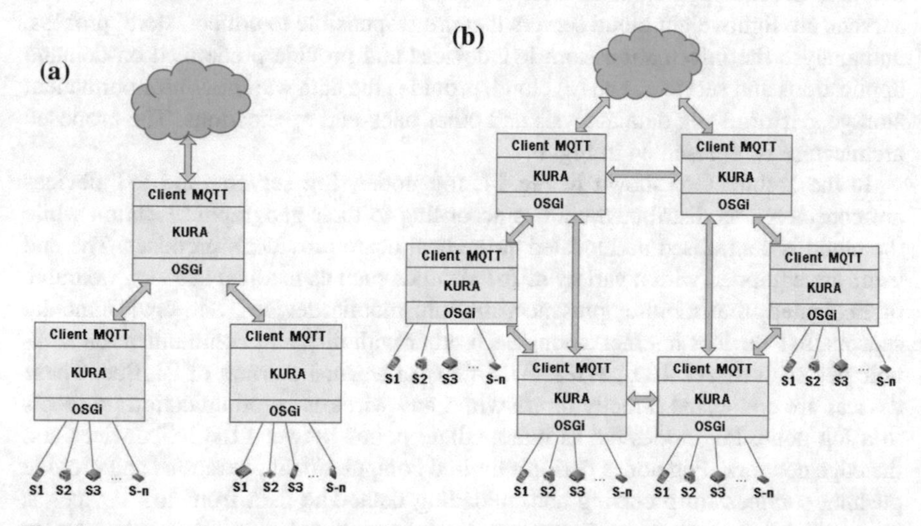


**Fig. 13** **a** Support for cluster organisation of IoT gateways, **b** support for mesh organisation of IoT gateways [44]

the authors [44], this extension to the fog architecture offers the following advantages: enabling hierarchical topologies, gateway-level MQTT message aggregation, real-time message delivery and reactions, actuation capacity and message priorities, locality awareness and locality-oriented optimisations, gateway-cloud connection optimisation. For more details the reader is referred to [44].

Two additional gateway extensions are proposed by Bellavista and Zanni [44], these include support for cluster and mesh gateway organisations of IoT gateways, shown in Fig. 13a, b, respectively

The most significant advantages associated with the support of cluster/mesh gateway organisations of IoT include: Kura gateway specialization, locality exploitation and data quality, geo-distribution, scalability, security and privacy.

The authors report that with the extensions proposed good scalability and limited overhead can be coupled, via proper configuration tuning and implementation optimisations, with the significant advantages of containerisation in terms of flexibility and easy deployment, also when working on top of existing, off-the-shelf, and limited-cost gateway nodes [44].

### 3.2.3 Fog Computing for Healthcare Service Delivery

Another interesting approach to fog computing developed in the context of healthcare service provisioning is presented in [48]. The architecture proposed contain three main components: (a) fog nodes, that support multiple communication protocols in order to aggregate data from various heterogeneous IoT devices; (b) fog servers, are lightweight cloud servers that are responsible to collect, store, process, and analyse the information from IoT devices and provide predefined on-demand applications and services, and (c) cloud, provides the data warehouse for permanent storage, performs big data analysis and other back-end applications. The proposed architecture is represented in Fig. 14.

In the architecture shown in Fig. 14, fog nodes, fog servers, and IoT devices are considered as distributed entities according to their geographic location while the cloud is centralised and located at the healthcare provider's premises. The end users are equipped with a variety of IoT devices such as medical sensors, wearable or implanted, that monitor presence, motion, mobile devices, and environmental sensors. IoT devices interact among each other with different communication technologies such as 3G, LTE, WiFi, WiMAX, 6Lowpan, Ethernet, or ZigBee. These devices are connected directly or via wired and wireless communication protocols to a fog node. Fog nodes act as intermediate points between the IoT devices and the edge network. Fog nodes perform limited computational operations and provide pushing services for receiving and uploading data. The data from IoT devices is transmitted to the fog server. Fog servers perform protocol conversion, data storage, processing, filtering, and analysis. Fog servers are able to make decisions and provide services to the end users based on predefined policies and rules without the necessity to interact with the cloud server [48]. However, it is assumed that fog servers have limited hardware capabilities and cannot fully support the creation of new services and applications. Hence, fog servers have to interact with the cloud in order to provide prediction models, enable big data analysis as well as new service execution and orchestration. This means that the fog cannot replace the cloud but they have to cooperate in order to provide timely value-added services to the end users. For example, data which has been processed, analysed, and temporarily stored in fog servers can be transmitted to the cloud for permanent storage and further analysis or, if there is not any necessity to be stored, these data can be removed from fog servers.

In order to illustrate the benefits that the architecture integrating IoT, fog, and cloud computing offers, in their work the authors presented two use cases scenarios, the first one is related to daily monitoring for provisioning healthcare services; the second one refers to an eCall service system to provide immediate assistance to people involved

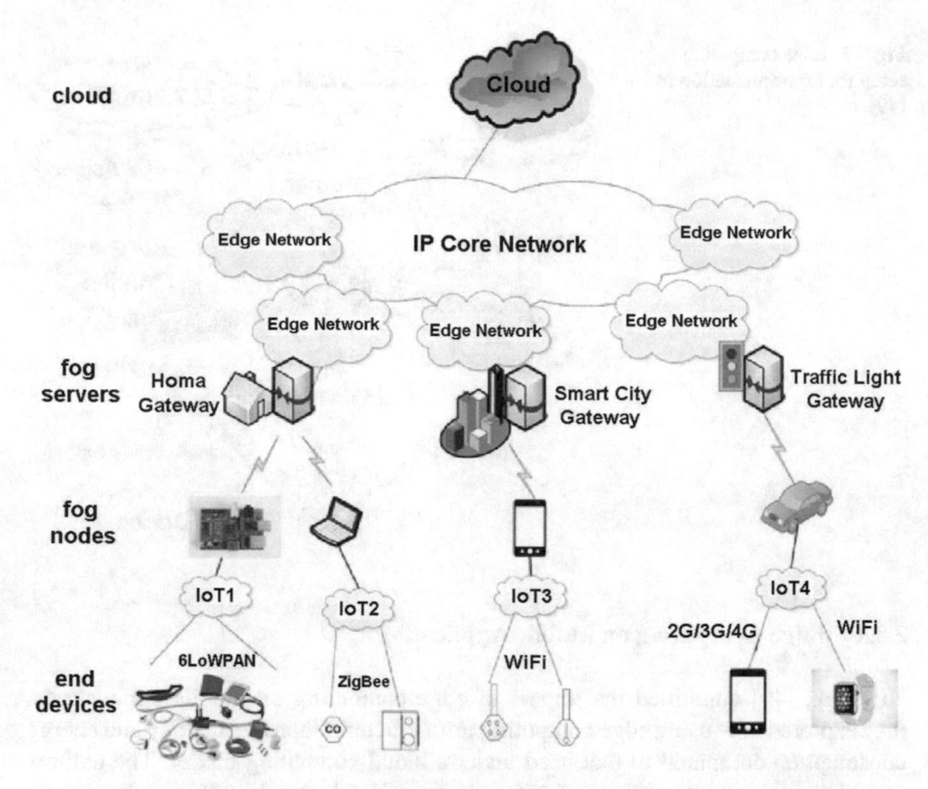


Fig. 14 Architecture integrating IoT, fog and cloud computing for healthcare service provisioning [48]

in a road collision accident. The main advantages, found as possible benefits of using such architecture, are a faster and accurate treatment delivery, reduction of medical costs, improvement of doctor–patient relationships, and the delivery of personalised treatment oriented to users' needs and preferences, for more details the reader is referred to [48].

## 3.3 Edge Computing Implementation Review

As described previously, in this work edge computing refers to placing applications, data and processing capabilities at the edge of the Internet. Therefore, in this section a review of various implementations of edge computing is presented.

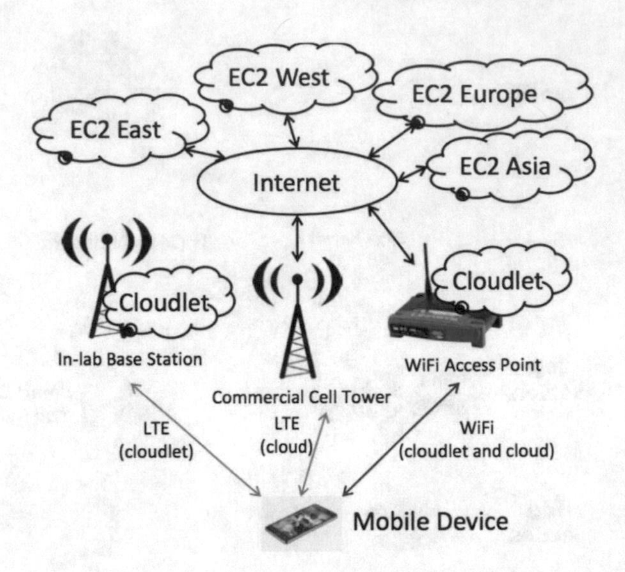


**Fig. 15** Edge computing setup for experimentation in [49]

### 3.3.1 Edge Computing on Mobile Applications

Hu et al., [49] quantified the impact of edge computing on mobile applications, they explored how using edge computing infrastructure improves latency and energy consumption compared to that used under a cloud computing model. The authors considered five configurations of network: No offload, cloud by WiFi, cloudlet by WiFi, cloudlet by LTE and cloud by LTE, with three statically prepartitoned applications from existing research, face recognition application (FACE), augmented reality application (MAR) and physics-based computer graphics example (FLUID). The proposed experimental setup is shown in Fig. 15. The results presented show that edge computing improves response time and energy consumption significantly for mobile devices on WiFi and LTE networks. Hence local processing consumes less power and is faster than sending the data to a distant cloud. Also results show that offloading computation blindly to the cloud can be a losing strategy. Offloading to a distant cloud can result in lower performance and higher energy costs than running locally on mobile devices. Edge computing achieves performance and energy improvements through offloading computing services at the edge of the Internet for mobile devices [49].

### 3.3.2 Femto Cloud System

An interesting approach is the proposed by Habak et al. [50], which consider how a collection of co-located devices can be orchestrated to provide a cloud service at the edge. As an answer to this question, the authors propose the FemtoCloud system, which provides a dynamic, self-configuring and multiple mobile devices

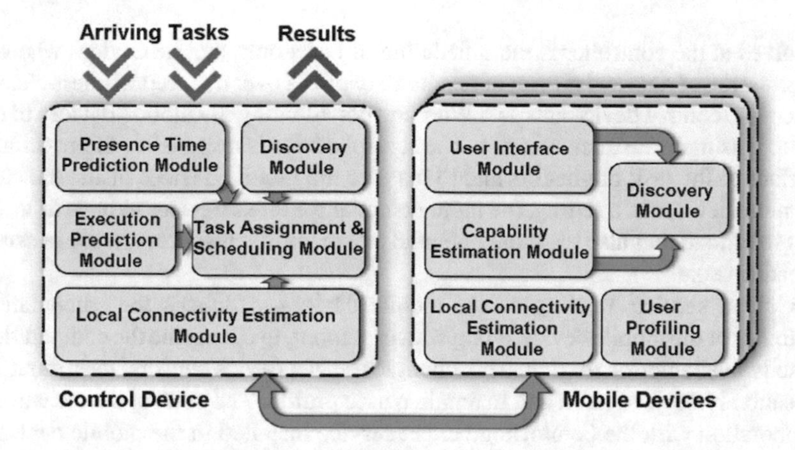


**Fig. 16** The FemtoCloud system architecture [50]

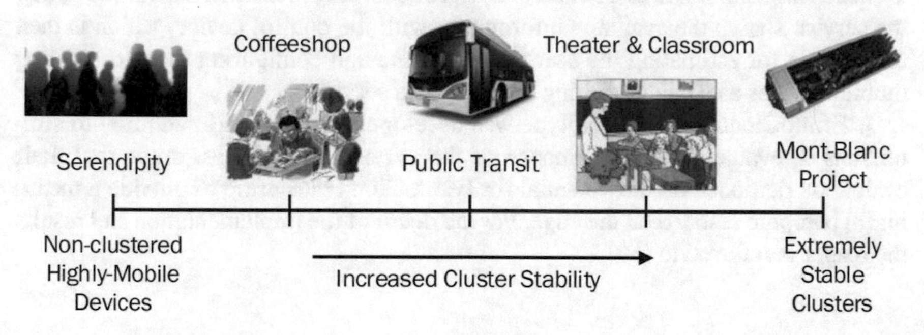


**Fig. 17** Mobile cluster stability representation [50]

cloudlet by coordinating a cluster of mobile devices. The general FemtoCloud system architecture is shown in Fig. 16.

The FenmtoCloud architecture consists of two main components, thc control device and the co-located mobile devices, as shown in Fig. 16. The co-located devices can appear in different scenarios including: passengers with mobile devices using public transit services, students in classrooms and groups of people sitting in a coffee shop, for example. It is assumed, however, that the devices are in a cluster stability, meaning that there is predictability in the duration of time that a given device is available for use in the FemtoCloud. A representation of cluster stability in different settings is shown in Fig. 17.

The control device is assumed as being provided as part of the infrastructure of the cluster scenario environment, this is the coffee shop, the university, etc. The architecture identifies the functionality to be realised in the controller and in the mobile devices, and the information to communicate within and between the two. Therefore, the FemtoCloud controller is responsible for deciding which mobile devices will be added to the compute cluster in the current environment. A critical problem that must be solved at the controller is the scheduling of tasks onto mobile devices where the transmission of data and receipt of results all happens over a shared wireless channel. Hence, the control device acts as a WiFi hotspot allowing the mobile devices to connect to it using infrastructure mode. The control device is responsible of providing an interface to the task originators and to manage the mobile devices inside the cloud. The more devices in a setting, the more potential for those devices to have tasks they want to offload, but also the more potential exists for idle resources to use in serving offloaded tasks.

A client service, running on the mobile devices, estimates the computational capability of the mobile device, along with user input, to determine the computational capacity available for sharing. This client leverages device sensors, user input, and utilisation history, to build and maintain a user profile. The control device works in collaboration with the FemtoCloud client service installed in the mobile devices to acquire information about the device characteristics and user profiles. The service uses such information to assign task to devices according to a heuristic. Subsequently, the service shares the available information with the control device, which is then responsible for estimating the user presence time and configuring the participating mobile devices as a cloud offering compute as a service.

A FemtoCloud system prototype was developed and used it in addition to simulations to evaluate the performance of the system. The authors argue that their evaluation demonstrate the potential for FemtoCloud clustering to provide a meaningful compute resource at the edge. For the detail of the implementation and results the reader is referred to [50].

## 4 Conclusions

In a traditional cloud-based architecture application intelligence and storage are centralised in server wire centers. It was believed that this architecture would satisfy the needs of most of the IoT applications. However, as IoT applications started demanding real-time response (for control loops, or health care monitoring applications, for example), generating an increasing amount of data and consuming cloud services overstretching the network and cloud infrastructure, added to a limited network bandwidth, this idea started to break down. In order to meet the network strains of data transmissions to the cloud, different approaches for providing computational, networking, and storage capabilities closer to the end users have been proposed. Among them, there can be distinguished the fog and edge computing approaches as the most relevant ones.

Fog computing has been proposed as a system-level architecture that can optimise the distribution of computational, networking, and storage capabilities in a hierarchy of levels at the continuum network from cloud to IoT end devices. It seeks to provide the exact balance of capacity among the three basic capabilities, computational, networking, and storage, at the precise level of the network where they are the most optimally located. Nevertheless, fog computing should be considered not as replacement of the cloud, but as a supplement to the cloud for the most critical aspects of network operations. Fog also supplements and greatly expands the capabilities of intelligent endpoint devices and other smart objects.

Fog computing can provide enhanced features such as user mobility support, location awareness, dense geographical distribution, low latency, and delays that could not be supported inherently by the cloud computing approach. These characteristics are significant for provisioning delay-sensitive services such as healthcare and emergency services.

The aim of edge computing is to place computing and storage resources at the edge of the Internet, in close proximity to mobile devices or sensors. Edge nodes include base stations, routers and switches that route network traffic. Placing data and data-intensive applications at the edge reduces the volume and distance that data must be moved. Hence, computing on edge nodes closer to application users is a platform for application providers to improve their service.

In this chapter, the two main paradigms that evolved from the initial concept of cloud computing have been presented and some application implementations have been reviewed. The idea was to define every concept from the point of view of decentralisation of data processing, networking and storage capabilities. As seen in the applications reviewed, the utilisation of a given paradigm for a particular application is problem specific, which needs an analysis according with the requirements and characteristics of each approach.

**Acknowledgements** The authors acknowledge the support to carry out this work from Instituto Politécnico Nacional under grant SIP-1894.

## References

1. Zhou, H.: The Internet of Things in the Cloud: A Middleware Perspective. CRC Press, Boca Raton (2013)
2. Weiser, M., Gold, R., Brown, J.S.: The origins of ubiquitous computing research at PARC in the late 1980s. IBM Syst. J. **38**(4), 693–696 (1999)
3. Ashton, K.: That "Internet of things" thing. RFiD J. **22**(7), 97–114 (2009)
4. Uckelmann, D., Harrison, M., Michahelles, F.: An architectural approach towards the future Internet of things. In: Uckelmann, D., Harrison, M., Michahelles, F. (eds.) Architecting the Internet of Things, pp. 1–24. Springer, Berlin (2011)
5. Gubbi, J., Buyya, R., Marusic, S., Palaniswami, M.: Internet of things (IoT): a vision, architectural elements, and future directions. Future Gener. Comput. Syst. **29**(7), 1645–1660 (2013)
6. Kotis, K., Katasonov, A.: Semantic interoperability on the web of things: the semantic smart gateway framework. In: Proceedings of the IEEE Sixth International Conference on Complex, Intelligent and Software Intensive Systems (CISIS), pp. 630–635 (2012)
7. Mazhelis, O., Warma, H., Leminen, S., Ahokangas, P., Pussinen, P., Rajahonka, M., Siuruainen, R., Okkonen, H., Shveykovskiy, A., Myllykoski, J.: Internet-of-things market, value networks, and business models: State of the art report. University of Jyvaskyla. http://internetofthings.fi/extras/IoTSOTAReport2013.pdf. Accessed 23 February 2017 (2013)
8. McFadin, P.: Internet of things: where does the data go? WIRED. Accessed 15 Jan 2017 (2015). https://www.wired.com/insights/2015/03/internet-things-data-go/

9. Dey, S., Mukherjee, A., Paul, H.S., Pal, A.: Challenges of using edge devices in IoT computation grids. In: Porceedings of IEEE 2013 International Conference on Parallel and Distributed Systems (ICPADS), pp. 564–569 (2013)
10. MQTT. Accessed 17 April 2017. http://mqtt.org/documentation
11. Krawiec, P., Sosnowski, M., Batalla, J.M., Mavromoustakis, C.X., Mastorakis, G., Pallis, E.: Survey on technologies for enabling real-time communication in the web of things. In: Batalla, J.M. et al. (eds.) Beyond the Internet of Things, pp. 323–339. Springer International Publishing, Switzerland (2017)
12. CoAP. Accessed 17 April 2017. http://coap.technology/
13. Patierno, P.: Hybrid IoT: On fog computing, gateways, and protocol translation. DZone/IoT Zone. Accessed 19 Dec 2016 (2016). https://dzone.com/articles/the-hybrid-internet-of-things-1
14. Cox, P.A.: Mobile cloud computing devices, trends, issues, and the enabling technologies, developerWorks, IBM. Accessed 20 Dec 2016 (2011). https://www.ibm.com/developerworks/cloud/library/cl-mobilecloudcomputing/cl-mobilecloudcomputing-pdf.pdf
15. IBM Watson.: The power of analytics at the edge. IBM. Accessed 15 November 2016 (2016). http://www-01.ibm.com/common/ssi/cgi-bin/ssialias?htmlfid=WWS12351USEN
16. Rayes, A., Salam, S.: Fog computing. In: Internet of Things — From Hype to Reality, pp. 139–164 . Springer International Publishing AG. (2017)
17. CISCO: Connections counter: The Internet of everything in motion. Accessed 3 March 2017 (2013). http://newsroom.cisco.com/feature-content?articleId=1208342
18. Nielsen, J.: Nielsen's law of Internet bandwidth. Accessed 3 March 2017 (1998). https://www.nngroup.com/articles/law-of-bandwidth/
19. Byers, C.C., Wetterwald, P.: Fog computing distributing data and intelligence for resiliency and scale necessary for IoT: the internet of things (Ubiquity symposium). Ubiquity **2015**, 4:1–4:12 (2015)
20. Internet Edge Solution Overview. Accessed 5 February 2017 (2010). http://www.cisco.com/c/en/us/td/docs/solutions/Enterprise/WAN_and_MAN/Internet_Edge/InterEdgeOver.pdf
21. Biron, J., Follett, J.: Foundational Elements of an IoT Solution. O'Reilly Media Inc, Sebastopol (2016)
22. Mell, P., Grance, T.: The NIST Definition of Cloud Computing, pp. 800–145. NIST Special Publication (2011)
23. Armbrust, M., Fox, A., Griffith, R., Joseph, A.D., Katz, R., Konwinski, A., Lee, G., Patterson, D., Rabkin, A., Stoica, I., Zaharia, M.: A view of cloud computing. Commun. ACM **53**(4), 50–58 (2010)
24. What is cloud computing? IBM. Accessed 3 March 2017 (2017). https://www.ibm.com/cloud-computing/learn-more/what-is-cloud-computing/
25. Vaquero, L.M., Rodero-Merino, L.: Finding your way in the fog: towards a comprehensive definition of fog computing. ACM SIGCOMM Comput. Commun. Rev. **44**(5), 27–32 (2014)
26. Dinh, H.T., Lee, C., Niyato, D., Wang, P.: A survey of mobile cloud computing: architecture, applications, and approaches. Wireless communications and mobile computing **13**(18), 1587–1611 (2013)
27. Tordera, E. M., Masip-Bruin, X., Garcia-Alminana, J., Jukan, A., Ren, G. J., Zhu, J., Farre, J.: What is a Fog Node: a tutorial on current concepts towards a common definition (2016). arXiv:1611.09193
28. Bonomi, F., Milito, R., Zhu, J., Addepalli, S.: Fog computing and its role in the Internet of things. In: Proceedings of the First Edition of the MCC Workshop on Mobile Cloud Computing, pp. 13–16. Helsinki, Finland, (2012)
29. Varghese, B., Wang, N., Nikolopoulos, D.S., Buyya, R.: Feasibility of fog computing (2017). arXiv:1701.05451
30. OpenFog reference architecture for fog computing, OpenFog Consortium. OPFRA001.020817. Accessed 3 March 2017 (2017). https://www.openfogconsortium.org/ra/
31. Stojmenovic, I.: Fog computing: a cloud to the ground support for smart things and machine-to-machine networks. In: IEEE Australasian Telecommunication Networks and Applications Conference (ATNAC), pp. 117–122 (2014)

32. Yi, S., Li, C., Li, Q.: A survey of fog computing: concepts, applications and issues. In: Proceedings of the 2015 Workshop on Mobile Big Data. ACM (2015)
33. Garcia-Lopez, P., Montresor, A., Epema, D., Datta, A., Higashino, T., Iamnitchi, A., Barcellos, M., Felber, P., Riviere, E.: Edge-centric computing: vision and challenges. ACM SIGCOMM Comput. Commun. Rev. **45**(5), 37–42 (2015)
34. Reinders, J.: Intel Threading Building Blocks: Outfitting C++ for Multi-core Processor Parallelism. O'Reilly, Sebastopol (2007)
35. Varghese, B., Wang, N., Barbhuiya, S., Kilpatrick, P., Nikolopoulos, D.S.: Challenges and opportunities in edge computing. In: IEEE International Conference on Smart Cloud (SmartCloud), pp. 20–26 (2016)
36. Mobile edge computing - Introductory technical white paper. ETSI. https://portal.etsi.org/portals/0/tbpages/mec/docs/mobile-edge_computing_-_introductory_technical_white_paper_v1%2018-09-14.pdf. Accessed 3 March 2017 (2014)
37. Beck, M.T., Werner, M., Feld, S., Schimper, S.: Mobile edge computing: a taxonomy. In: Proceedings of the Sixth International Conference on Advances in Future Internet (2014)
38. Ahmed, A., Ahmed, E.: A survey on mobile edge computing. In: IEEE 10th International Conference on Intelligent Systems and Control (ISCO), pp. 1–8 (2016)
39. Satyanarayanan, M., Bahl, P., Caceres, R., Davies, N.: The case for VM-based cloudlets in mobile computing. IEEE Pervasive Comput. **8**(4), 2–11 (2009)
40. Satyanarayanan, M., Chen, Z., Ha, K., Hu, W., Richter, W., Pillai, P.: Cloudlets: at the leading edge of mobile-cloud convergence. In: Proceedings of IEEE 6th International Conference on Mobile Computing, Applications and Services (MobiCASE), pp. 1–9 (2014)
41. Gao, L., Luan, T.H., Liu, B., Zhou, W., Yu, S.: Fog computing and its applications in 5G. In: 5G Mobile Communications, pp. 571–593. Springer International Publishing, Switzerland (2017)
42. Abawajy, J.H., Hassan, M.M.: Federated Internet of things and cloud computing pervasive patient health monitoring system. IEEE Commun. Mag. **55**(1), 48–53 (2017)
43. Yang, C., Yu, M., Hu, F., Jiang, Y., Li, Y.: Utilizing cloud computing to address big geospatial data challenges. Comput. Environ. Urban Syst. **61**, 120–128 (2017)
44. Bellavista, P., Zanni, A.: Feasibility of fog computing deployment based on Docker containerization over RaspberryPi. In: Proceedings of the ACM 18th International Conference on Distributed Computing and Networking. Hyderabad, India (2017)
45. Kura. Accessed 3 March 2017. https://eclipse.org/kura
46. Docker. Accessed 3 March 2017. https://www.docker.io
47. Raspberry Pi. Accessed 17 May 2017. https://www.raspberrypi.org/
48. Andriopoulou, F., Dagiuklas, T., Orphanoudakis, T.: Integrating IoT and fog computing for healthcare service delivery. In: Keramidas, G. et al. (eds.) Components and Services for IoT Platforms, pp. 213–232. Springer International Publishing, Switzerland (2017)
49. Hu, W., Gao, Y., Ha, K., Wang, J., Amos, B., Chen, Z., Pillai, P., Satyanarayanan, M.: Quantifying the impact of edge computing on mobile applications. In: Proceedings of the 7th ACM SIGOPS Asia-Pacific Workshop on Systems. Hong Kong, China (2016)
50. Habak, K., Ammar, M., Harras, K.A., Zegura, E.: Femto clouds: leveraging mobile devices to provide cloud service at the edge. In: IEEE 8th International Conference on Cloud Computing (CLOUD), pp. 9–16 (2015)

# Part II
# Search, Optimization and Hybrid Algorithms

# Integer Programming Models and Heuristics for Non-crossing Euclidean 3-Matchings

**Rodrigo Alexander Castro Campos, Marco Antonio Heredia Velasco, Gualberto Vazquez Casas and Francisco Javier Zaragoza Martínez**

**Abstract** Given a set $P$ of $3k$ points in general position in the plane, a Euclidean 3-matching is a partition of $P$ into $k$ triplets, such that the cost of each triplet $(u, v, w)$ is the sum of the lengths of the segments $\overline{uv}$ and $\overline{wv}$, and the cost of the 3-matching is the sum of the costs of its triplets. We are interested in finding non-crossing Euclidean 3-matchings of minimum and maximum cost. As these are hard combinatorial problems, we present and evaluate three integer programming models and three heuristics for them.

**Keywords** Euclidean 3-matching · Non-crossing · Integer programming · Linear relaxation · Heuristics

## 1 Introduction

Let $P$ be a set of $n = 3k$ points in the Euclidean plane in general position (i.e., no three points of $P$ are collinear). A 3-matching is a partition of $P$ into $k$ disjoint subsets of 3 points each, called *triplets*. There are several ways of assigning a cost to a triplet, for example the perimeter or the area of the corresponding triangle [3, 10]. In our case, we represent a triplet $(u, v, w)$ by the segments $\overline{uv}$ and $\overline{wv}$ and its

R.A. Castro Campos · G. Vazquez Casas (✉)
Posgrado en Optimización, Universidad Autónoma Metropolitana
Unidad Azcapotzalco, Mexico City, Mexico
e-mail: gvc@correo.azc.uam.mx

R.A. Castro Campos
e-mail: racc@correo.azc.uam.mx

M.A. Heredia Velasco · F.J. Zaragoza Martínez
Departamento de Sistemas, Universidad Autónoma Metropolitana
Unidad Azcapotzalco, Mexico City, Mexico
e-mail: hvma@correo.azc.uam.mx

F.J. Zaragoza Martínez
e-mail: franz@correo.azc.uam.mx

© Springer International Publishing AG 2018

Y. Maldonado et al. (eds.), *NEO 2016*, Studies in Computational Intelligence 731,
https://doi.org/10.1007/978-3-319-64063-1_5

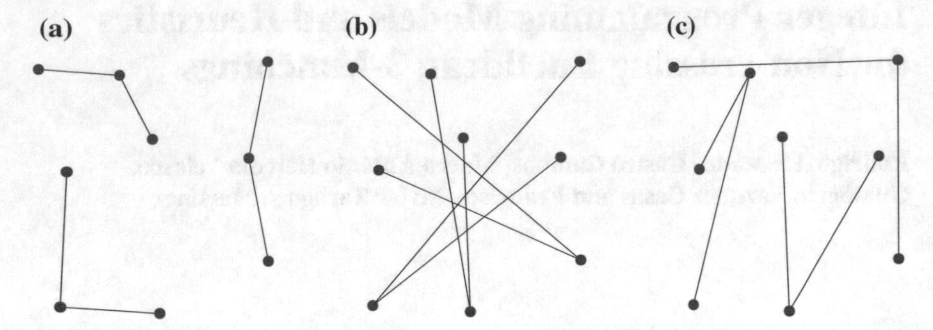


**Fig. 1** Three Euclidean 3-matchings of the same point set $P$: **a** of minimum cost, **b** of maximum cost, and **c** non-crossing of maximum cost

cost is given by the sum of the lengths of those segments [3, 7, 8, 11]. The cost of a 3-matching of $P$ is the sum of the costs of its chosen triplets.

In the usual formulation of the Euclidean 3-matching problem we need to find a minimum cost 3-matching of $P$. See Fig. 1a. This problem has several applications, especially in the insertion of components on a printed circuit (PC) board [2, 6, 12]. Johnsson, Magyar, and Nevalainen [7, 8] introduced two integer programming formulations for this problem, and proved that its decision version is NP-complete, if each triplet has an arbitrary positive cost (i.e., not necessarily Euclidean). The problem remains NP-complete even if the points of $P$ correspond to vertices of a unit distance graph [13] (a metric cost function).

Besides the minimization problem, we are interested in a similar maximization problem: finding a maximum cost non-crossing Euclidean 3-matching of $P$. We say that two segments *cross* each other if they intersect in a common interior point. It is easy to see that, in a minimum cost Euclidean 3-matching of $P$, any two representing segments of triplets do not cross, while in a maximum cost Euclidean 3-matching there could be many crossings. See Fig. 1.

Both problems, minimum cost and maximum cost non-crossing, are challenging, and we believe that both are NP-hard. Similar non-crossing maximization problems have been studied before [1, 4]. Exact solutions to both problems can be attained through Integer Programming, however, in order to obtain good solutions in feasible times, we fix our attention to heuristics.

The rest of this paper is organized as follows: In Sect. 2 we present two well-known integer programming formulations of the minimum Euclidean 3-matching problem and we introduce three new integer programming formulations that use half as many variables. We also show how to adapt these formulations to avoid crossings in the maximum non-crossing Euclidean 3-matching problem. Using both the old and new formulations, we solve to optimality some benchmark instances and we also obtain their linear programming relaxations. Section 3 presents three heuristics specially designed for our problems and compare their solutions to the optimal ones. For this, we use the same benchmark instances as before. Finally, in Sect. 4, we present our conclusions and our ideas for future work.

## 2 Integer Programming Formulations

In this section we study several integer programming formulations for the minimum cost 3-matching problem, three of them new. Later on, we discuss how to adapt these formulations to solve the maximum cost non-crossing problem.

For the following models, the cost $c_{rs}$ is assumed to be the Euclidean length of the segment $\overline{rs}$, for any two points $r, s \in P$.

### *2.1 Models for the Minimum Cost Problem*

#### 2.1.1 First Integer Programming Formulation

In 1998, Johnsson, Magyar, and Nevalainen gave their first integer programming formulation for the minimum cost 3-matching of $P$ [7]. Given $P$, let $D = (P, A)$ be the directed complete graph, so that for every $u, v \in P$ with $u \neq v$, we have the two arcs $uv, vu \in A$. For each arc $uv \in A$, let $x_{uv}$ be a binary variable representing whether arc $uv$ has been chosen ($x_{uv} = 1$) or not ($x_{uv} = 0$) as part of the solution. The main idea is that, the segments $\overline{su}$ and $\overline{vu}$ of a chosen triplet $(s, u, v)$ are represented by the two arcs $su$, $vu$ (emphasizing that $u$ is the central point of the triplet). In the integer program, Eq. (1) implies that, for each $s \in P$, either the first term is 1 and the second is 0 or vice versa. In the former case, two arcs point toward $s$, while in the latter case one arc points away from $s$. In other words, either $s$ is the central point of the path or $s$ is an end of the path. See Fig. 2.

$$\min z = \sum_{rs\in A} c_{rs}x_{rs}$$

subject to

$$\frac{1}{2}\sum_{rs\in A} x_{rs} + \sum_{sk\in A} x_{sk} = 1 \ \forall s \in P \tag{1}$$

$$x_{rs} \in \{0, 1\} \ \forall rs \in A. \tag{2}$$

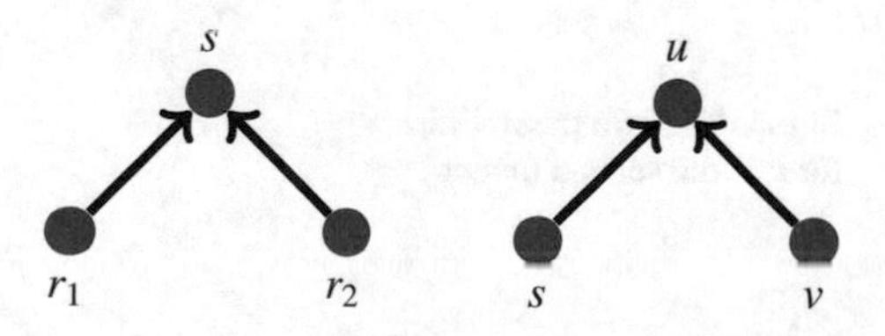


**Fig. 2** The two possibilities for $s$ in this model

### 2.1.2 Second Integer Programming Formulation

In 1999, Johnsson, Magyar, and Nevalainen gave their second integer programming formulation for the minimum cost problem [8]. As before, let $D = (P, A)$ be a directed graph, so that for every two different points $u, v \in P$, we have the two arcs $uv, vu \in A$, additionally, for each $v \in P$ we have a loop from $v$ to $v$. For each arc $uv \in A$, let $x_{uv}$ be a binary variable representing whether arc $uv$ has been chosen ($x_{uv} = 1$) or not ($x_{uv} = 0$) as part of the solution. The main idea is to represent a chosen triplet $(s, u, v)$ by the three arcs $us, uu, uv$.

$$\min z = \sum_{rs \in A, r \neq s} c_{rs} x_{rs}$$

subject to

$$\sum_{rs \in A} x_{rs} = 1 \; \forall s \in P \tag{3}$$

$$\sum_{s \in P} x_{ss} = \frac{1}{3}|P| \tag{4}$$

$$\sum_{sr \in A, r \neq s} x_{sr} = 2x_{ss} \; \forall s \in P \tag{5}$$

$$x_{rs} \in \{0, 1\} \; \forall rs \in A. \tag{6}$$

Equation (3) implies that exactly one arc enters each point. Equation (4) implies that exactly one third of the points are chosen as centers. Equation (5) implies that if $s$ is a center, then exactly two arcs leave $s$. See Fig. 3.

### 2.1.3 Triplet Integer Programming Formulation

The previous two models have the possible disadvantage of requiring two variables for each pair of points of $P$, whereas the following *triplet* model has only one variable for each pair of points (and no variables associated to the central points of a triplet). Let $U = (P, E)$ be the geometric complete graph, so that for every two different points $u, v \in P$ we have the line segment $\overline{uv} \in E$. For each $S \subseteq P$, let $\delta(S)$ be the

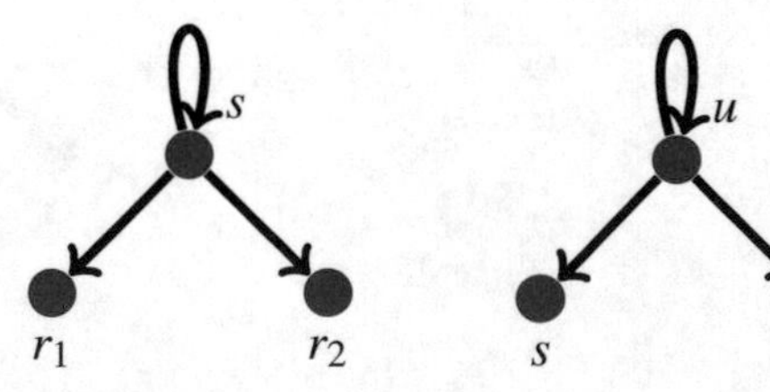


**Fig. 3** The two possibilities for $s$ in the second model

set of edges with one end point in $S$ and the other in $P \setminus S$ and let $\gamma(S)$ be the set of edges with both end points in $S$. For each edge $\overline{uv} \in E$, let $x_{uv}$ be a binary variable representing whether such edge has been chosen ($x_{uv} = 1$) or not ($x_{uv} = 0$). For each $F \subset E$, let $x(F) = \sum_{\overline{uv} \in F} x_{uv}$. Consider now the integer program [13]:

$$\min z = \sum_{\overline{rs} \in E} c_{rs} x_{rs}$$

subject to

$$x(\delta(v)) \geq 1 \; \forall v \in P \tag{7}$$

$$x(\delta(r,s,t)) \leq 3(2 - x(\gamma(r,s,t))) \; \forall r,s,t \in P \tag{8}$$

$$x(\delta(r,s,t)) \geq \frac{3}{2}(2 - x(\gamma(r,s,t))) \; \forall r,s,t \in P \tag{9}$$

$$x_{rs} \in \{0,1\} \; \forall \overline{rs} \in E. \tag{10}$$

Constraint (7) implies that each point must be incident to at least one chosen edge. Now consider the point set $\{r, s, t\}$. If a triplet of this point set is chosen, then $x(\gamma(r,s,t)) = 2$ and $x(\delta(r,s,t)) = 0$. If instead only one edge in $\gamma(r,s,t)$ is chosen, then $x(\gamma(r,s,t)) = 1$ and $2 \leq x(\delta(r,s,t)) \leq 3$. Finally, if no edge in $\gamma(r,s,t)$ is chosen, then $x(\gamma(r,s,t)) = 0$ and $3 \leq x(\delta(r,s,t)) \leq 6$. Note that all this cases are valid for constraints (8) and (9). See Fig. 4.

Conversely, the three disallowed cases are: paths with more than two edges, points with more than two edges, and isolated edges. First, let $r, s, t, u$ be consecutive points in a path of three (or more) edges. If $u = r$, then we have $x(\gamma(r,s,t)) = 3$, violating (8). If $u \neq r$, then $x(\gamma(r,s,t)) = 2$ and $x(\delta(r,s,t)) \geq 1$, also violating (8). Second, let $r$ be a point with at least three neighbors $s, t, u$. Then $x(\gamma(r,s,t)) \geq 2$ and $x(\delta(r,s,t)) \geq 1$, also violating (8). Finally, let $\overline{rs}$ be an isolated edge and let $t \in P \setminus \{r, s\}$. Constraints (8) and (9) imply that $t$ has degree either 2 or 3. Since the latter is impossible, it follows that $t$ has degree 2 and, therefore, we have selected a set of cycles. This is a contradiction, because paths with three or more edges are impossible. See Fig. 5.

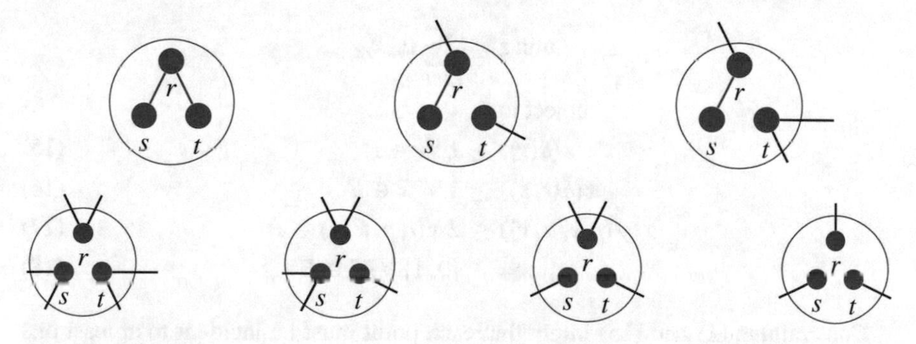

**Fig. 4** All valid cases for our model

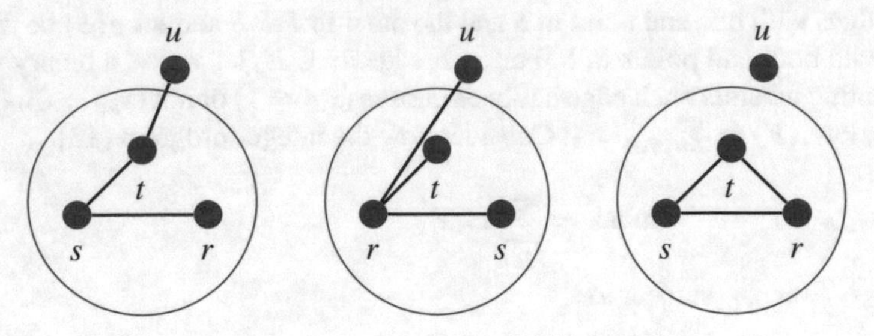


**Fig. 5** Invalid cases for our model

### 2.1.4 New Integer Programming Formulations

Here we introduce two new models for the minimum cost problem. We call them *Pair integer programming formulation* and *Quad integer programming formulation*, due to the maximum number of points involved in each of their constraints. Because of their similarities, both models are described together.

We take the geometric graph $U = (P, E)$, and the functions $\delta(S)$, $\gamma(S)$ and $x(F)$, as they were defined in Sect. 2.1.3.

The Pair integer program is defined as:

$$\min z = \sum_{\overline{rs} \in E} c_{rs} x_{rs}$$

subject to

$$x(\delta(v)) \geq 1 \; \forall v \in P \tag{11}$$

$$x(\delta(r)) + x(\delta(s)) \geq 2 + x_{rs} \; \forall \overline{rs} \in E \tag{12}$$

$$x(\delta(r)) + x(\delta(s)) \leq 4 - x_{rs} \; , \forall \overline{rs} \in E \tag{13}$$

$$x_{rs} \in \{0, 1\} \; \forall \overline{rs} \in E \tag{14}$$

The Quad integer program is defined as:

$$\min z = \sum_{\overline{rs} \in E} c_{rs} x_{rs}$$

subject to

$$x(\delta(v)) \geq 1 \; \forall v \in P \tag{15}$$

$$x(\delta(r, s)) \geq 1 \; \forall \overline{rs} \in E \tag{16}$$

$$x(\gamma(r, s, u, v)) \leq 2 \; \forall (r, s, u, v) \in P \tag{17}$$

$$x_{rs} \in \{0, 1\} \; \forall \overline{rs} \in E \tag{18}$$

Constraints (11) and (15) imply that each point must be incident to at least one chosen edge. Now consider the point set $\{r, s, t\}$, and suppose that a triplet of this

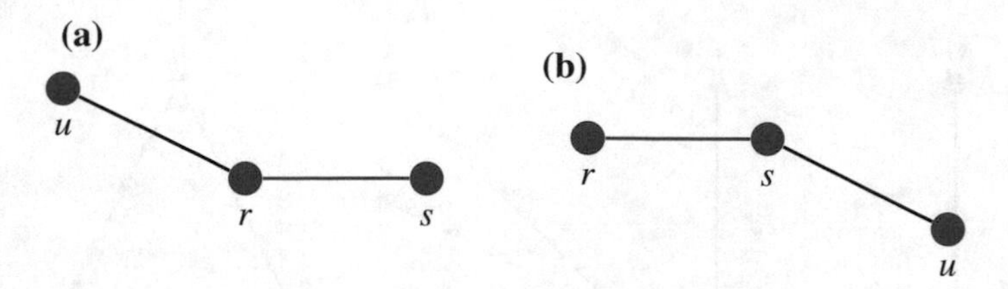


**Fig. 6** Valid cases for both models

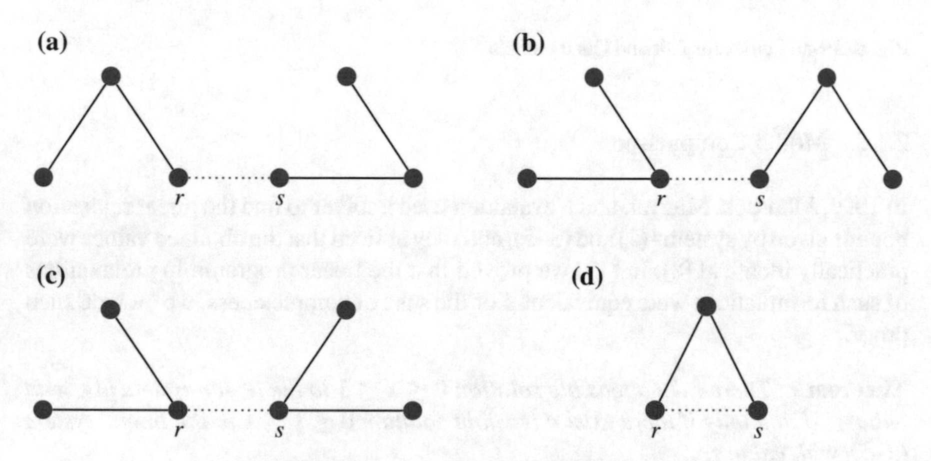


**Fig. 7** More valid cases for both models

point set is chosen. We have two cases, either $r$ is the central point of the triplet or it is not. In both cases $2 + 1 \leq x(\delta(r)) + x(\delta(s)) \leq 4 - 1$ and $x(\gamma(r, s, u, v)) \leq 2$; those values are valid for the constraints of the two models. See Fig. 6.

Suppose now that some segment, $\overline{rs}$, was not chosen. As $r$ and $s$ must belong to some triplet, we can only have the cases depicted in Fig. 7. In all cases $2 + 0 \leq x(\delta(r)) + x(\delta(s)) \leq 4 - 0$ and $x(\delta(r, s)) \geq 2$; those values are also valid for the constraints of the two models.

Conversely, the disallowed cases are: paths with more than two edges, points with more than two edges, triangles and isolated edges. First, let $u, r, s, v$ be consecutive points in a path of three (or more) edges. Either if $u = v$ of $u \neq v$, it is true that $x(\delta(r)) + x(\delta(s)) \nleq 4 - 1$ and $x(\gamma(r, s, u, v)) \nleq 2$, violating (13) and (17). The same inequalities hold for the case when $r$ is touched by two edges (or more). Finally, constraints (12) and (16) exclude isolated edges. See Fig. 8.

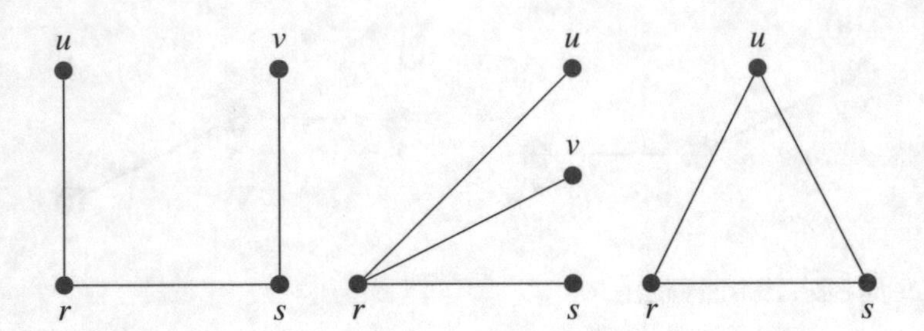


**Fig. 8** Invalid cases for Pair and Quad models

### 2.1.5 Model Comparison

In 1999, Johnsson, Magyar, and Nevalainen used a solver to find the linear relaxation bounds given by systems (1) and (3–5), and they noticed that the obtained values were practically identical [8]. In [13] we proved that the linear programming relaxations of such formulations were equivalent. For the sake of completeness, we include such proof.

**Theorem 1** *There exist a feasible solution* $0 \le x \le 1$ *to the linear system (1), with value z, if and only if there exist a feasible solution* $0 \le y \le 1$ *to the linear system (3–5), with value z.*

*Proof* Both formulations use similar variable names; for the sake of clarity, the term $y_{rs}$ is used to refer to the variable $x_{rs}$ of the linear system (3–5), for all $r, s \in P$.

Let $\mathbf{0} \le y \le \mathbf{1}$ be a feasible solution to the linear system (3–5). Let $x$ be the vector given by $x_{rs} = y_{sr}$ for all $r, s \in P$ with $r \ne s$. Clearly $\mathbf{0} \le x \le \mathbf{1}$. For each $s \in P$ we have

$$\frac{1}{2}\sum_{rs\in A} x_{rs} + \sum_{sk\in A} x_{sk} = \frac{1}{2}\sum_{sr\in A, r\ne s} y_{sr} + \sum_{ks\in A} y_{ks} \tag{19}$$

$$= \frac{1}{2}(2y_{ss}) + (1 - y_{ss}) = 1, \tag{20}$$

that is, $x$ satisfies the linear system (1).

Conversely, let $\mathbf{0} \le x \le \mathbf{1}$ be a feasible solution to the linear system (1). Let $y$ be the vector given by $y_{rs} = x_{sr}$ for all $r, s \in P$ with $r \ne s$, and $y_{ss} = 1 - \sum_{rs\in A, r\ne s} y_{rs} \le 1$ for all $s \in P$. Also, for each $s \in P$ we have

$$y_{ss} = 1 - \sum_{rs\in A, r\ne s} y_{rs} \tag{21}$$

$$= 1 - \sum_{sk\in A} x_{sk} \tag{22}$$

$$= \frac{1}{2} \sum_{rs \in A} x_{rs} \geq 0, \tag{23}$$

and hence $\mathbf{0} \leq y \leq \mathbf{1}$. By definition, (3) holds. For $s \in P$

$$\sum_{sr \in A, r \neq s} y_{sr} = \sum_{rs \in A} x_{rs} \tag{24}$$

$$= 2\left(1 - \sum_{sr \in A} x_{sr}\right) \tag{25}$$

$$= 2\left(1 - \sum_{rs \in A, r \neq s} y_{rs}\right) \tag{26}$$

$$= 2y_{ss}, \tag{27}$$

and hence (5) holds. Finally, (3) and (5) imply (4). Adding $y_{ss}$ to both sides of (5) and summing over $s \in P$ we obtain

$$3 \sum_{s \in P} y_{ss} = \sum_{s \in P} \sum_{sr \in A} y_{sr} \tag{28}$$

$$= \sum_{r \in P} \sum_{rs \in A} y_{rs} \tag{29}$$

$$= \sum_{r \in P} 1 = |P|. \tag{30}$$

In other words, the constraints (4) are implied by the constraints (3) and (5) and therefore are unnecessary.

Observe that in both directions of the proof $x_{sr} = y_{rs}$ and hence the cost of both solutions is the same. □

Likewise, in [13] we compared the linear programming relaxations of the first two models and that of the Triplet model. Additionally, using the state-of-the-art MIP solver Gurobi [5], we solved to optimality all instances that were used as tests in [7, 8], in the benchmark available at http://www.cs.utu.fi/research/projects/3mp/. Each instance in that benchmark is a point set in the Euclidean plane, of size divisible by 3.

Table 1 contains, for each instance, the relaxation value and integrality gap of all previous models. The integrality gap is taken as the ratio between the optimal value and the relaxation value. Naturally, the linear relaxation values differ between models.

In our experiments the Quad model had better relaxation than the rest of the models, we conjecture that this is always the case.

Another observation comes from drawing the fractional solutions of the linear programming relaxations (edges are colored black if chosen, white if not chosen, and in various shades of gray if chosen fractionally). As the first two models (Sects. 2.1.1

**Table 1** Relaxation values and integrality gaps of all models. Tighter relaxations are shown in bold

| Case | $n$ | Minimum value | Relaxation 1998/1999 | Integrality gap | Relaxation pair | Integrality gap | Relaxation triplet | Integrality gap | Relaxation Quad | Integrality gap |
|---|---|---|---|---|---|---|---|---|---|---|
| f21 | 21 | 159.7289 | 146.4494 | 1.0907 | 156.5128 | 1.0205 | 141.2287 | 1.1310 | 156.6416 | **1.0197** |
| f27 | 27 | 8011.0786 | 7379.4580 | 1.0856 | 7791.3999 | 1.0282 | 7182.3311 | 1.1154 | 7837.2612 | **1.0222** |
| f33 | 33 | 255.6944 | 236.9880 | 1.0789 | 255.6944 | **1.0000** | 232.2087 | 1.1011 | 255.6944 | **1.0000** |
| f39 | 39 | 282.3146 | 237.7430 | 1.1875 | 251.3423 | 1.1232 | 237.4670 | 1.1889 | 257.4570 | **1.0966** |
| 39a | 39 | 783.9551 | 705.7473 | 1.1108 | 769.1493 | 1.0192 | 718.6500 | 1.0909 | 775.6470 | **1.0107** |
| 39b | 39 | 826.1430 | 685.4586 | 1.2052 | 764.0848 | 1.0812 | 680.2161 | 1.2145 | 781.6820 | **1.0569** |
| 39c | 39 | 959.3756 | 865.2224 | 1.1088 | 944.3328 | 1.0159 | 866.2079 | 1.1076 | 946.0240 | **1.0141** |
| 39d | 39 | 781.2205 | 699.9003 | 1.1162 | 766.1649 | 1.0197 | 709.9151 | 1.1004 | 781.2205 | **1.0000** |
| 39e | 39 | 872.3539 | 758.9849 | 1.1494 | 813.9905 | 1.0717 | 739.8327 | 1.1791 | 831.5713 | **1.0490** |
| 42a | 42 | 949.0908 | 805.3827 | 1.1784 | 882.7597 | 1.0751 | 810.1607 | 1.1715 | 903.2889 | **1.0507** |
| 42b | 42 | 860.5917 | 727.7668 | 1.1825 | 806.2082 | 1.0675 | 735.9717 | 1.1693 | 814.2614 | **1.0569** |
| 45a | 45 | 1013.1434 | 902.5522 | 1.1225 | 1003.4940 | 1.0096 | 905.4010 | 1.1190 | 1013.1430 | **1.0000** |
| 45b | 45 | 985.6019 | 785.1058 | 1.2554 | 940.5861 | 1.0479 | 810.2633 | 1.2164 | 962.0711 | **1.0245** |
| 48a | 48 | 996.6130 | 945.5751 | 1.0540 | 996.3883 | 1.0002 | 936.0416 | 1.0647 | 996.6131 | **1.0000** |
| 48b | 48 | 967.8344 | 817.5964 | 1.1838 | 961.9909 | 1.0061 | 832.4138 | 1.1627 | 967.8344 | **1.0000** |
| 51a | 51 | 983.5503 | 843.9761 | 1.1654 | 962.6867 | 1.0217 | 850.3052 | 1.1567 | 979.7631 | **1.0039** |
| 51b | 51 | 1003.8185 | 892.6889 | 1.1245 | 971.0931 | 1.0337 | 883.1294 | 1.1367 | 994.0304 | **1.0098** |
| h01 | 51 | 265.6100 | 243.4752 | 1.0909 | 255.2249 | 1.0407 | 232.7736 | 1.1411 | 255.8900 | **1.0380** |
| h02 | 84 | 1007.1086 | 876.5153 | 1.1490 | 947.6500 | 1.0627 | 870.1146 | 1.1574 | 964.0787 | **1.0446** |
| man | 84 | 1007.1086 | 876.5153 | 1.1490 | 947.6500 | 1.0627 | 870.1146 | 1.1574 | 964.0787 | **1.0446** |
| h03 | 99 | 751.5259 | 684.7681 | 1.0975 | 736.4505 | 1.0205 | 666.2244 | 1.1280 | 737.9276 | **1.0184** |
| f99 | 99 | 386.2318 | 363.1123 | 1.0637 | 377.9442 | 1.0219 | 351.9001 | 1.0976 | 379.3111 | **1.0182** |
| rat99 | 99 | 751.5259 | 684.7681 | 1.0975 | 736.4505 | 1.0205 | 666.2244 | 1.1280 | 737.9276 | **1.0184** |
| 120a | 120 | 1508.7162 | 1229.2390 | 1.2274 | 1430.6870 | 1.0545 | 1251.5170 | 1.2055 | 1466.5570 | **1.0287** |

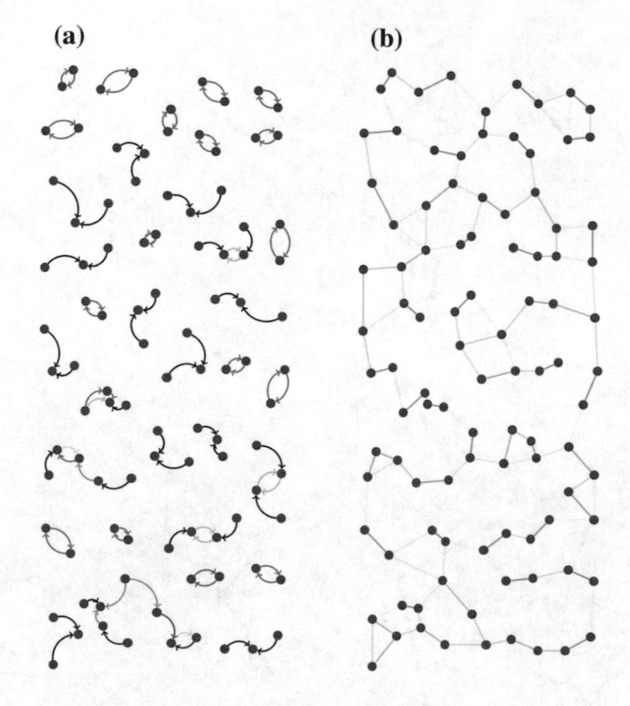


**Fig. 9** Relaxation for instance h03: **a** in the 1998 model and **b** in the Triplet model

and 2.1.2) have equivalent linear programming relaxations, the drawings of the solutions look the same for each instance, as expected. However, it seems that there are no edges with positive value that cross (as in Fig. 9a). Since this happened for each experiment, we also conjecture that this is always the case. The same property was seen in each experiment of the Triplet model (as in Fig. 9b), again we conjecture that this is always the case. See Fig. 9.

Conversely, we know that such property is not true for the Pair model, since instance 39b has at least one crossing. See Fig. 10.

## 2.2 *Models for the Maximum Cost Problem*

Simply changing the objective function from minimization to maximization does not work: The resulting maximum cost 3-matching will almost certainly contain many crosses (as in Fig. 1b). Therefore, in order to obtain an integer programming formulation for the maximum cost non-crossing 3-matching problem, we need to add a family of constraints to deal with crossings.

Note that a crossing between $\overline{ik}$ and $\overline{jl}$ occurs if and only if $i, j, k, l$ are the vertices of a convex quadrangle ($\overline{ik}$ and $\overline{jl}$ would be the diagonals of this quadrangle). Hence, we can add these constraints to the first two models, with $i, j, k, l \in P$:

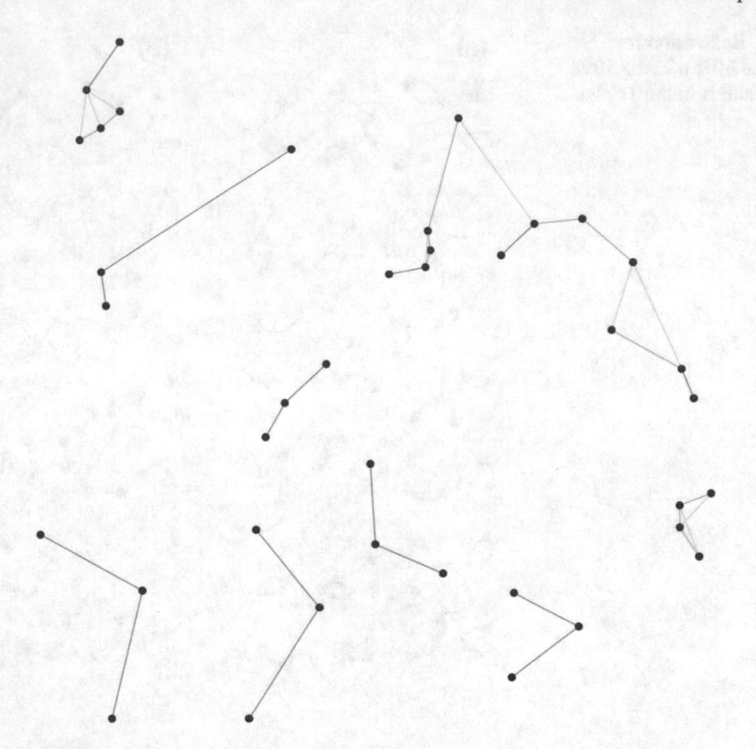

**Fig. 10** Relaxation for instance 39b in the Pair model

$$x_{ik} + x_{ki} + x_{jl} + x_{lj} \leq 1 \quad \text{if } i, j, k, \text{ and } l \text{ form a convex quadrangle.} \tag{31}$$

Or we can add these other constraints to the last three models, with $i, j, k, l \in P$:

$$x_{ik} + x_{jl} \leq 1 \quad \text{if } i, j, k, \text{ and } l \text{ form a convex quadrangle.} \tag{32}$$

In either case, these $O(n^4)$ constraints imply that we can only select one of the two diagonals of each convex quadrangle, therefore avoiding crossings.

## 3 Heuristics

In this section we introduce three heuristics to compute 3-matchings of $P$. All three are used for both: minimization and maximization problems; with slight modifications. We use the benchmark mentioned in Sect. 2.1.5 to test these strategies.

## 3.1 Statements of the Heuristics

The following statements are for the minimization problem. For the maximization problem, the necessary modifications are explained in parenthesis.

- *Windrose*: Sort the points of $P$ along the $x$-axis. Using this ordering, partition $P$ in subsets of 3 consecutive points. For each of these subsets consider the permutation of its elements $(u, v, w)$ that minimizes (resp. maximizes) the sum of the lengths of $\overline{uv}$ and $\overline{wv}$, and add this triplet to the solution of this direction. Repeat this process along the $y$-axis, along a 45° line and along a $-45°$ line. Among the four solutions, select that of minimum cost (resp. maximum cost). See Fig. 11.
- *ConvHull*: At each step: compute conv$(P)$, the convex hull of $P$ (see Fig. 12a); search for two consecutive segments in conv$(P)$ whose sum of lengths is minimum (resp. maximum); add the induced triplet of points to the solution and delete them

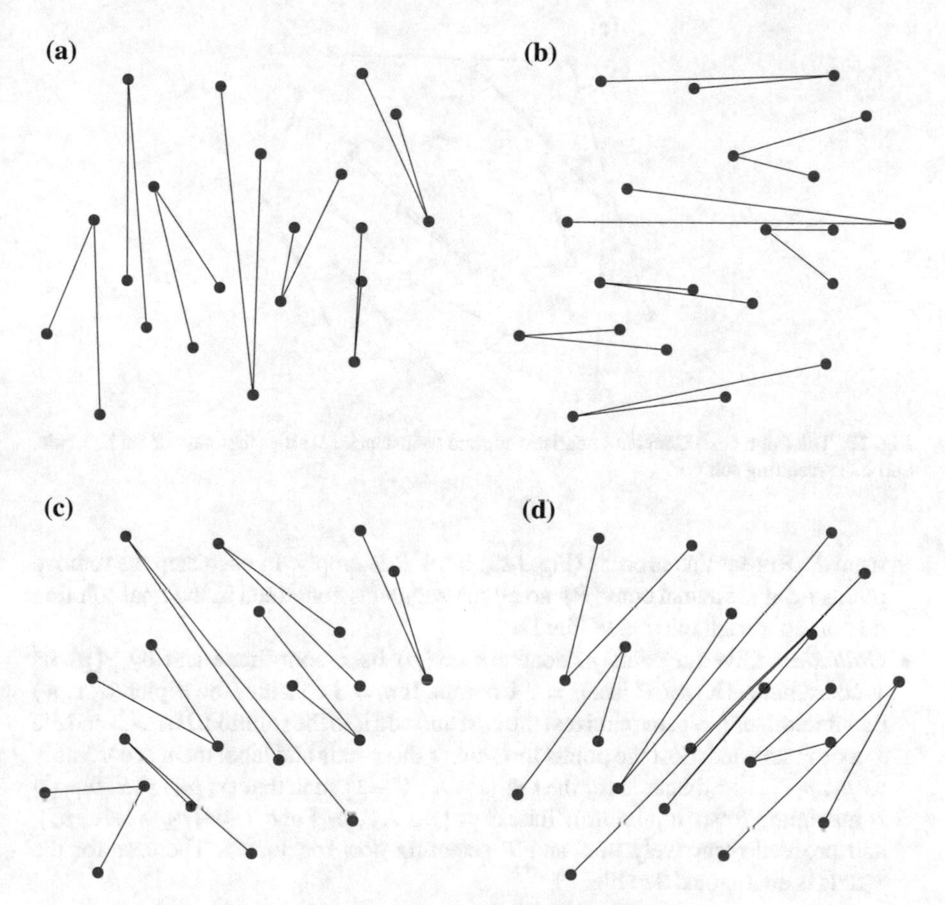


**Fig. 11** The four stages of (high cost) Windrose heuristic for instance f21: **a** along the $x$-axis, **b** along the $y$-axis, **c** along a 45° *line*, and **d** along a $-45°$ line

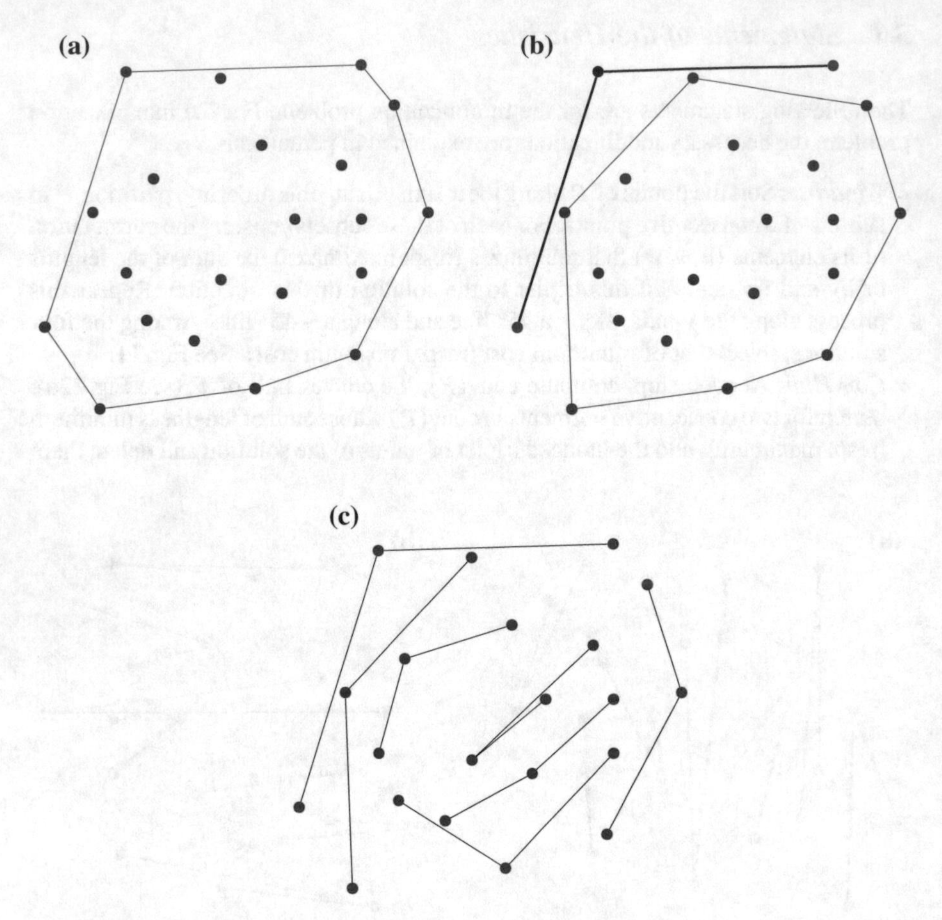


**Fig. 12** The (high cost) ConvHull heuristic applied to instance f21: **a–b** first pass of the heuristic, and **c** the resulting solution

from $P$. Repeat this process (Fig. 12b) until $P$ is empty. In each step we remove points from the actual $\text{conv}(P)$, so all the segments contained in the final solution do not cross each other. See Fig. 12.

- *Guillotine*: Given a point $p$, denote by $x(p)$ its $x$-coordinate and by $y(p)$ its $y$-coordinate. The set $P$ has $n = 3k$ points. If $n = 3$ consider the triplet $(u, v, w)$ that minimizes (resp. maximizes) the cost and add it to the solution. If $n > 3$ and the $x$-axis is selected: Sort the points in $P$ along the $x$-axis and label them accordingly as $p_1, p_2, \ldots, p_n$. Search for the $i \in \{1, \ldots, k-1\}$ such that $|x(p_{3i}) - x(p_{3i+1})|$ is maximum (resp. minimum). Take $S = \{p_1, \ldots, p_{3i}\}$ and $T = \{p_{3i+1}, \ldots, p_n\}$, and proceed recursively in $S$ and $T$ selecting now the $y$-axis. The case for the $y$-axis is analogous. See Fig. 13.

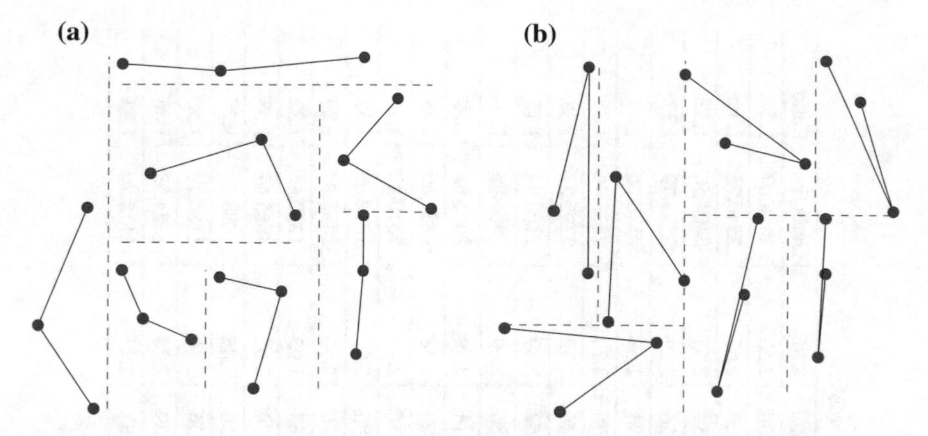


**Fig. 13** Guillotine heuristic applied to instance f21: **a** *low* cost and **b** *high* cost

## 3.2 Experimental Results

Tables 2, 3 and 4 summarize the results obtained by our heuristics, compared against the optimal values. The optimal values were computed with the MIP solver Gurobi [5]. For the minimum cost 3-matching of $P$ we used the integer programming formulation of Sect. 2.1.1, and for the maximum cost we adapted that same formulation in accordance to Sect. 2.2.

In Table 3 we see first that it is feasible to obtain the corresponding minimum value using Gurobi even with sets of about 200 points, usually in less than 30 seconds. Then we report the results obtained in 2000 by Johnsson, Magyar, and Nevalainen [9] using several heuristic strategies, primarily genetic algorithms. Their results are very nearly optimal, usually within 1% of the minimum. However, they needed about one hour of time in a Pentium 133 to obtain them. Even considering that our hardware is about 50 times faster, their heuristic needed more time to obtain an approximate solution than Gurobi needed to obtain the optimal solution. The rest of the columns show the results obtained by our heuristics. All our heuristics are very fast (the whole set of tests ran in only a few seconds), but only the Guillotine heuristic gave reasonably good solutions. It is likely that this heuristic is a 2-approximation algorithm for the minimization problem. This intuition comes from reading the ratio column, that is, the quotient cost of the solution obtained by the heuristic divided by the cost of the optimal solution. Clearly such quotient is $\geq 1$ for a minimization problem, and the closer it is to 1, the better.

As we can see in Table 4, even with small sets of points ($n \leq 51$), the computation of some maximization instances took several hours (even a few days) on a computer with two AMD Opteron 6174 at 2.2 GHz and 128GB RAM; whilst our heuristics took an negligible amount of time (thus not listed in the tables) for the same instances. Again, the performance of our heuristics can be read from the ratio column. In this case, the quotient is $\leq 1$, but again the closer it is to 1, the better. Table 4 highlights in

**Table 2** Comparison of the average best results in [9] (among 20 runs) and our heuristics, for the minimum cost problem

| Case | $n$ | Minimum | $\text{Time}_{[s]}$ | [9] | Ratio | Windrose | Ratio | ConvHull | Ratio | Guillotine | Ratio |
|---|---|---|---|---|---|---|---|---|---|---|---|
| p01 | 201 | 1862.29 | 10.74 | 1869.63 | 1.00 | 9812.23 | 5.27 | 4715.08 | 2.53 | 2318.07 | 1.24 |
| p02 | 201 | 1826.01 | 21.01 | 1835.51 | 1.01 | 9662.64 | 5.29 | 4871.67 | 2.67 | 2540.78 | 1.39 |
| p03 | 201 | 1812.18 | 7.20 | 1815.13 | 1.00 | 9040.09 | 4.99 | 4629.54 | 2.55 | 2404.50 | 1.33 |
| p04 | 201 | 1839.27 | 28.50 | 1844.78 | 1.00 | 8621.45 | 4.69 | 4214.26 | 2.29 | 2513.79 | 1.37 |
| p05 | 201 | 1856.08 | 8.62 | 1872.04 | 1.01 | 9611.48 | 5.18 | 4999.93 | 2.69 | 2598.20 | 1.40 |
| p06 | 201 | 1775.64 | 19.49 | 1786.95 | 1.01 | 9048.62 | 5.10 | 4838.06 | 2.72 | 2333.94 | 1.31 |
| p07 | 201 | 1802.04 | 10.13 | 1816.51 | 1.01 | 8880.21 | 4.93 | 4907.46 | 2.72 | 2392.24 | 1.33 |
| p08 | 201 | 1822.25 | 36.71 | 1838.47 | 1.01 | 8794.89 | 4.83 | 4904.52 | 2.69 | 2100.94 | 1.15 |
| p09 | 201 | 1846.29 | 19.91 | 1851.40 | 1.00 | 9459.49 | 5.12 | 5171.86 | 2.80 | 2506.06 | 1.36 |
| p10 | 201 | 1913.10 | 8.20 | 1919.95 | 1.00 | 9194.90 | 4.81 | 5342.14 | 2.79 | 2653.62 | 1.39 |
| p11 | 150 | 15227.38 | 5.32 | 15228.28 | 1.00 | 51179.90 | 3.36 | 36207.00 | 2.38 | 19487.60 | 1.28 |
| p12 | 150 | 3908.57 | 4.49 | 3922.06 | 1.00 | 15401.40 | 3.94 | 9430.38 | 2.41 | 5310.42 | 1.36 |
| p13 | 195 | 1438.88 | 6.61 | 1447.40 | 1.01 | 4581.78 | 3.18 | 2885.85 | 2.01 | 1972.89 | 1.37 |
| p14 | 159 | 25536.29 | 13.74 | 25585.65 | 1.00 | 80710.80 | 3.16 | 93374.70 | 3.66 | 32855.20 | 1.29 |
| p15 | 204 | 1062.44 | 9.42 | 1078.29 | 1.01 | 5166.97 | 4.86 | 2070.83 | 1.95 | 1292.43 | 1.22 |
| p16 | 129 | 3526.94 | 3.54 | 3526.94 | 1.00 | 12970.40 | 3.68 | 9201.30 | 2.61 | 4865.67 | 1.38 |
| p17 | 99 | 386.23 | 1.53 | 386.26 | 1.00 | 1043.88 | 2.70 | 713.50 | 1.85 | 488.35 | 1.26 |
| p18 | 99 | 751.53 | 1.90 | 751.53 | 1.00 | 1651.83 | 2.20 | 1395.56 | 1.86 | 1186.79 | 1.58 |
| p19 | 201 | 1344.08 | 9.04 | 1351.63 | 1.01 | 5397.97 | 4.02 | 3063.39 | 2.28 | 1824.53 | 1.36 |
| p20 | 222 | 1590.82 | 13.46 | 1602.03 | 1.01 | 7549.48 | 4.75 | 3614.87 | 2.27 | 2118.70 | 1.33 |

**Table 3** Minimum cost and heuristics

| Case | $n$ | Minimum | $Time_{[s]}$ | Windrose | Ratio | ConvHull | Ratio | Guillotine | Ratio |
|---|---|---|---|---|---|---|---|---|---|
| f21 | 21 | 159.73 | 0.07 | 212.40 | 1.330 | 221.28 | 1.385 | **182.64** | 1.143 |
| f27 | 27 | 8011.08 | 0.08 | 10326.00 | 1.289 | 13758.40 | 1.717 | **9578.44** | 1.196 |
| f33 | 33 | 255.69 | 0.09 | 459.28 | 1.796 | 350.74 | 1.372 | **299.38** | 1.171 |
| 39a | 39 | 783.96 | 0.31 | 1629.76 | 2.079 | 1705.26 | 2.175 | **903.88** | 1.153 |
| 39b | 39 | 826.14 | 0.51 | 1651.56 | 1.999 | 1283.06 | 1.553 | **1002.43** | 1.213 |
| 39c | 39 | 959.38 | 0.10 | 1670.49 | 1.741 | 1531.89 | 1.597 | **1192.85** | 1.243 |
| 39d | 39 | 781.22 | 0.29 | 1832.70 | 2.346 | 1303.39 | 1.668 | **845.48** | 1.082 |
| 39e | 39 | 872.35 | 0.70 | 1427.32 | 1.636 | 1471.62 | 1.687 | **948.66** | 1.087 |
| f39 | 39 | 282.31 | 0.92 | **292.98** | 1.038 | 454.32 | 1.609 | 489.56 | 1.734 |
| 42a | 42 | 949.09 | 0.43 | 2092.11 | 2.204 | 1563.40 | 1.647 | **1022.72** | 1.078 |
| 42b | 42 | 860.59 | 0.63 | 1697.77 | 1.973 | 1644.05 | 1.910 | **951.38** | 1.105 |
| 45a | 45 | 1013.14 | 0.19 | 2298.06 | 2.268 | 1775.77 | 1.753 | **1319.36** | 1.302 |
| 45b | 45 | 985.60 | 0.78 | 1790.62 | 1.817 | 1847.84 | 1.875 | **1153.54** | 1.170 |
| 48a | 48 | 996.61 | 0.10 | 2254.78 | 2.262 | 1609.26 | 1.615 | **1220.06** | 1.224 |
| 48b | 48 | 967.83 | 0.25 | 2155.98 | 2.228 | 2093.32 | 2.163 | **1381.74** | 1.428 |

(continued)

**Table 3** (continued)

| Case | $n$ | Minimum | $\text{Time}_{[s]}$ | Windrose | Ratio | ConvHull | Ratio | Guillotine | Ratio |
|---|---|---|---|---|---|---|---|---|---|
| 51a | 51 | 983.55 | 0.51 | 2550.64 | 2.593 | 1866.47 | 1.898 | **1131.93** | 1.151 |
| 51b | 51 | 1003.82 | 0.60 | 2140.50 | 2.132 | 1927.80 | 1.920 | **1029.04** | 1.025 |
| eil51 | 51 | 265.61 | 0.68 | 532.57 | 2.005 | 421.18 | 1.586 | **323.10** | 1.216 |
| man | 84 | 1007.11 | 2.53 | 2693.77 | 2.675 | 2858.21 | 2.838 | **1309.44** | 1.300 |
| f99 | 99 | 386.23 | 1.58 | 1043.88 | 2.703 | 713.49 | 1.847 | **488.35** | 1.264 |
| 120a | 120 | 1508.72 | 10.52 | 5125.51 | 3.397 | 3336.75 | 2.212 | **1926.33** | 1.277 |
| h04 | 129 | 3526.94 | 3.52 | 12970.40 | 3.678 | 9201.30 | 2.609 | **4865.67** | 1.380 |
| 201a | 201 | 1945.09 | 14.23 | 9164.72 | 4.712 | 4327.80 | 2.225 | **2424.20** | 1.246 |
| 240a | 240 | 2068.67 | 12.99 | 11089.60 | 5.361 | 5760.70 | 2.785 | **2871.38** | 1.388 |
| 300a | 300 | 2202.32 | 56.24 | 13835.30 | 6.282 | 5790.96 | 2.629 | **3082.12** | 1.399 |
| 360a | 360 | 2520.41 | 281.49 | 17297.90 | 6.863 | 7458.42 | 2.959 | **3700.60** | 1.468 |
| h07 | 441 | 30979.16 | 116.22 | 127301.00 | 4.109 | 104987.00 | 3.389 | **49480.10** | 1.597 |
| 510a | 510 | 2999.25 | 341.95 | 23933.90 | 7.980 | 9170.59 | 3.058 | **4588.09** | 1.530 |
| h08 | 573 | 4106.92 | 153.54 | 23507.60 | 5.724 | 9835.20 | 2.395 | **6314.24** | 1.537 |
| rat783 | 783 | 5269.62 | 558.24 | 36178.10 | 6.865 | 15074.40 | 2.861 | **8320.26** | 1.579 |
| h10 | 1002 | 148206.63 | 8746.71 | 1133310.00 | 7.647 | 553990.00 | 3.738 | **210241.00** | 1.419 |

**Table 4** Maximum cost and heuristics

| Case | $n$ | Maximum | Time$_{[hr]}$ | Windrose | Ratio | ConvHull | Ratio | Guillotine | Ratio |
|---|---|---|---|---|---|---|---|---|---|
| f21 | 21 | 492.19 | 0.01 | **418.00** | 0.849 | 307.54 | 0.625 | 277.77 | 0.564 |
| f27 | 27 | 39330.22 | 0.03 | **35834.30** | 0.911 | 24311.40 | 0.618 | 18078.50 | 0.460 |
| f33 | 33 | 1136.34 | 0.32 | **887.77** | 0.781 | 712.82 | 0.627 | 601.63 | 0.529 |
| 39a | 39 | 4997.58 | 2.41 | **4107.73** | 0.822 | 2774.23 | 0.555 | 2172.61 | 0.435 |
| 39b | 39 | 4548.58 | 2.31 | **3077.47** | 0.677 | 2496.36 | 0.549 | 1632.18 | 0.359 |
| 39c | 39 | 4410.10 | 1.75 | **3726.14** | 0.845 | 2725.35 | 0.618 | 1887.40 | 0.428 |
| 39d | 39 | 4728.09 | 3.06 | **3728.20** | 0.789 | 2974.18 | 0.629 | 2330.61 | 0.493 |
| 39e | 39 | 4804.63 | 2.07 | **4205.22** | 0.875 | 3068.20 | 0.639 | 1783.95 | 0.371 |
| f39 | 39 | 3804.50 | 0.54 | **3320.16** | 0.873 | 1998.54 | 0.525 | 994.54 | 0.261 |
| 42a | 42 | 5026.50 | 8.71 | **3818.60** | 0.760 | 3225.15 | 0.642 | 2588.07 | 0.515 |
| 42b | 42 | 4918.73 | 8.69 | **3524.74** | 0.717 | 3433.42 | 0.698 | 2502.66 | 0.509 |
| 45a | 45 | 5710.43 | 12.29 | **5093.17** | 0.892 | 3319.70 | 0.581 | 2416.68 | 0.423 |
| 45b | 45 | 6009.25 | 8.68 | **4676.96** | 0.778 | 3772.16 | 0.628 | 2255.34 | 0.375 |
| 48a | 48 | 5861.28 | 57.81 | **5097.48** | 0.870 | 3665.41 | 0.625 | 2608.46 | 0.445 |
| 48b | 48 | 5932.78 | 38.58 | **4947.42** | 0.834 | 3583.77 | 0.604 | 2484.23 | 0.419 |
| 51a | 51 | 6266.31 | 52.05 | **5150.67** | 0.822 | 4042.01 | 0.645 | 2004.47 | 0.320 |
| 51b | 51 | 6276.55 | 48.88 | **4489.25** | 0.715 | 3770.45 | 0.601 | 2497.22 | 0.398 |
| eil51 | 51 | 1247.13 | 89.29 | **098.85** | 0.881 | 749.35 | 0.601 | 545.16 | 0.437 |

bold the best results obtained by our heuristics, always from the Windrose heuristic. It is not clear whether any of our three heuristics is in fact an approximation algorithm.

## 4 Conclusions

In this paper we have approached both the minimization and maximization versions of the non-crossing Euclidean 3-matching problem. Both of them are believed to be NP-hard problems because their combinatorial counterparts are known to be NP-complete even for metric costs. However, the non-crossing maximization version seems to be harder than the minimization version. This appears to be true for many geometric network optimization problems.

Our first approach was based on proposing integer programming formulations for the minimization version. There were two previously known formulations and we have proposed three new formulations. The main difference is that our formulations use one variable per edge, while the previously known formulations used two variables per edge. We also studied their linear programming relaxations. A previous result of ours was that the linear programming relaxations of the two previously known formulations are equivalent, i.e., they give the same minimum value. We conjecture that at least two of our formulations (Pair and Quad) give a tighter bound. We also conjecture that the Triple linear programming relaxation gives a planar graph. In order to obtain models for the maximization version, we had to add constraints to avoid crossings.

Our second approach was based on proposing heuristics for our two problems. The main reason for this was that solving our models to optimality is very time consuming, particularly so for the maximization version. We have proposed three heuristics specific to these problems, each of them running in fractions of a second. Although our heuristics do not give the best possible results, the time needed to run previously known heuristics exceeds the time to obtain the optimal values using state-of-the-art solvers. Furthermore, our heuristics were easily adapted for both minimization and maximization versions.

Our future work on these problems shall include proving the conjectures mentioned above, as well as designing approximation algorithms for our problems.

## References

1. Alon, N., Rajagopalan, S., Suri, S.: Long non-crossing configurations in the plane. In: Proceedings of the ninth annual symposium on computational geometry, SCG '93, pages 257–263. New York, USA, ACM (1993)
2. Crama, Y., Oerlemans, A.G., Spieksma, F.: ProDuction Planning In Automated Manufacturing. Springer, Berlin (1994)
3. Crama, Y., Spieksma, F.C.: Approximation algorithms for three-dimensional assignment problems with triangle inequalities. Eur. J. Oper. Res. **60**(3), 273–279 (1992)

4. Dumitrescu, A., Tóth, C.D.: Long non-crossing configurations in the plane. Discret. Comput. Geom. **44**(4), 727–752 (2010)
5. Gurobi Optimization. Gurobi optimizer reference manual. http://www.gurobi.com/. June 2015
6. Johnsson, M., Leipl, T., Nevalainen, O.: Determining the manual setting order of components on PC boards. J. Manuf. Syst. **15**(3), 155–163 (1996)
7. Johnsson, M., Magyar, G., Nevalainen, O.: On the Euclidean 3-matching problem. Nord. J. Comput. **5**(2), 143–171 (1998)
8. Magyar, G., Johnsson, M., Nevalainen, O.: On the exact solution of the Euclidean three-matching problem. Acta Cybern. **14**(2), 357376 (1999)
9. Magyar, G., Johnsson, M., Nevalainen, O.: An adaptive hybrid genetic algorithm for the three-matching problem. IEEE Trans. Evol. Comput. **4**(2), 135–146 (2000)
10. Spieksma, F.C., Woeginger, G.J.: Geometric three-dimensional assignment problems. Eur. J. Oper. Res. **91**(3), 611–618 (1996)
11. Tanahashi, R., Zhi-Zhong, C.: A deterministic approximation algorithm for maximum 2-path packing. IEICE Trans. Inf. Syst. **93**(2), 241–249 (2010)
12. Van Laarhoven, P.J.M., Zijm, W.H.M.: Production preparation and numerical control in PCB assembly. Int. J. Flex. Manuf. Syst. **5**(3), 187–207 (1993)
13. Vazquez Casas, G., Castro Campos, R.A., Heredia Velasco, M.A., Zaragoza Martínez, F.J.: A triplet integer programming model for the Euclidean 3-matching problem. In: 12th International Conference on Electrical Engineering, Computing Science and Automatic Control (CCE 2015), 1–4 Oct 2015

# A Multi-objective Robust Ellipse Fitting Algorithm

**Heriberto Cruz Hernández and Luis Gerardo de la Fraga**

**Abstract** The ellipse fitting problem is very common in science and industry. There exists a broad research on this problem, and currently we can find deterministic as well as stochastic methods in the specialized literature. Within the deterministic methods we find the algebraic approaches that are time and memory efficient, but which are, on the other hand, very sensitive to noise and outliers. Stochastic methods aim to find good solutions by exploring a reduced number of the possible ellipses. Two of the main characteristics of stochastic methods is that they are very robust to outliers and noise while they have a higher cost, in terms of functions evaluations, compared to the algebraic methods (with closed form solutions). In this paper, we propose to approach the ellipse fitting problem via using a multi-objective genetic algorithm. We simultaneously optimize two contradictory fitting functions, as this approach helps to enhance the exploration of the genetic algorithm. From the solution set we introduce another function that selects the correct solution. To validate our approach we detect ellipses from data sets with high amounts of noise (up to 100%), and we compare our results to others obtained via a classical mono-objective approach.

**Keywords** Ellipse fitting · Multi-objective optimization · Genetic algorithm · Robust optimization · Computer vision

## 1 Introduction

Ellipse fitting is a highly investigated problem with many applications in science and industry. Ellipse fitting is currently present in many fields like biology (cells counting, detection and segmentation [4, 24]), computer vision (camera calibration [2, 5], head tracking), pattern recognition (shape analysis [19, 26, 31]), medicine (eye tracking and ultrasonic medical imaging [7, 8, 32]), and autonomous robotics [20, 26, 27].

---

H. Cruz Hernández · L.G. de la Fraga (✉)
Computer Science Department, Cinvestav, Av. IPN 2508, 07360 Mexico City, México
e-mail: fraga@cs.cinvestav.mx

© Springer International Publishing AG 2018

Y. Maldonado et al. (eds.), *NEO 2016*, Studies in Computational Intelligence 731,
https://doi.org/10.1007/978-3-319-64063-1_6

There are two main ways to define an ellipse: (i) the algebraic form shown in Eq. (1) and (ii) the parametric form presented in Eq. (2). The algebraic ellipse equation is given by:

$$f(\boldsymbol{a}, \boldsymbol{p}) = \boldsymbol{a}^{\mathrm{T}}\boldsymbol{x} = ax^2 + bxy + cy^2 + dx + ey + g = 0, \tag{1}$$

where $\boldsymbol{a} = [a, b, c, d, e, g]^{\mathrm{T}}$ and $\boldsymbol{x} = [x^2, xy, y^2, x, y, 1]^{\mathrm{T}}$; $[x, y]^{\mathrm{T}}$ represents a point over the ellipse perimeter. This is also the general conic equation and represents an ellipse when $ac - b^2 < 0$.

The parametric ellipse form is given by:

$$\begin{bmatrix} x(t) \\ y(t) \end{bmatrix} = \begin{bmatrix} \cos\theta & -\sin\theta \\ \sin\theta & \cos\theta \end{bmatrix} \begin{bmatrix} a_1 \cos t \\ b_1 \sin t \end{bmatrix} + \begin{bmatrix} x_0 \\ y_0 \end{bmatrix}, \tag{2}$$

where the ellipse parameters can be expressed by vector $\boldsymbol{b} = [a_1, b_1, x_0, y_0, \theta]^{\mathrm{T}}$. $a_1$ is called the semi mayor ellipse axis, $b_1$ the semi minor ellipse axis, $[x_0, y_0]^{\mathrm{T}}$ the ellipse center, and $\theta$ the rotation of the ellipse with respect to axis $x$. All these five parameters have the geometric meaning shown in Fig. 1.

The ellipse fitting problem is defined as the task to find the set of ellipse parameters **a** or **b** (both can be transformed into each other [12]) from a set of at least five points in $\mathbb{R}^2$.

Currently, there exist many solution approaches for this problem which can be divided into deterministic and stochastic methods.

Deterministic methods allow to get a unique solution from a determined set of observations (or points). In this classification we find algebraic methods [1, 15], and the ones based on the Hough transform [25].

Two of the most relevant algebraic methods in literature are the ones proposed by Fitzgibbon et al. [16] and Ahn et al. [1]. Both approaches minimize an error function based on the distance of the observations considering the algebraic distance in Eq. (1). The approach of Ahn et al. [1] considers orthogonal Euclidean distances, i.e., the magnitude of the line perpendicular to the ellipse tangent, that passes through a determined observation point. An overview of the most relevant state of the art

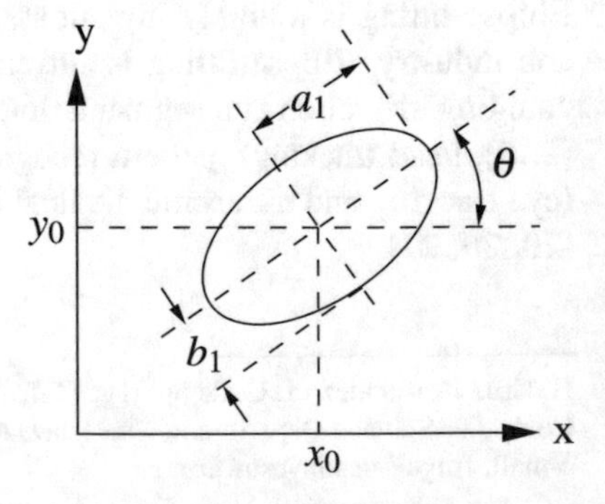


**Fig. 1** Geometric meaning of ellipse parameters

deterministic methods can be found in Kanatani et al. [21]. The Hough transform is a voting scheme that identifies the salient geometrical entities, in this case ellipses, by counting the points that belong to the entity, the ellipses with more votes are selected as the solution.

The main characteristic of algebraic methods is their low computational costs. These methods describe the problem in such a way that the solution can be obtained by solving a closed-form expression or by a numerical method within very few iterations. However, they have the disadvantage to be very sensible to outliers what makes them unsuitable for real world applications where automatic detection is required. Hough transform is robust to outliers but its cost (in terms of both memory and time) increases exponentially with the number of parameters to find and it is also sensitive to noise.

The stochastic methods perform solution space exploration avoiding the exploration of all possible solutions by randomized or evolutionary strategies. These methods can be considered as heuristics and their aim is to find good solutions without assuring optimality. The most relevant work in this classification is Random Sample Consensus (RANSAC) [15, 28, 35] but other evolutionary approaches like genetic algorithms or differential evolution have been also proposed [9, 13].

Two of the main characteristics of stochastic methods in ellipse fitting is that they are very robust to outliers but their computational cost (in terms of memory and time) is higher compared to the algebraic methods. Since stochastic methods performance is an aspect that can still be improved, and since the development of evolutionary optimization heuristics is an active research area that is showing good results, in this paper we propose to solve the ellipse (and multiple ellipse) fitting from very noisy data sets with a multi-objective approach together with a novel ellipse selection function. We believe that the multi-objectivization of ellipse fitting problem increases the exploration characteristic of the heuristic such as the successful application of multi-objectivization to other problems do [22, 33, 34].

This paper is organized as follows. In Sect. 2, we present some of the relevant works in the state of the art. In Sect. 3, we formalize our problem and present the background for this paper. In Sect. 4 we describe our proposal for the single ellipse, as well as the understanding of multi-ellipse fitting. In Sect. 5, we describe all the experiments we performed. In Sect. 6 we present the results of our experiments. In Sect. 7, we present the discussion for the results and finally, in Sect. 8, we present the conclusions and future work.

## 2 Previous Work

In the specialized literature we find that both new algebraic and stochastic methods are being proposed. Recently, Yu et al. [36] proposed an objective function based on a geometric definition that exploits the fact that for each point on an ellipse, the sum of distances from the point to the foci is constant. The authors propose to minimize the objective function by descendant gradient, and their method results are robust to noise, specially with ellipses with high eccentricity.

Arellano and Dahyot [3] proposed an ellipse fitting method based on the minimization of the Euclidean distance between tuned(GMM) [29]. The authors construct one GMM from the observations and another GMM from an initial set of parameters. The initial set of parameters is refined by an annealing approach that minimizes the Euclidean distance between the GMMs, giving as result the refining of the ellipse parameters. The method shows to be suitable for real world applications but has the disadvantage to require an initial set of parameters which has to be fine-tuned.

Mulleti and Seelamantula [30] proposed to model ellipses as analog signals, specifically, as finite-rate-of-innovation (FRI) signals [14]. The authors study the robust ellipse fitting specially for estimating the ellipse parameters from noisy ellipse arcs. This case is of special interest since traditionally methods tend to generate small ellipses around the arc in such cases.

Grbić et al. [18] proposed a hybrid method that uses RANSAC and a least squares method. The approach first performs a variation of the k-means algorithm to find circles as first solution for the ellipses. Those circles define partitions on the image. For each partition, the circle is refined to get an ellipse through singular value decomposition (SVD) and RANSAC. The authors showed that their approach achieved good results in synthetic data (multiple ellipses with noisy edges) and real images. This method requires to know an apriori estimation of the initial parameters or to have an initial partition of the image.

Fornaciari et al. [17] proposed a method that exploits the presence of arcs in the input data set. The algorithm performs first a digital image processing to find edges and then arcs from the edges. The approach classifies the detected arcs based on their convexity and a quadrant analysis in order to select the ones which could belong to ellipses. Finally, the selected arcs are processed through the Hough transform. The method results robust to noise and outliers since it start performing the arc detection (a medium level characteristic). It performs well for images with preferable complete ellipses and is not suitable to detect very segmented ellipses or point clouds.

Cruz-Díaz et al. [9] proposed to use a genetic algorithm (GA) to perform the ellipse fitting from a given point cloud $C$. Each individual in the algorithm encodes an integer vector of size five, which are the indexes of five points in $C$. Each individual is used to fit an ellipse using the orthogonal distance least squares method of [1]. The estimated ellipse is evaluated by automatic computing indicators based on the number of points near the boundaries of the estimated ellipse. The method is very robust to outliers but has the disadvantage of requiring more iterations than other methods.

Even though the efficiency of robust ellipse fitting and detection has been improved in recent years, most improvements have been achieved in noisy conditions without considering high amounts of outliers. Nonetheless, considering the results in [9], it is possible to further improve the ellipse detection from data sets with high ratio of noise. With this goal, this work explores the use of a multi-objective genetic algorithm in the multi-ellipse robust detection and fitting.

## 3 Background

In the following paragraphs will be defined several terms needed to understands this work.

**Outliers**. From the probabilistic point of view, an outlier is an observation that is inconsistent with the rest of the observations of a determined data set. Outliers are frequent in real data and are often not deleted from data sets since they are processed by automatic computational systems and algorithms. Outliers are of special interest in this paper since traditional deterministic methods can totally fail even in the presence of a single outlier. In our problem, the characteristic that group several points is the geometrical model of the ellipse in (1) or (2). Thus, an outlier for a given ellipse is a point that belongs to another ellipse. Or in other words, the points that belong to an ellipse are outliers for the other ellipses in a given data set.

**Noise**. It is possible to distinguish two kinds of noise: (1) random noise added to the location of each ellipse point, and (2) a set of points which are not generated from a recognized geometrical model. Noise of kind (1) is our work is negligible. It can be seen also as the truncation of the point positions to nearest integer values, as pixels positions within an digital image. Noise of kind (2) in our work are points with positions randonly generated, with an uniform distribution, within the borders of an image.

**Multi-objective optimization** is a numerical optimization tool suitable for those problems that consider more than one objective function (often in conflict). This kind of problems is very common since, in real world, often there exist more than one aspect to optimize.

In general form, a multi-objective problem (MOP) is defined as [10]:

$$\min_{\boldsymbol{x} \in D} \boldsymbol{f}(\boldsymbol{x}), \quad \boldsymbol{f} : D \subset \mathbb{R}^m \rightarrow \mathbb{R}^l, \tag{3}$$

where $\boldsymbol{f}(\boldsymbol{x}) = [f_1(\boldsymbol{x}), f_2(\boldsymbol{x}), \ldots, f_l(\boldsymbol{x})]^{\mathrm{T}}$, $l$ is the number of objective functions to minimize with $l \geq 2$ and $\boldsymbol{x} = [x_1, x_2, \ldots, x_m]^{\mathrm{T}}$ is the set of decision variables in the feasible solutions set (or domain) $D$.

Multi-objective problems are characterized themselves for not having a unique solution. The multiple objective functions lead to trade-offs that result in a set of solutions. A feasible solution $\boldsymbol{x}^* \in D$ is that where there not exist other solution $\boldsymbol{x}$ that dominates $\boldsymbol{x}^*$. A solution $\boldsymbol{y} = [y_1, y_2, \ldots, y_m]^{\mathrm{T}}$ dominates [10] another solution $\boldsymbol{z} = [z_1, z_2, \ldots, z_m]^{\mathrm{T}}$, considering minimization, iff $\forall i \in [1, \ldots, l]$, $f_i(\boldsymbol{y}) \leq f_i(\boldsymbol{z})$, and $\exists i \in [1, \ldots, l]$ such that $f_i(\boldsymbol{y}) < f_i(\boldsymbol{z})$. The set of all non-dominated solutions is formally called the Pareto optimal set (PO). A Pareto set is the set $X$ of point solutions $\boldsymbol{x}_i$ corresponding to the found optimal trade-offs. The evaluation of the Pareto set forms a plot in the objectives space which is called the Pareto front.

# 4 Proposed Approach

Our proposal is inspired by the fact that multi-objective optimization algorithms allow to get a set of solutions instead of a unique one. We exploit this fact to first find a reduced number of possible ellipses to later select one of the ellipses as final solution. To find the set of ellipses we apply the multi-objective optimization algorithm to get a finite size approximation of the Pareto set. Then, given the finite approximation of the Pareto set, in a similar task of a decision maker, we choose the final solution through a different fitness function, i.e., the final ellipse. This process is illustrated in Fig. 2.

## 4.1 *Input Point Cloud*

We consider as input a point cloud $C$ in $\mathbb{R}^2$. Each point $\boldsymbol{p}_i = [x, y]^{\mathrm{T}} \in C$ can be a point belonging to one or more ellipses, and $\boldsymbol{p}_i$ can be affected by noise or even it can be an outlier.

## 4.2 *Multi-objective Ellipse Fitting Formulation*

We use the so-called Non-dominated Sorting Genetic Algorithm II (NSGAII) [11]. NSGAII is a very popular multi-objective genetic algorithm that is known to have good performance for two objectives. We define the decision variable vector as $\boldsymbol{x} = [i_1, i_2]^{\mathrm{T}}$, where $i_1$ and $i_2$ are integer indexes of two points, $\boldsymbol{p}_{i_1}$ and $\boldsymbol{p}_{i_2}$ in $C$. We expect the genetic algorithm to choose $i_1$ and $i_2$ as points of an ellipse.

As objectives we have two contradictory fitness functions. These two fitness functions evaluate in different aspects a given ellipse $E$. The first function $f_1$ evaluates

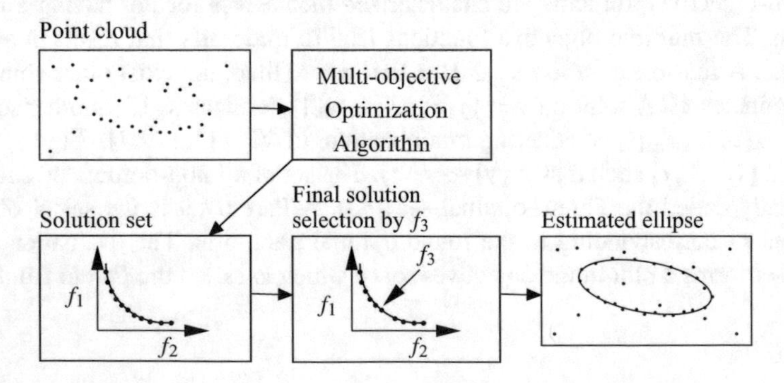


**Fig. 2** General idea for the proposed approach

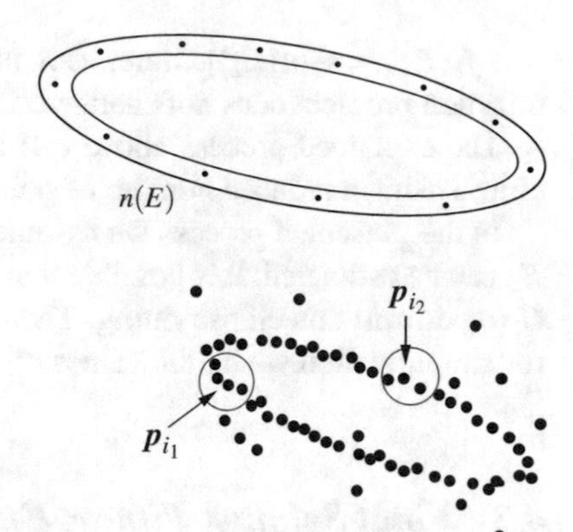


**Fig. 3** Points form the ellipse $E$, and the two ellipses form its neighborhood $n(E)$

**Fig. 4** Two centered circles at $\boldsymbol{p}_{i_1}$ and $\boldsymbol{p}_{i_2}$ with radius $r$. Four points are selected from the *left* circle and three points from the *right* one

the area of the ellipse and the second $f_2$ is counting the points in $C$ that belong to the ellipse neighborhood. We simultaneously minimize $f_1$ and $f_2$. The ellipse neighborhood, as defined in [9], is the set of points inside a threshold of orthogonal distance from an ellipse. A point $\boldsymbol{p}$ is in the ellipse $E$ neighborhood, denoted as $n(E)$, if $d_{\text{orthogonal}}(\boldsymbol{p}, E) < T$, where $d(\cdot)$ is the Euclidean distance from the point to other point on the ellipse, which its line that joint both points is also orthogonal to the ellipse perimeter, and $T$ is a threshold. An illustration for the ellipse neighborhood is shown in Fig. 3.

#### 4.2.1 Individuals Evaluation Process

The process for evaluating a solution $\boldsymbol{x} = [i_1, i_2]^{\mathrm{T}}$ is the following.

We first select the points with indexes $i_1$ and $i_2$ from $C$, i.e., $\boldsymbol{p}_{i_1}$ and $\boldsymbol{p}_{i_2}$. Then two circles with centers $\boldsymbol{p}_{i_1}$, $\boldsymbol{p}_{i_1}$ and radius $r$ are defined, and all points inside the circles are selected. A point $\boldsymbol{p}$ is inside the circle $k$ if

$$|\boldsymbol{p}_{i_k} - \boldsymbol{p}| \leq r, \text{ for } k = \{1, 2\}.$$

This step is illustrated in Fig. 4.

Using the points inside the two circles we estimate an ellipse $E_1$ through the method in [12] which simplifies the calculations of algebraic method proposed in Fitzgibbon et al. [16]. Then we identify all those points inside the $E_1$ neighborhood, $n(E_1)$.

Using the points in $n(E_1)$ we estimate a second ellipse $E_2$. For the estimation of $E_2$ we use again the algebraic method. $E_2$ is the found ellipse from the points $\boldsymbol{p}_{i_1}$ and $\boldsymbol{p}_{i_2}$.

We evaluate our two fitness functions using $E_2$. The value of $f_1$ and $f_2$ are computed as the area and the cardinality of the neighborhood of $E_2$, i.e.: $f_1(E_2) = \pi ab$

and $f_2(E_2) = -|n(E_2)|$. Here, as $\pi$ is a constant in the calculation of $f_1$, the optimization problem does not change if it is used $f_1(E) = ab$.

The explained process above can be seen as a local search. $E_1$ is a first fitted ellipse using a reduced number of points and $E_2$ as a refined version of $E_1$.

In the presented process we assume that $C$ has such a configuration that $E_1$ and $E_2$ can be estimated. It is possible that there cannot be found five such points within $C$ to perform this ellipse fitting. Then, those cases do not represent an ellipse, and for simplification we fix the values of $f_1$ and $f_2$ equal to zero.

## 4.3 Final Solution Fitness Function

From the application of the multi-objective genetic algorithm along the process in Sect. 4.2.1 we get a solution set $X = \{\boldsymbol{x}_1, \boldsymbol{x}_2, \ldots, \boldsymbol{x}_m\}$ and, their associated ellipses, as the example given in Fig. 5.

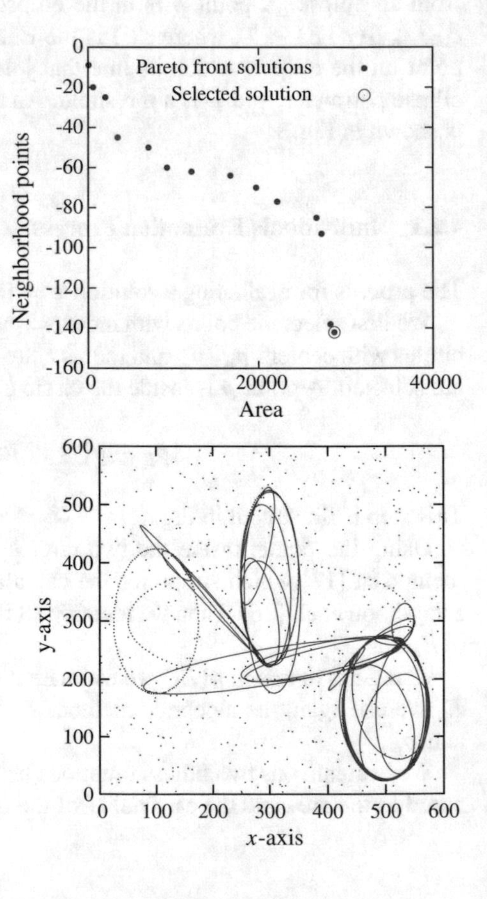


**Fig. 5** *Top* Pareto front output for a run of our approach. *Bottom* associated ellipses to the Pareto front in the *top* and the selected final solution

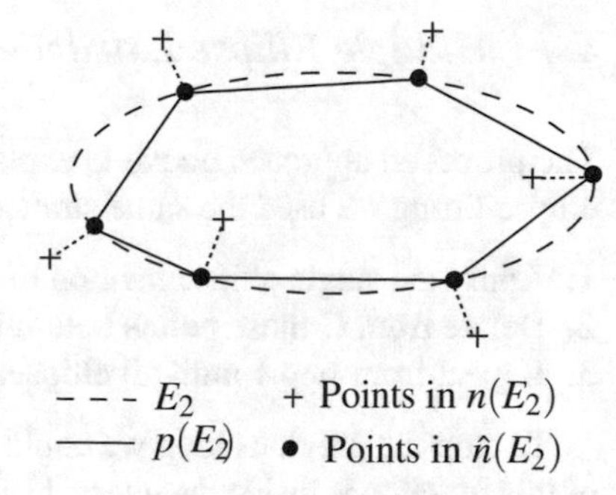


**Fig. 6** Polygon estimation from the ellipse $E_2$

We choose the final found ellipse $E_{\text{final}}$, out of the computed Pareto set, with a third fitness function $f_3$. We propose $f_3$ as an indicator of the distribution of points on the found ellipses. The ellipse those points yield the best distribution will be chosen as the final solution.

For evaluating $f_3$ we apply the procedure in Sect. 4.2.1 to get an ellipse $E_2$ and its neighborhood set $n(E_2)$. Then each element in a new set $\hat{n}(E_2)$ is computed. Cardinality of $\hat{n}(E_2)$ is equal to the cardinality of $n(E_2)$. Each element in $\hat{n}(E_2)$ is the corresponding point in $E_2$ of each element (point) in $n(E_2)$. Each element in $\hat{n}(E_2)$ is the nearest point on $E_2$ to its corresponding point in $E_2$ with is also orthogonal to the perimeter ellipse to the line that joints the corresponding points in $n(E_2)$ and $\hat{n}(E_2)$. Both sets $n(E_2)$ and $\hat{n}(E_2)$ as shown with one example in Fig. 6.

The points in $\hat{n}(E_2)$ define a irregular polygon of $m$ edges. We denote this polygon as $p(E_2)$. Our purpose with this polygon is illustrated in Fig. 6. Notice that when the number of points in $\hat{n}(E_2)$ tends to infinite, the sum of polygon edges will become similar to the ellipse $E_2$ perimeter. With this idea we compute $f_3$ as the reason of the polygon perimeter and the ellipse perimeter as shown in Eq. (4). $f_3$ takes higher values when the polygon has more and well distributed points on $E_2$.

$$f_3 = \frac{\text{polygon perimeter} * \text{number of points}}{\text{ellipse perimeter}}. \tag{4}$$

To compute the ellipse perimeter there exist exact methods based on infinite series or integrals [6], however since these methods are not suitable for real world applications without avoid truncation, we used the Ramanujan perimeter approximation in Eq. (5).

$$P_E \approx \pi \left[ 3(a+b) - \sqrt{(a+3b)(3a+b)} \right]. \tag{5}$$

As final solution we choose the ellipse with the highest $f_3$ value.

### 4.4 Multiple Ellipse Extraction

The proposed approach above is explained for a single ellipse fitting. For multiple ellipse fitting we used the same strategy used in [9], i.e.:

1. Apply the single ellipse method to detect an ellipse $E$ from a set of points $C$.
2. Delete from $C$ those points belonging to $E$.
3. Repeat from step 1 until all ellipses are extracted.

To stop the previous loop we could use the value of function $f_3$ in (4). The value of this function is inside the interval $[0, n]$, where $n$ is the number of points that form the found ellipse. Then a threshold value can be used to stop the loop, where this value represents when a set of points is consider an ellipse according to its number of points.

## 5 Experiments

To validate our proposal we performed various experiments in different conditions. We adopted the same benchmark used in [9] in which different configurations of four ellipses are used to test the method.

### 5.1 Data Sets

We generated multiple data sets with three configurations of four ellipses point clouds. We considered the three cases in [9], i.e., disjoint, overlapped, and nested ellipses. These configurations are shown in Fig. 7. We varied the amount of noise from 0% to 100% in steps of 20%. In total we generated 18 different point sets. Details for the point sets are further detailed in Table 1.

Each data set, at the beginning of the detection process, has four ellipses, then for any ellipse the points of the other three ellipses are outliers ($126 \cdot 3 = 378$ points). The percentage of noise is calculated as the random position points against the total number of points for the four ellipses ($126 \cdot 4 = 504$ points). Each time an ellipse is found, its points are erased from the data set, then the number of outliers decreases, and the percentage of noise (with respect to the points that belong to ellipses) increases, as is depicted in Table 2.

### 5.2 Experiments Settings

For each data set described in Sect. 5.1 we performed 200 executions. For all our experiments we fixed population size, crossover and mutation probabilities and we

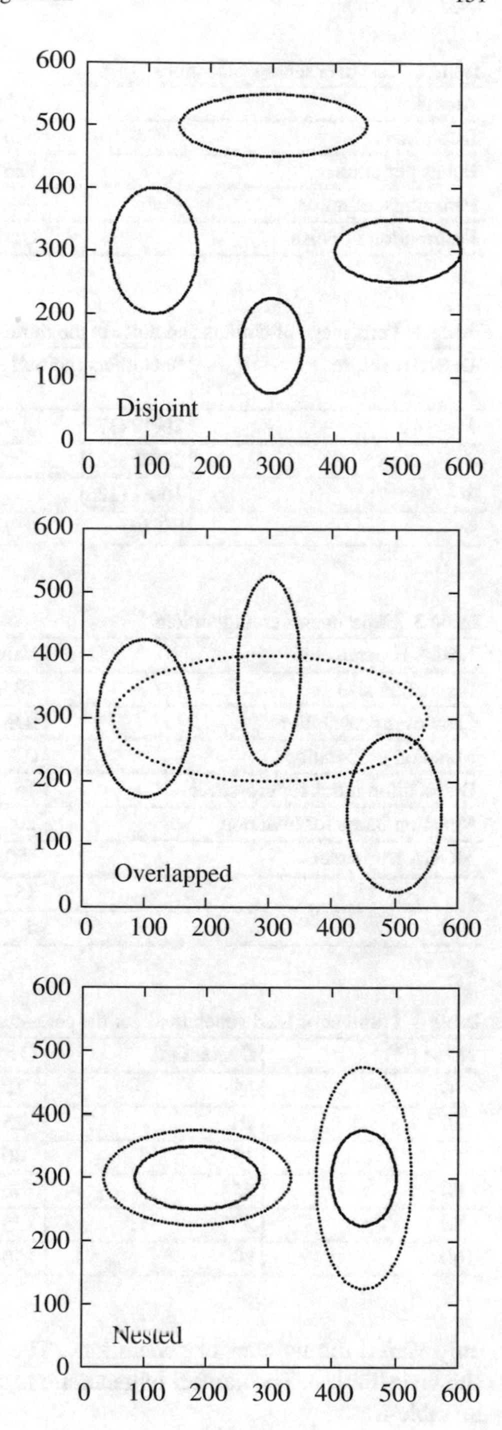


**Fig. 7** Test cases for our experiments

**Table 1** Test data sets specifications

| Aspect | Values |
|---|---|
| Image size | 600 × 600 pixels |
| Points per ellipse | 126 points for each ellipse |
| Percentage of noise | 0%, 20%, 40%, 60%, 80%, and 100% |
| Distribution of noise | Uniform |

**Table 2** Percentage of outliers and noise in the multiple ellipse detection problem

| Detected ellipse | % Outliers (points) | % Noise (noise points/ellipses points) |
|---|---|---|
| 1 | 300% (378) | 100% (504/504) |
| 2 | 200% (252) | 133% (504/378) |
| 3 | 100% (126) | 200% (504/252) |
| 4 | 0% (0) | 400% (504/126) |

**Table 3** Experiments configurations

| NSGA-II parameter | Value |
|---|---|
| Population size | 20 individuals |
| Crossover probability | 0.8 |
| Mutation probability | 0.2 |
| Distribution index for crossover | 15 |
| Mutation index for mutation | 20 |
| MOGA parameters | Value |
| $r$ | 15 |
| $T$ | 5 |

**Table 4** Number of used generations for the percentage of noise and each case of study

| Noise (%) | Disjoint | Overlapped | Nested |
|---|---|---|---|
| 0 | 5 | 10 | 5 |
| 20 | 10 | 25 | 10 |
| 40 | 15 | 40 | 30 |
| 60 | 25 | 80 | 50 |
| 80 | 35 | 125 | 55 |
| 100 | 80 | 150 | 95 |

only varied the number of generations. The list of NSGA-II parameters used are shown in Table 3. The number of generations used in for each configuration is shown in Table 4.

# 6 Results

For each test data set described in Sect. 5 we executed the proposed approach to detect four ellipses, and we counted the number of correctly detected ellipses. An ellipse is considered as detected correctly if it has 107 or more points in its neighborhood, i.e., the 85% of the ground-truth conforming points. We computed the mean $\bar{x}$, the minimum $m$, and the maximum $M$ number of detected ellipses for each group of executions. We show the number of ellipses and the number of evaluations for each case in Tables 5, 6, and 7. For comparison we also show the results of [9]. Our approach is labeled as MOGA while the results in [9] as GA. Underlined numbers represent a better result of an approach over the other. Some instances of the results obtained for the three cases with 100% of noise are shown in Fig. 8.

**Table 5** Results for the disjoint case

| Noise (%) | MOGA | | | | GA | | | |
|---|---|---|---|---|---|---|---|---|
| | m | M | $\bar{x}$ | Eval. | m | M | $\bar{x}$ | Eval. |
| 0 | 4 | 4 | 4.00 | 100 | 4 | 4 | 4.00 | 3710 |
| 20 | 3 | 4 | 3.99 | 200 | 4 | 4 | 3.99 | 5180 |
| 40 | 3 | 4 | 3.99 | 300 | 3 | 4 | 3.92 | 5530 |
| 60 | 4 | 4 | 4.00 | 500 | 2 | 4 | 3.95 | 5518 |
| 80 | 3 | 4 | 3.96 | 700 | 2 | 4 | 3.87 | 6230 |
| 100 | 3 | 4 | 3.94 | 1600 | 2 | 4 | 3.84 | 6300 |

**Table 6** Results for the overlapped case

| Noise (%) | MOGA | | | | GA | | | |
|---|---|---|---|---|---|---|---|---|
| | m | M | $\bar{x}$ | Eval. | m | M | $\bar{x}$ | Eval. |
| 0 | 4 | 4 | 4.00 | 200 | 4 | 4 | 4.00 | 5040 |
| 20 | 2 | 4 | 3.98 | 500 | 3 | 4 | 3.97 | 5040 |
| 40 | 3 | 4 | 3.99 | 800 | 3 | 4 | 3.97 | 5460 |
| 60 | 3 | 4 | 3.97 | 1600 | 3 | 4 | 3.94 | 5530 |
| 80 | 2 | 4 | 3.98 | 2500 | 3 | 4 | 3.97 | 5810 |
| 100 | 2 | 4 | 3.85 | 3000 | 2 | 4 | 3.82 | 6160 |

**Table 7** Results for the nested case

| Noise (%) | MOGA | | | | GA | | | |
|---|---|---|---|---|---|---|---|---|
| | m | M | $\bar{x}$ | Eval. | m | M | $\bar{x}$ | Eval. |
| 0 | 4 | 4 | 4.00 | 100 | 4 | 4 | 4.00 | 6300 |
| 20 | 4 | 4 | 4.00 | 200 | 4 | 4 | 4.00 | 5250 |
| 40 | 4 | 4 | 4.00 | 600 | 4 | 4 | 4.00 | 5460 |
| 60 | 3 | 4 | 3.98 | 1000 | 3 | 4 | 3.96 | 5600 |
| 80 | 2 | 4 | 3.90 | 1100 | 2 | 4 | 3.88 | 5740 |
| 100 | 2 | 4 | 3.91 | 1900 | 3 | 4 | 3.94 | 6160 |

## 7 Discussion

Our approach shows an important reduction in the number of evaluations required to solve the multiple ellipse fitting problem. For small percentage of noise, for instance, for 0% of noise (but very high number of outliers) we observe in Tables 5, 6 and 7 that our approach requires hundreds of evaluations against thousands that require the GA (i.e. in Table 5, MOGA requires 100 and GA requires 3710 evaluations of the fitness function. And the same applies for the other two cases in Tables 6 and 7). Thus, the results in Sect. 6 of our novel approach are very robust and also efficient for multiple ellipse detection. This is a main contribution of this paper, since other methods like RANSAC, Hough transform or a mono-objective GA [9] require thousands of evaluations.

Our approach also showed better results in the number of correctly found ellipses. We see an improvement in the mean of found ellipse in all cases in Tables 5, 6 and 7. Only in a single case the GA is slightly better that our approach: with 100% in the nested case, in the last row of Table 7, 3.91 versus 3.94 average ellipses found by MOGA and GA, respectively Our approach obtains better results in 17 of the 18 used data set. And our approach always use less number of evaluations of the fitness function.

The most difficult case for our approach is the overlapped case. This is because some points can be considered part of more than one ellipse. When a first ellipse is detected and its points deleted from the data set, the other ellipses lose some of their points.

We associate the good performance of our approach with three main aspects:

- The genetic algorithm individuals have only two values (two points) instead of the five variable values that represent an ellipse. This increases the probability of finding common ellipse points at the beginning of the evolutionary process.
- The multi-objective approach increases the exploration of all possible ellipses and makes the final ellipse selection simpler.

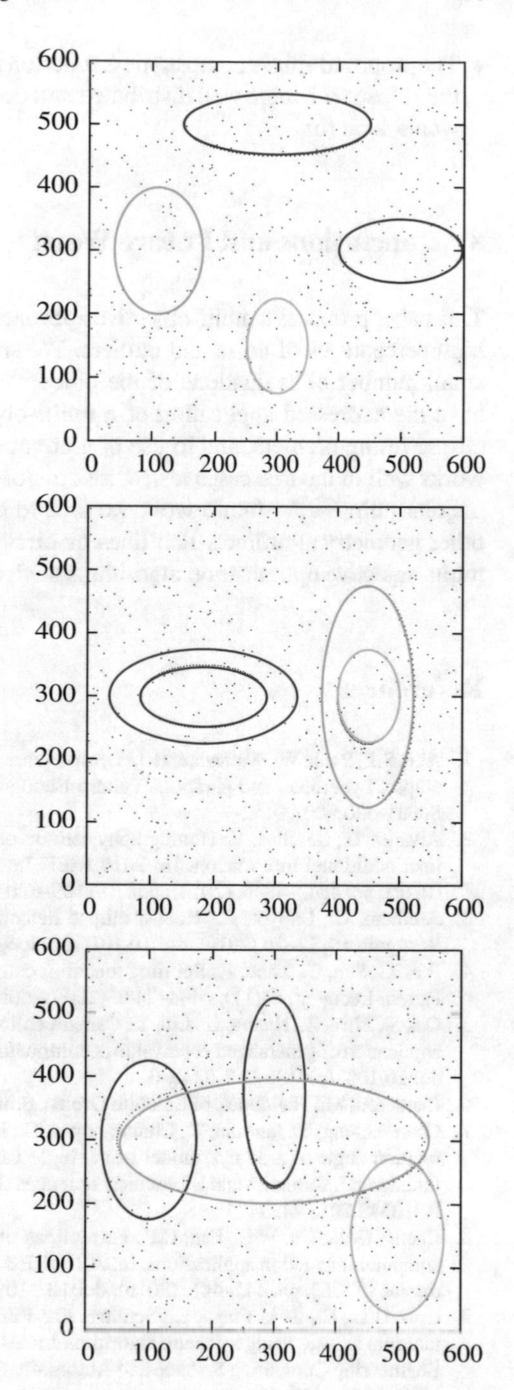


**Fig. 8** Final results for every case with 100% of noise

- The proposed ellipse completeness function is a geometric approach that will return the ellipse with more well distributed and dense points. This is what a human agent would look for.

## 8 Conclusions and Future Work

This paper proposes a multi-objective approach for ellipse fitting from data sets with high percentages of noise and outliers. We successfully solved the problem with a small number of evaluations of the objective function. As main contributions we have the addressed application of a multi-objective optimization algorithm to the ellipse fitting problem, and to use of a novel ellipse fitness function. Our approach works well in the tree cases tested, i.e., disjoint, overlapped (the most difficult) and disjoint ellipses. As future work we aim to explore similar approaches for fitting other geometrical artifacts like lines or circles, and to use other more specialized multi-objective optimization algorithms such as HCS [23].

## References

1. Ahn, S.J., Rauh, W., Warnecke, H.J.: Least-squares orthogonal distances fitting of circle, sphere, ellipse, hyperbola, and parabola. Pattern Recognit. **34**(12), 2283–2303 (2001). doi:10.1016/S0031-3203(00)00152-7
2. Alvarez, L., Caselles, V.: Homography estimation using one ellipse correspondence and minimal additional information. In: 2014 IEEE International Conference on Image Processing (ICIP), pp. 4842–4846. (2014). doi:10.1109/ICIP.2014.7025981
3. Arellano, C., Dahyot, R.: Robust ellipse detection with Gaussian mixture models. Pattern Recognit. **58**, 12–26 (2016). doi:10.1016/j.patcog.2016.01.017
4. Bai, X., Sun, C., Zhou, F.: Splitting touching cells based on concave points and ellipse fitting. Pattern Recognit. **42**(11), 2434–2446 (2009). doi:10.1016/j.patcog.2009.04.003
5. Cai, S., Zhao, Z., Huang, L., Liu, Y.: Camera calibration with enclosing ellipses by an extended application of generalized eigenvalue decomposition. Mach. Vis. Appl. **24**(3), 513–520 (2013). doi:10.1007/s00138-012-0446-0
6. Capderou, M.: Handbook of Satellite Orbits: From Kepler to GPS. Springer, Berlin (2014)
7. Charoenpong, T., Jantima, T., Chianrabupra, C., Mahasitthiwat, V.: A new method to estimate rotation angle of a 3d eye model from single camera. In: 2015 International Conference on Intelligent Informatics and Biomedical Sciences (ICIIBMS), pp. 398–402 (2015). doi:10.1109/ICIIBMS.2015.7439474
8. Cheng, C.W., Ou, W.L., Fan, C.P.: Fast ellipse fitting based pupil tracking design for human-computer interaction applications. In: 2016 IEEE International Conference on Consumer Electronics (ICCE), pp. 445–446. (2016). doi:10.1109/ICCE.2016.7430685
9. Cruz-Díaz, C., de la Fraga, L., Schütze, O.: Fitness function evaluation for the detection of multiple ellipses using a genetic algorithm. In: 2011 8th International Conference on Electrical Engineering Computing Science and Automatic Control (CCE), pp. 1–6. (2011). doi:10.1109/ICEEE.2011.6106652
10. Deb, K.: Multi-Objective Optimization Using Evolutionary Algorithms. Wiley Interscience Series in Systems and Optimization. Wiley, New York (2001)

11. Deb, K., Pratap, A., Agarwal, S., Meyarivan, T.: A fast and elitist multiobjective genetic algorithm: NSGA-II. IEEE Trans. Evol.Comput. **6**(2), 182–197 (2002). doi:10.1109/4235.996017
12. de la Fraga, L., Cruz Díaz, C.: Fitting an ellipse is equivalent to find the roots of a cubic equation. In: 2011 8th International Conference on Electrical Engineering Computer Science and Automatic Control, IEEE pp. 1–4. (2011)
13. de la Fraga, L., Vite-Silva, I., Cruz-Cortes, N.: Euclidean distance fit of conics using differential evolution. Evolutionary Image Analysis and Signal Processing. Studies in Computational Intelligence, pp. 171–184. Springer, Berlin (2009)
14. Eldar, Y., Kutyniok, G.: Compressed Sensing: Theory and Applications. Cambridge University Press, Cambridge (2012)
15. Fischler, M.A., Bolles, R.C.: Random sample consensus: a paradigm for model fitting with applications to image analysis and automated cartography. Commun. ACM **24**(6), 381–395 (1981). doi:10.1145/358669.358692
16. Fitzgibbon, A., Pilu, M., Fisher, R.B.: Direct least square fitting of ellipses. IEEE Trans. Pattern Anal. Mach. Intell. **21**(5), 476–480 (1999). doi:10.1109/34.765658
17. Fornaciari, M., Prati, A., Cucchiara, R.: A fast and effective ellipse detector for embedded vision applications. Pattern Recognit. **47**(11), 3693–3708 (2014). doi:10.1016/j.patcog.2014.05.012
18. Grbić, R., Grahovac, D., Scitovski, R.: A method for solving the multiple ellipses detection problem. Pattern Recognit. **60**, 824–834 (2016). doi:10.1016/j.patcog.2016.06.031
19. Johansson, E., Johansson, D., Skog, J., Fredriksson, M.: Automated knot detection for high speed computed tomography on Pinus sylvestris L. and Picea abies (L.) Karst. using ellipse fitting in concentric surfaces. Comput. Electron. Agric. **96**, 238–245 (2013). doi:10.1016/j.compag.2013.06.003
20. Jung, Y., Lee, D., Bang, H.: Study on ellipse fitting problem for vision-based autonomous landing of an UAV. In: 2014 14th International Conference on Control, Automation and Systems (ICCAS), pp. 1631–1634. (2014). doi:10.1109/ICCAS.2014.6987819
21. Kanatani, K., Sugaya, Y., Kanazawa, Y.: Advances in Computer Vision and Pattern Recognition. Springer, Berlin (2016)
22. Knowles, J.D., Watson, R.A., Corne, D.W.: Reducing local optima in single-objective problems by multi-objectivization. In: Zitzler, E., Deb, K., Thiele, L., Coello, C.A.C., Corne, D. (eds.) Proceedings of the First International Conference on Evolutionary Multi-Criterion Optimization (EMO 2001). LNCS, vol. 1993, pp. 269–283. Springer, Berlin (2001)
23. Lara, A., Sanchez, G., Coello Coello, C., Schütze, O.: HCS: a new local search strategy for memetic multiobjective evolutionary algorithms. IEEE Trans. Evol. Comput. **14**(1), 112–132 (2010)
24. Liao, M., Zhao, Y.Q., Li, X.H., Dai, P.S., Xu, X.W., Zhang, J.K., Zou, B.J.: Automatic segmentation for cell images based on bottleneck detection and ellipse fitting. Neurocomputing **173**, 615–622 (2016). doi:10.1016/j.neucom.2015.08.006
25. Lu, W., Tan, J.: Detection of incomplete ellipse in images with strong noise by iterative randomized hough transform (irht). Pattern Recognit. **41**(4), 1268–1279 (2008). doi:10.1016/j.patcog.2007.09.006, http://www.sciencedirect.com/science/article/pii/S0031320307004128
26. Ma, Z., Ho, K.C.: Asymptotically efficient estimators for the fittings of coupled circles and ellipses. Digit. Signal Process. Rev. J. **25**(1), 28–40 (2014). doi:10.1016/j.dsp.2013.10.022
27. Masuzaki, T., Sugaya, Y., Kanatani, K.: Floor-wall boundary estimation by ellipse fitting. In: 2015 IEEE 7th International Conference on Cybernetics and Intelligent Systems (CIS) and IEEE Conference on Robotics, Automation and Mechatronics (RAM), pp. 30–35. (2015). doi:10.1109/ICCIS.2015.7274592
28. Matas, J., Chum, O.: Randomized RANSAC with Td, d test. Image Vis. Comput. **22**(10), 837–842 (2004). doi:10.1016/j.imavis.2004.02.009. British Machine Vision Computing 2002
29. McNicholas, P.: Mixture Model-Based Classification. CRC Press, Boca Raton (2016)
30. Mulleti, S., Seelamantula, C.S.: Innovation Sampling Principle **25**(3), 1451–1464 (2016)
31. Panagiotakis, C., Argyros, A.: Parameter-free modelling of 2D shapes with ellipses. Pattern Recognit. **53**, 259–275 (2016). doi:10.1016/j.patcog.2015.11.004

32. Rueda, S., Knight, C.L., Papageorghiou, A.T., Noble, J.A.: Oriented feature-based coupled ellipse fitting for soft tissue quantification in ultrasound images. In: 2013 IEEE 10th International Symposium on Biomedical Imaging, pp. 1014–1017. (2013). doi:10.1109/ISBI.2013.6556649
33. Segura, C., Coello Coello, C., Miranda, G., Leon, C.: Using multi-objective evolutionary algorithms for single-objective optimization. 4OR **11**(3), 201–228 (2013)
34. Vite Silva, I., Cruz Cortés, N., Toscano Pulido, G., de la Fraga, L.: Optimal triangulation in 3D computer vision using a multi-objective evolutionary algorithm. In: EvoWorkshops 2007, Lecture Notes in Computer Science 4448, pp. 330–339. (2007)
35. Wang, Y., Zheng, J., Xu, Q.Z., Li, B., Hu, H.M.: An improved RANSAC based on the scale variation homogeneity. J. Vis. Commun. Image Represent. B **40**, 751–764 (2016). doi:10.1016/j.jvcir.2016.08.019
36. Yu, J., Kulkarni, S.R., Poor, H.V.: Robust ellipse and spheroid fitting. Pattern Recognit. Lett. **33**(5), 492–499 (2012). doi:10.1016/j.patrec.2011.11.025

# Gradient-Based Multiobjective Optimization with Uncertainties

**Sebastian Peitz and Michael Dellnitz**

**Abstract** In this article we develop a gradient-based algorithm for the solution of multiobjective optimization problems with uncertainties. To this end, an additional condition is derived for the descent direction in order to account for inaccuracies in the gradients and then incorporated into a subdivision algorithm for the computation of global solutions to multiobjective optimization problems. Convergence to a superset of the Pareto set is proved and an upper bound for the maximal distance to the set of substationary points is given. Besides the applicability to problems with uncertainties, the algorithm is developed with the intention to use it in combination with model order reduction techniques in order to efficiently solve PDE-constrained multiobjective optimization problems.

## 1 Introduction

In many applications from industry and economy, one is interested in simultaneously optimizing several criteria. For example, in transportation one wants to reach a destination as fast as possible while minimizing the energy consumption. This example illustrates that in general, the different objectives contradict each other. Therefore, the task of computing the set of optimal compromises between the conflicting objectives, the so-called *Pareto set*, arises. This leads to a multiobjective optimization problem (MOP). Based on the knowledge of the Pareto set, a *decision maker* can use this information either for improved system design or for changing parameters during operation, as a reaction on external influences or changes in the system state itself.

Multiobjective optimization is an active area of research. Different methods exist to address MOPs, e.g. deterministic approaches [10, 21], where ideas from scalar optimization theory are extended to the multiobjective situation. In many cases, the resulting solution method involves solving multiple scalar optimization

S. Peitz (✉) · M. Dellnitz
Department of Mathematics, Paderborn University, Warburger Str. 100,
33098 Paderborn, Germany
e-mail: speitz@math.upb.de

© Springer International Publishing AG 2018

Y. Maldonado et al. (eds.), *NEO 2016*, Studies in Computational Intelligence 731,
https://doi.org/10.1007/978-3-319-64063-1_7

problems consecutively. Continuation methods make use of the fact that under certain smoothness assumptions, the Pareto set is a manifold that can be approximated by continuation methods known from dynamical systems theory [15]. Another prominent approach is based on evolutionary algorithms [4], where the underlying idea is to evolve an entire set of solutions (population) during the optimization process. Set oriented methods provide an alternative deterministic approach to the solution of MOPs. Utilizing subdivision techniques (cf. [8, 17, 29]), the desired Pareto set is approximated by a nested sequence of increasingly refined box coverings. In the latter approach, gradients are evaluated to determine a descent direction for all objectives.

Many solution approaches are gradient-free since the benefit of derivatives is less established than in single objective optimization [3]. Exceptions are scalarization methods since the algorithms used in this context mainly stem from scalar optimization. Moreover, in so-called *memetic algorithms* [22], evolutionary algorithms are combined with local search strategies, where gradients are utilized (see e.g. [20, 30, 31]). Finally, several authors also develop gradient-based methods for MOPs directly. In [9, 12, 26], algorithms are developed where a single descent direction is computed in which all objectives decrease. In [3], a method is presented by which the entire set of descent directions can be determined, an extension of Newton's method to MOPs with quadratic convergence is presented in [13].

Many real world problems possess uncertainties for various reasons such as the ignorance of the exact underlying dynamical system or unknown material properties. The use of reduced order models in order to decrease the computational effort also introduces errors which can often be quantified [34]. A similar concept exists in the context of evolutionary computation, where *surrogate-assisted evolutionary computation* (see e.g. [18] for a survey) is often applied in situations where the exact model is too costly to solve. In this setting, the set of *almost Pareto optimal points* has to be computed (see also [35], where the idea of $\varepsilon$ *efficiency* was introduced). Many researchers are investigating problems related to uncertainty quantification and several authors have addressed multiobjective optimization problems with uncertainties. In [16, 33], probabilistic approaches to multiobjective optimization problems with uncertainties were derived independently. In [2, 5, 6, 32], evolutionary algorithms were developed for problems with uncertain or noisy data in order to compute robust approximations of the Pareto set. In [27], stochastic search was used, cell mapping techniques were applied in [11, 14], the weighted sum method was extended to uncertainties.

In this article, we present extensions to the gradient-free and gradient-based global subdivision algorithms for unconstrained MOPs developed in [8] which take into account inexactness in both the function values and the gradients. The algorithms compute a set of solutions which is a superset of the Pareto set with an upper bound for the distance to the Pareto set. The remainder of the article is organized in the following manner. In Sect. 2, we give a short introduction to multiobjective optimization in general and gradient-based descent directions for MOPs. In Sect. 3, a descent direction for all objectives under inexact gradient information is developed and an upper bound for the distance to a Pareto optimal point is given. In Sect. 4, the subdivision algorithm presented in [8] is extended to inexact function and gradient values before we present our results in Sect. 5 and draw a conclusion in Sect. 6.

## 2 Multiobjective Optimization

Consider the continuous, unconstrained multiobjective optimization problem

$$\min_{x \in \mathbb{R}^n} F(x) = \min_{x \in \mathbb{R}^n} \begin{pmatrix} f_1(x) \\ \vdots \\ f_k(x) \end{pmatrix}, \tag{MOP}$$

where $F : \mathbb{R}^n \to \mathbb{R}^k$ is a vector valued objective function with continuously differentiable objective functions $f_i : \mathbb{R}^n \to \mathbb{R}$, $i = 1, \ldots, k$. The space of the parameters $x$ is called the *decision space* and the function $F$ is a mapping to the $k$-dimensional *objective space*. In contrast to single objective optimization problems, there exists no total order of the objective function values in $\mathbb{R}^k$, $k \geq 2$ (unless the objectives are not conflicting). Therefore, the comparison of values is defined in the following way [21]:

**Definition 1** Let $v, w \in \mathbb{R}^k$. The vector $v$ is *less than* $w$ ($v <_p w$), if $v_i < w_i$ for all $i \in \{1, \ldots, k\}$. The relation $\leq_p$ is defined in an analogous way.

A consequence of the lack of a total order is that we cannot expect to find isolated optimal points. Instead, the solution of (MOP) is the set of optimal compromises, the so-called *Pareto set* named after Vilfredo Pareto:

**Definition 2**

(a) A point $x^* \in \mathbb{R}^n$ *dominates* a point $x \in \mathbb{R}^n$, if $F(x^*) \leq_p F(x)$ and $F(x^*) \neq F(x)$.
(b) A point $x^* \in \mathbb{R}^n$ is called *(globally) Pareto optimal* if there exists no point $x \in \mathbb{R}^n$ dominating $x^*$. The image $F(x^*)$ of a (globally) Pareto optimal point $x^*$ is called a *(globally) Pareto optimal value*.
(c) The set of non-dominated points is called the *Pareto set* $\mathcal{P}_S$, its image the *Pareto front* $\mathcal{P}_F$.

Consequently, for each solution that is contained in the Pareto set, one can only improve one objective by accepting a trade-off in at least one other objective. That is, roughly speaking, in a two-dimensional problem, we are interested in finding the "lower left" boundary of the reachable set in objective space (cf. Fig. 1b). A more detailed introduction to multiobjective optimization can be found in e.g. [10, 21].

Similar to single objective optimization, a necessary condition for optimality is based on the gradients of the objective functions. In the multiobjective situation, the corresponding Karush–Kuhn–Tucker (KKT) condition is as follows:

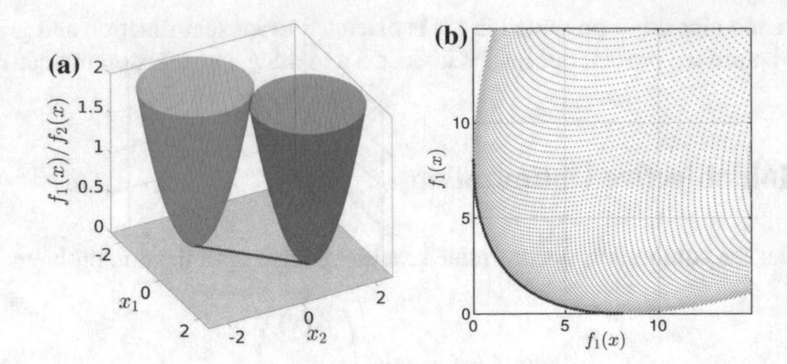


**Fig. 1** The *red lines* depict the Pareto set (**a**) and Pareto front (**b**) of an exemplary multiobjective optimization problem (two paraboloids) of the form (MOP) with $n = 2$ and $k = 2$

**Theorem 1** ([19]) *Let $x^*$ be a Pareto point of* (MOP). *Then, there exist nonnegative scalars $\alpha_1, \ldots, \alpha_k \geq 0$ such that*

$$\sum_{i=1}^{k} \alpha_i = 1 \text{ and } \sum_{i=1}^{k} \alpha_i \nabla f_i(x^*) = 0. \tag{1}$$

Observe that (1) is only a necessary condition for a point $x^*$ to be a Pareto point and the set of points satisfying (1) is called the set of *substationary points* $\mathcal{P}_{S,sub}$. Under additional smoothness assumptions (see [15]) $\mathcal{P}_{S,sub}$ is locally a $(k-1)$-dimensional manifold. Obviously $\mathcal{P}_{S,sub}$ is a superset of the Pareto set $\mathcal{P}_S$.

If $x \notin \mathcal{P}_{S,sub}$ then the KKT conditions can be utilized in order to identify a descent direction $q(x)$ for which all objectives are non-increasing, i.e.:

$$-\nabla f_i(x) \cdot q(x) \geq 0, \quad i = 1, \ldots, k. \tag{2}$$

One way to compute a descent direction satisfying (2) is to solve the following auxiliary optimization problem [26]:

$$\min_{\alpha \in \mathbb{R}^k} \left\{ \left\| \sum_{i=1}^{k} \alpha_i \nabla f_i(x) \right\|_2^2 \;\middle|\; \alpha_i \geq 0,\ i = 1, \ldots, k, \sum_{i=1}^{k} \alpha_i = 1 \right\}. \tag{QOP}$$

Using (QOP), we obtain the following result:

**Theorem 2** ([26]) *Define $q : \mathbb{R}^n \to \mathbb{R}^n$ by*

$$q(x) = -\sum_{i=1}^{k} \widehat{\alpha}_i \nabla f_i(x), \tag{3}$$

*where $\widehat{\alpha}$ is a solution of* (QOP). *Then either $q(x) = 0$ and $x$ satisfies* (1), *or $q(x)$ is a descent direction for all objectives $f_1(x), \ldots, f_k(x)$ in $x$. Moreover, $q(x)$ is locally Lipschitz continuous.*

*Remark 1* As in the classical case of scalar optimization there exist in general infinitely many valid descent directions $q(x)$ and using the result from Theorem 2 yields one particular direction. As already stated in the introduction, there are alternative ways to compute such a direction, see e.g. [12] for the computation of a single direction or [3], where the entire set of descent directions is determined.

## 3 Multiobjective Optimization with Inexact Gradients

Suppose now that we only have approximations $\widetilde{f}_i(x)$, $\nabla \widetilde{f}_i(x)$ of the objectives $f_i(x)$ and their gradients $\nabla f_i(x)$, $i = 1, \ldots, k$, respectively. To be more precise we assume that

$$\widetilde{f}_i(x) = f_i(x) + \bar{\xi}_i, \qquad \|\widetilde{f}_i(x) - f_i(x)\|_2 = \|\bar{\xi}_i\|_2 \leq \xi_i, \tag{4}$$

$$\nabla \widetilde{f}_i(x) = \nabla f_i(x) + \bar{\varepsilon}_i, \qquad \|\nabla \widetilde{f}_i(x) - \nabla f_i(x)\|_2 = \|\bar{\varepsilon}_i\|_2 \leq \varepsilon_i, \tag{5}$$

where the upper bounds $\xi_i$, $\varepsilon_i$ are given. In the following, when computing descent directions we will assume that

$$\varepsilon_i \leq \|\nabla f_i(x)\|_2 \quad \text{for all } x,$$

since otherwise, we are already in the vicinity of the set of stationary points as will be shown in Lemma 1. Using (5) we can derive an upper bound for the angle between the exact and the inexact gradient by elementary geometrical considerations (cf. Fig. 2a):

$$\sphericalangle(\nabla f_i(x), \nabla \widetilde{f}_i(x)) = \arcsin\left(\frac{\|\bar{\varepsilon}_i\|_2}{\|\nabla f_i(x)\|_2}\right) \leq \arcsin\left(\frac{\varepsilon_i}{\|\nabla f_i(x)\|_2}\right) =: \varphi_i. \tag{6}$$

Here we denote the largest possible angle (i.e. the "worst case") by $\varphi_i$. Based on this angle, one can easily define a condition for the inexact gradients such that the exact gradients could satisfy (1) for the first time. This is precisely the case if each inexact gradient deviates from the hyperplane defined by (1) at most by $\varphi_i$, see Fig. 2b. This motivates the definition of an *inexact descent direction*:

**Definition 3** A direction analog to (3) but based on inexact gradients, i.e.

$$q_u(x) = -\sum_{i=1}^{k} \widehat{\alpha}_i \nabla \widetilde{f}_i(x) \quad \text{with } \widehat{\alpha}_i \geq 0 \text{ for } i = 1, \ldots, k \text{ and } \sum_{i=1}^{k} \widehat{\alpha}_i = 1, \tag{7}$$

is called *inexact descent direction*.

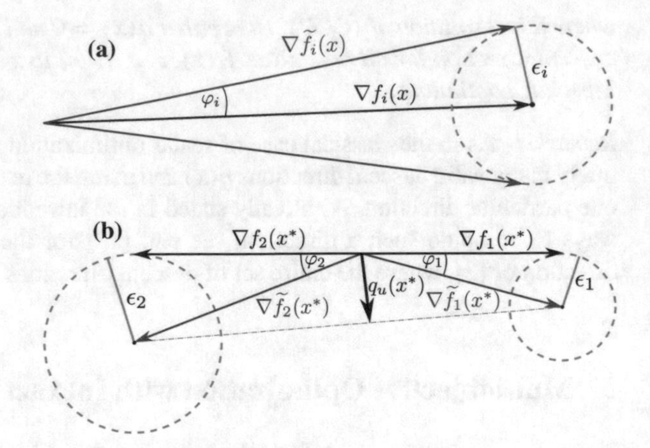


**Fig. 2** **a** Maximal angle between the exact and the inexact gradient in dependence on the error $\varepsilon_i$. **b** Maximal angle between the inexact gradients in the situation where $x^*$ satisfies (1) ($\pi - (\varphi_1 + \varphi_2)$ in the 2D case)

We can prove an upper bound for the norm of $q_u(x^*)$ when $x^*$ satisfies the KKT condition of the exact problem:

**Lemma 1** *Consider the multiobjective optimization problem* (MOP) *with inexact gradient information according to* (5). *Let $x^*$ be a point satisfying the KKT conditions* (1) *for the exact problem. Then, the inexact descent direction $q_u(x^*)$ is bounded by*

$$\|q_u(x^*)\|_2 \leq \|\varepsilon\|_\infty.$$

*Proof* Since $x^*$ satisfies (1), we have $\sum_{i=1}^{k} \widehat{\alpha}_i \nabla f_i(x^*) = 0$. Consequently,

$$\sum_{i=1}^{k} \widehat{\alpha}_i \nabla \widetilde{f}_i(x) = \sum_{i=1}^{k} \widehat{\alpha}_i \left(\nabla f_i(x) + \bar{\varepsilon}_i\right) = \sum_{i=1}^{k} \widehat{\alpha}_i \bar{\varepsilon}_i,$$

and thus,

$$\begin{aligned} \left\| \sum_{i=1}^{k} \widehat{\alpha}_i \nabla \widetilde{f}_i(x) \right\|_2 &= \left\| \sum_{i=1}^{k} \widehat{\alpha}_i \bar{\varepsilon}_i \right\|_2 \leq \sum_{i=1}^{k} \widehat{\alpha}_i \|\bar{\varepsilon}_i\|_2 \leq \sum_{i=1}^{k} \widehat{\alpha}_i \varepsilon_i \\ &\leq \sum_{i=1}^{k} \widehat{\alpha}_i \|\varepsilon\|_\infty = \|\varepsilon\|_\infty \sum_{i=1}^{k} \widehat{\alpha}_i = \|\varepsilon\|_\infty. \end{aligned}$$

□

A consequence of Lemma 1 is that in the presence of inexactness, we cannot compute the set of points satisfying (1) exactly. At best, we can compute the set of points determined by $\|q_u(x)\|_2 \leq \|\varepsilon\|_\infty$. In the following section we will derive a criterion for the inexact descent direction $q_u(x)$ which guarantees that it is also a descent direction for the exact problem when $x$ is sufficiently far away from the set of substationary points.

### 3.1 Descent Directions in the Presence of Inexactness

The set of valid descent directions for (MOP) is a cone defined by the intersection of all half-spaces orthogonal to the gradients $\nabla f_1(x), \ldots, \nabla f_k(x)$ (cf. Fig. 3a), i.e. it consists of all directions $q(x)$ satisfying

$$\sphericalangle(q(x), -\nabla f_i(x)) \leq \frac{\pi}{2}, \quad i = 1, \ldots, k. \tag{8}$$

This fact is well known from scalar optimization theory, see e.g. [23]. Observe that here we allow directions also to be valid for which all objectives are at least non-decreasing. In terms of the angle $\gamma_i \in [0, \pi/2]$ between the descent direction $q(x)$ and the hyperplane orthogonal to the gradient of the $i$th objective, this can be expressed as

$$\gamma_i = \frac{\pi}{2} - \arccos\left(\frac{q(x) \cdot (-\nabla f_i(x))}{\|q(x)\|_2 \cdot \|\nabla f_i(x)\|_2}\right) \geq 0, \quad i = 1, \ldots, k. \tag{9}$$

*Remark 2* Note that $\gamma_i$ is equal to $\pi/2$ when $q(x) = \nabla f_i(x)$ and approaches zero when a point $x$ approaches the set of substationary points, i.e.

$$\lim_{\|q(x)\|_2 \to 0} \gamma_i \to 0 \quad \text{for } i = 1, \ldots, k.$$

We call the set of all descent directions satisfying (8) the *exact cone* $\mathcal{Q}$. If we consider inexact gradients according to (5) then $\mathcal{Q}$ is reduced to the *inexact cone* $\mathcal{Q}_u$ by the respective upper bounds for the angular deviation $\varphi_i$:

$$\gamma_i \geq \varphi_i \geq 0, \quad i = 1, \ldots, k. \tag{10}$$

This means that we require the angle between the descent direction $q(x)$ and the $i$th hyperplane to be at least as large as the maximum deviation between the exact and the inexact gradient (cf. Fig. 3b). Thus, if an inexact descent direction $q_u(x)$ satisfies (10), then it is also a descent direction for the exact problem.

We would like to derive an algorithm by which we can determine an inexact descent direction $q_u(x)$ in such a way that it is also valid for the exact problem. To this end, we derive an additional criterion which is equivalent to (10). Concretely, we prove the following lemma:

**Lemma 2** *Consider the multiobjective optimization problem* (MOP) *with inexact gradient information according to* (5). *Let $q_u(x)$ be an inexact descent direction according to Definition 3. We assume $\|q_u(x)\|_2 \neq 0$, $\|\nabla \tilde{f}_i(x)\|_2 \neq 0, i = 1, \ldots, k$. Then* (10) *is equivalent to*

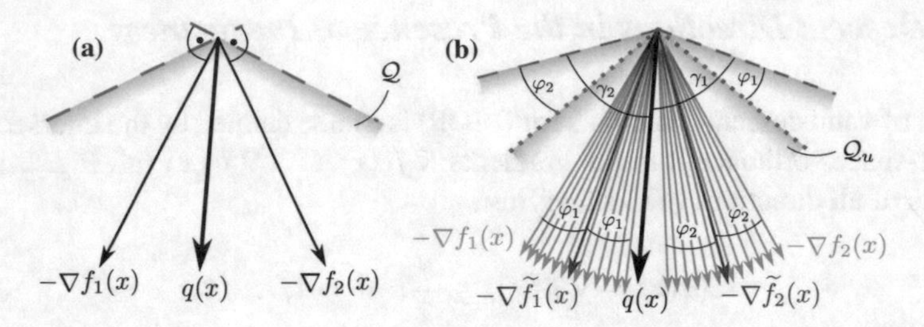


**Fig. 3** **a** Set of valid descent directions (the exact cone $\mathcal{Q}$ bounded by the *dashed lines*) determined by the intersection of half-spaces defined by the negative gradients. **b** Reduction of the set of valid descent directions in dependence on the errors $\varepsilon_i$ (the inexact cone $\mathcal{Q}_u$ bounded by the *dotted lines*). The *gray vectors* represent the set of possible values of the exact gradients $\nabla f_i(x)$ and the inexact cone $\mathcal{Q}_u$ is defined by the "most aligned" (here the uppermost) realizations of $\nabla f_i(x)$

$$\widehat{\alpha}_i \geq \frac{1}{\|\nabla \widetilde{f}_i(x)\|_2^2} \left( \|q_u(x)\|_2 \varepsilon_i - \sum_{\substack{j=1 \\ j \neq i}}^{k} \widehat{\alpha}_j \left( \nabla \widetilde{f}_j(x) \cdot \nabla \widetilde{f}_i(x) \right) \right), \quad i = 1, \ldots, k. \tag{11}$$

*In particular, $q_u(x)$ is a descent direction for all objective functions $f_i(x)$ if (11) is satisfied.*

*Proof* Inserting the expression for $\gamma_i$ in (9) into (10) yields

$$\frac{q_u(x) \cdot \left(-\nabla \widetilde{f}_i(x)\right)}{\|q_u(x)\|_2 \cdot \|\nabla \widetilde{f}_i(x)\|_2} \geq \cos\left(\frac{\pi}{2} - \varphi_i\right) = \sin\left(\varphi_i\right) = \frac{\varepsilon_i}{\|\nabla \widetilde{f}_i(x)\|_2}. \tag{12}$$

Using the definition of $q_u(x)$ this is equivalent to

$$\begin{aligned}
\frac{\left(\sum_{j=1}^{k} \widehat{\alpha}_j \nabla \widetilde{f}_j(x)\right) \cdot \nabla \widetilde{f}_i(x)}{\|q_u(x)\|_2 \, \|\nabla \widetilde{f}_i(x)\|_2} &= \frac{\widehat{\alpha}_i \|\nabla \widetilde{f}_i(x)\|_2^2 + \sum_{j=1, j \neq i}^{k} \widehat{\alpha}_j \nabla \widetilde{f}_j(x) \cdot \nabla \widetilde{f}_i(x)}{\|q_u(x)\|_2 \, \|\nabla \widetilde{f}_i(x)\|_2} \\
&= \frac{\|\nabla \widetilde{f}_i(x)\|_2}{\|q_u(x)\|_2} \widehat{\alpha}_i + \frac{\sum_{j=1, j \neq i}^{k} \widehat{\alpha}_j \nabla \widetilde{f}_j(x) \cdot \nabla \widetilde{f}_i(x)}{\|q_u(x)\|_2 \, \|\nabla \widetilde{f}_i(x)\|_2} \\
&\geq \frac{\varepsilon_i}{\|\nabla \widetilde{f}_i(x)\|_2}
\end{aligned}$$

$$\iff \widehat{\alpha}_i \geq \frac{1}{\|\nabla \widetilde{f}_i(x)\|_2^2} \left( \|q_u(x)\|_2 \varepsilon_i - \sum_{\substack{j=1 \\ j \neq i}}^{k} \widehat{\alpha}_j \left( \nabla \widetilde{f}_j(x) \cdot \nabla \widetilde{f}_i(x) \right) \right).$$

□

*Remark 3* By setting $\varepsilon_i = 0$ (i.e. $\widetilde{f_i}(x) = f_i(x)$ for $i = 1, \dots, k$) in (11) and performing some elemental manipulations, we again obtain the condition (2) for an exact descent direction:

$$\begin{aligned} \widehat{\alpha}_i &\geq \frac{1}{\|\nabla f_i(x)\|_2^2} \left( - \sum_{\substack{j=1 \\ j\neq i}}^{k} \widehat{\alpha}_j \left(\nabla f_j(x) \cdot \nabla f_i(x)\right) \right) \\ \Leftrightarrow \widehat{\alpha}_i \left(\nabla f_i(x) \cdot \nabla f_i(x)\right) &\geq \left( - \sum_{\substack{j=1 \\ j\neq i}}^{k} \widehat{\alpha}_j \left(\nabla f_j(x) \cdot \nabla f_i(x)\right) \right) \\ \Leftrightarrow -\nabla f_i(x) \cdot q(x) &\geq 0. \end{aligned}$$

The condition (11) can be interpreted as a lower bound for the "impact" of a particular gradient on the descent direction induced by $\widehat{\alpha}_i$. The larger the error $\varepsilon_i$, the higher the impact of the corresponding gradient needs to be in order to increase the angle between the descent direction and the hyperplane normal to the gradient. The closer a point $x$ is to the Pareto front (i.e. for small values of $\|q(x)\|_2$), the more confined the region of possible descent directions becomes. Hence, the inaccuracies gain influence until it is no longer possible to guarantee the existence of a descent direction for every objective. This is the case when the sum over the lower bounds from (11) exceeds one as shown in part (a) of the following result:

**Theorem 3** *Consider the multiobjective optimization problem* (MOP) *and suppose that the assumptions in Lemma 2 hold. Let*

$$\widehat{\alpha}_{min,i} = \frac{1}{\|\nabla \widetilde{f_i}(x)\|_2^2} \left( \|q_u(x)\|_2 \varepsilon_i - \sum_{\substack{j=1 \\ j\neq i}}^{k} \widehat{\alpha}_j \left(\nabla \widetilde{f_j}(x) \cdot \nabla \widetilde{f_i}(x)\right) \right), \quad i = 1, \dots, k.$$

*Then the following statements are valid:*

*(a) If $\sum_{i=1}^{k} \widehat{\alpha}_{min,i} > 1$ then $\mathcal{Q}_u = \emptyset$ (see Fig. 3b), and therefore it cannot be guaranteed that there is a descent direction for all objective functions.*
*(b) All points $x$ with $\sum_{i=1}^{k} \widehat{\alpha}_{min,i} = 1$ are contained in the set*

$$\mathcal{P}_{S,\varepsilon} = \left\{ x \in \mathbb{R}^n \;\middle|\; \left\| \sum_{i=1}^{k} \widehat{\alpha}_i \nabla f_i(x) \right\|_2 \leq 2\|\varepsilon\|_\infty \right\}. \tag{13}$$

*Proof* For part (a), suppose that we have a descent direction $q_u(x)$ for which $\sum_{i=1}^{k} \widehat{\alpha}_{min,i} > 1$. Then by (11)

$$
\begin{aligned}
& \sum_{i=1}^{k}\left(\frac{\|q_u(x)\|_2}{\|\nabla \widetilde{f}_i(x)\|_2^2}\varepsilon_i - \frac{\sum_{j=1,j\neq i}^{k}\widehat{\alpha}_j \nabla \widetilde{f}_j(x)\cdot\nabla \widetilde{f}_i(x)}{\|\nabla \widetilde{f}_i(x)\|_2^2}\right) > 1 \\
\Leftrightarrow\ & \sum_{i=1}^{k}\left(\frac{\|q_u(x)\|_2}{\|\nabla \widetilde{f}_i(x)\|_2^2}\varepsilon_i - \frac{\left(-q_u(x)-\widehat{\alpha}_i \nabla \widetilde{f}_i(x)\right)\cdot\nabla \widetilde{f}_i(x)}{\|\nabla \widetilde{f}_i(x)\|_2^2}\right) > 1 \\
\Leftrightarrow\ & \sum_{i=1}^{k}\left(\frac{\|q_u(x)\|_2}{\|\nabla \widetilde{f}_i(x)\|_2^2}\varepsilon_i - \frac{-q_u(x)\cdot\nabla \widetilde{f}_i(x)}{\|\nabla \widetilde{f}_i(x)\|_2^2} + \widehat{\alpha}_i\right) > 1 \\
\Leftrightarrow\ & \sum_{i=1}^{k}\left(\frac{\|q_u(x)\|_2}{\|\nabla \widetilde{f}_i(x)\|_2^2}\varepsilon_i - \frac{-q_u(x)\cdot\nabla \widetilde{f}_i(x)}{\|\nabla \widetilde{f}_i(x)\|_2^2}\right) > 1 - \sum_{i=1}^{k}\widehat{\alpha}_i = 0 \\
\Leftrightarrow\ & \sum_{i=1}^{k}\left(\frac{\varepsilon_i}{\|\nabla \widetilde{f}_i(x)\|_2} - \frac{-q_u(x)\cdot\nabla \widetilde{f}_i(x)}{\|q_u(x)\|_2\cdot\|\nabla \widetilde{f}_i(x)\|_2}\right) > 0 \\
\Leftrightarrow\ & \sum_{i=1}^{k}\sin\varphi_i > \sum_{i=1}^{k}\sin\gamma_i.
\end{aligned}
$$

Since $\varphi_i, \gamma_i \in [0, \pi/2]$ for $i = 1, \ldots, k$, it follows that $\varphi_i > \gamma_i$ for at least one $i \in \{1, \ldots, k\}$. This is a contradiction to (10) yielding $\mathscr{Q}_u = \emptyset$.

For part (b), we repeat the calculation from part (a) with the distinction that $\sum_{i=1}^{k}\widehat{\alpha}_{min,i} = 1$, and obtain $\sum_{i=1}^{k}\sin\varphi_i = \sum_{i=1}^{k}\sin\gamma_i$. This implies $\varphi_i = \gamma_i$ for $i = 1, \ldots, k$, i.e. the set of descent directions is reduced to a single valid direction. This is a situation similar to the one described in Lemma 1. Having a single valid descent direction results in the fact that there is now a possible realization of the gradients $\nabla f_i(x)$ such that each one is orthogonal to $q_u(x)$. In this situation, $x$ would satisfy (1) and hence, $\|q_u(x)\|_2 \leq \|\varepsilon\|_\infty$ which leads to

$$
\begin{aligned}
\left\|\sum_{i=1}^{k}\widehat{\alpha}_i\nabla f_i(x)\right\|_2 &= \left\|\sum_{i=1}^{k}\widehat{\alpha}_i\left(\nabla \widetilde{f}_i(x) + (-\bar{\varepsilon}_i)\right)\right\|_2 = \left\|q_u(x) + \sum_{i=1}^{k}\widehat{\alpha}_i(-\bar{\varepsilon}_i)\right\|_2 \\
&\leq \left\|q_u(x)\right\|_2 + \left\|\sum_{i=1}^{k}\widehat{\alpha}_i\bar{\varepsilon}_i\right\|_2 \leq \left\|q_u(x)\right\|_2 + \left\|\sum_{i=1}^{k}\bar{\varepsilon}_i\right\|_2 \leq 2\|\varepsilon\|_\infty.
\end{aligned}
$$

□

**Corollary 1** *If $\sum_{i=1}^{k}\widehat{\alpha}_{min,i} > 1$, the angle of the cone spanned by the exact gradients $\nabla f_i(x)$ is larger than for $\sum_{i=1}^{k}\widehat{\alpha}_{min,i} \leq 1$ (cf. Fig. 4b). Moreover, if $\|q_u(x)\|_2$ is monotonically decreasing for decreasing distances of $x$ to the set of stationary points $\mathscr{P}_{S,sub}$, then the result from Theorem 3 (b) holds for $\sum_{i=1}^{k}\widehat{\alpha}_{min,i} > 1$.*

The results from Theorem 3 are visualized in Fig. 4. In the situation where the inexactness in the gradients $\nabla \widetilde{f}_i(x)$ permits the exact gradients $\nabla f_i(x)$ to satisfy the KKT condition (1), $\mathscr{Q}_u = \emptyset$ and a descent direction can no longer be computed.

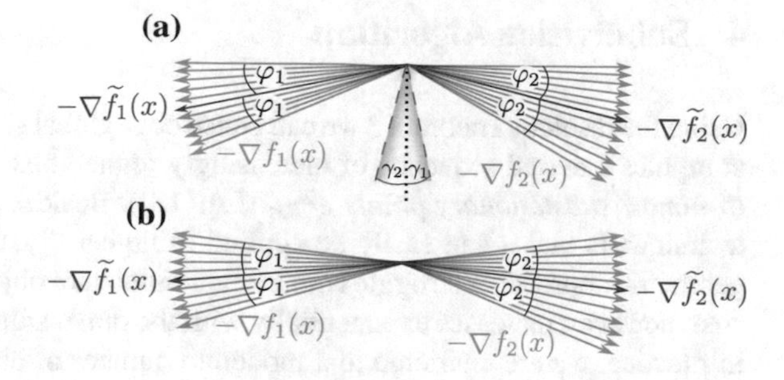


**Fig. 4** **a** Situation when a valid descent direction can no longer be computed. **b** Situation where $\sum_{i=1}^{k} \widehat{\alpha}_{min,i} > 1$

In order to numerically compute a valid descent direction, one can compute a direction by solving (QOP) and consequently verifying that $q_u(x)$ is indeed descending for all objectives by checking condition (11). If this is not the case, we can solve (QOP) again with adjusted lower bounds for $\widehat{\alpha}$ (cf. Algorithm 1). Alternatively, one can compute the entire set of descent directions [3] and chose a direction from this set which satisfies (10) or (11), respectively. In the examples we investigated, we observed that $q_u(x)$ is equal to $q(x)$ for the majority of points $x$. This is likely due to the fact that by solving (QOP), we obtain a steepest descent like direction which very often is relatively far away from the descent cone's boundaries. Close to the set of substationary points, we observe updates of the lower bounds $\widehat{\alpha}_{min,i}$, but in many cases, the line search strategy directly leads to points where $\mathcal{Q}_u = \emptyset$ without requiring adjusted bounds.

**Algorithm 1** (Descent direction for inexact gradients)

**Require:** Inexact gradients $\nabla\widetilde{f}_i(x)$, error bounds $\varepsilon_i$ and lower bounds $\widehat{\alpha}_{min,i} = 0, i = 1, \dots, k$
1: **loop**
2: Compute $\widehat{\alpha}_i$ by solving (QOP) with $\widehat{\alpha}_i \in [\widehat{\alpha}_{min,i}, 1], i = 1, \dots, k$
3: Evaluate the condition (11) and update $\widehat{\alpha}_{min,i}$
4: **if** $\sum_{i=1}^{k} \widehat{\alpha}_{min,i} \geq 1$ **then**
5: $q_u(x) = 0$ ($\mathcal{Q}_u = \emptyset$)
6: **STOP**
7: **else if** $\widehat{\alpha}_i \geq \widehat{\alpha}_{min,i}, i = 1, \dots, k$ **then**
8: $q_u(x) = -\sum_{i=1}^{k} \widehat{\alpha}_i \nabla\widetilde{f}_i(x)$
9: **STOP**
10: **end if**
11: **end loop**

The obtained descent direction can now be utilized to efficiently compute a point which is approximately Pareto optimal, utilizing the advantages of gradient-based methods. In order to compute the entire Pareto set for MOPs with inexact gradient information, we will combine the above result with the algorithm presented in [8] in the next section.

# 4 Subdivision Algorithm

Using the result of Theorem 3 we can construct a global subdivision algorithm which computes a nested sequence of increasingly refined box coverings of the entire set of *almost substationary points* $\mathcal{P}_{S,\varepsilon}$ (Eq. (13)). Besides globality, a benefit of this technique is that it can easily be applied to higher dimensions whereas especially geometric approaches struggle with a larger number of objectives. The computational cost, however, increases exponentially with the dimension of the Pareto set such that in practice, we are restricted to a moderate number of objectives. In the following, we recall the subdivision algorithm for the solution of multiobjective optimization problems with exact information before we proceed with our extension to inexact information. For details we refer to [8].

## 4.1 *Subdivision Algorithm with Exact Gradients*

In order to apply the subdivision algorithm to a multiobjective optimization problem, we first formulate a descent step of the optimization procedure using a line search approach, i.e.

$$x_{j+1} = g(x_j) = x_j + h_j q_j(x_j), \tag{14}$$

where $q_j(x_j)$ is the descent direction according to (3) and $h_j$ is an appropriately chosen step length (e.g. according to the Armijo rule [23] for all objectives). The subdivision algorithm was initially developed in the context of dynamical systems [7] in order to compute the global attractor of a dynamical system $g$ relative to a set $Q$, i.e. the set $A_Q$ such that $g(A_Q) = A_Q$. Using a multilevel subdivision scheme, the following algorithm yields an outer approximation of $A_Q$ in the form of a sequence of sets $\mathcal{B}_0, \mathcal{B}_1, \ldots$, where each $\mathcal{B}_s$ is a subset of $\mathcal{B}_{s-1}$ and consists of finitely many subsets $B$ (from now on referred to as *boxes*) of $Q$ covering the relative global attractor $A_Q$. For each set $\mathcal{B}_s$, we define a box diameter

$$\operatorname{diam}(\mathcal{B}_s) = \max_{B \in \mathcal{B}_s} \operatorname{diam}(B)$$

which tends to zero for $s \to \infty$ within Algorithm 2. Interpreting (14) as a dynamical system, the attractor is the set of points for which $q(x_j) = 0$, i.e. the set of points satisfying the KKT conditions. As a first step, we can prove that each accumulation point of the system is a substationary point for (MOP):

**Theorem 4** ([8]) *Suppose that $x^*$ is an accumulation point of the sequence $(x_j)_{j=0,1,\ldots}$ created by* (14). *Then, $x^*$ is a substationary point of* (MOP).

Using Algorithm 2, we can now compute an outer approximation of the attractor of the dynamical system (14) which contains all points satisfying (1). The attractor of a

**Algorithm 2** (Subdivision algorithm)

Let $\mathcal{B}_0$ be an initial collection of finitely many subsets of the compact set $Q$ such that $\bigcup_{B\in\mathcal{B}_0} B = Q$. Then, $\mathcal{B}_s$ is inductively obtained from $\mathcal{B}_{s-1}$ in two steps:

(i) **Subdivision**. Construct from $\mathcal{B}_{s-1}$ a new collection of subsets $\widehat{\mathcal{B}}_s$ such that

$$\bigcup_{B\in\widehat{\mathcal{B}}_s} B = \bigcup_{B\in\mathcal{B}_{s-1}} B,$$

$$\operatorname{diam}(\widehat{\mathcal{B}}_s) = \theta_s \operatorname{diam}(\mathcal{B}_{s-1}), \quad 0 < \theta_{min} \leq \theta_s \leq \theta_{max} < 1.$$

(ii) **Selection**. Define the new collection $\mathcal{B}_s$ by

$$\mathcal{B}_s = \left\{ B \in \widehat{\mathcal{B}}_s \;\middle|\; \exists \widehat{B} \in \widehat{\mathcal{B}}_s \text{ such that } g^{-1}(B) \cap \widehat{B} \neq \emptyset \right\}.$$

dynamical system is always connected, which is not necessarily the case for $\mathcal{P}_{S,sub}$. In this situation, the attractor is a superset of $\mathcal{P}_{S,sub}$. However, if $\mathcal{P}_{S,sub}$ is bounded and connected, it coincides with the attractor of (14), which is stated in the following theorem:

**Theorem 5** ([8]) *Suppose that the set $\mathcal{P}_{S,sub}$ of points $x \in \mathbb{R}^n$ satisfying* (1) *is bounded and connected. Let $Q$ be a compact neighborhood of $\mathcal{P}_{S,sub}$. Then, an application of Algorithm 2 to $Q$ with respect to the iteration scheme* (14) *leads to a sequence of coverings $\mathcal{B}_s$ which converges to the entire set $\mathcal{P}_{S,sub}$, that is,*

$$d_h(\mathcal{P}_{S,sub}, \mathcal{B}_s) \to 0, \quad \text{for } s = 0, 1, 2, \ldots,$$

*where $d_h$ denotes the Hausdorff distance.*

The concept of the subdivision algorithm is illustrated in Fig. 5, where in addition a numerical realization has been introduced (cf. Sect. 4.2.1 for details).

## 4.2 Subdivision Algorithm with Inexact Function and Gradient Values

In this section, we combine the results from [8] with the results from Sect. 3 in order to devise an algorithm for the approximation of the set of substationary points of (MOP) with inexact function and gradient information. So far, we have only considered inexactness in the descent direction where we only need to consider errors in the gradients. For the computation of a descent step, we further require a step length strategy [23] where we additionally need to consider errors in the function values. For this purpose, we extend the concept of *non-dominance* (2) to inexact function values:

**Definition 4** Consider the multiobjective optimization problem (MOP), where the objective functions $f_i(x), i = 1, \dots, k$, are only known approximately according to (4). Then

(a) a point $x^* \in \mathbb{R}^n$ *confidently dominates* a point $x \in \mathbb{R}^n$, if $\widetilde{f}_i(x^*) + \xi_i \leq \widetilde{f}_i(x) - \xi_i$ for $i = 1, \dots, k$ and $\widetilde{f}_i(x^*) + \xi_i < \widetilde{f}_i(x) - \xi_i$ for at least one $i \in 1, \dots, k$.
(b) a set $\mathscr{B}^* \subset \mathbb{R}^n$ *confidently dominates* a set $\mathscr{B} \subset \mathbb{R}^n$ if for every point $x \in \mathscr{B}$ there exists at least one point $x^* \in \mathscr{B}^*$ dominating $x$.
(c) The *set of almost non-dominated points* which is a superset of the Pareto set $\mathscr{P}_S$ is defined as:

$$\mathscr{P}_{S,\xi} = \left\{ x^* \in \mathbb{R}^n \middle| \nexists x \in \mathbb{R}^n \text{ with } \widetilde{f}_i(x) + \xi_i \leq \widetilde{f}_i(x^*) - \xi_i,\ i = 1, \dots, k \right\}. \tag{15}$$

Note that the same definition was also introduced in [28] in order to increase the number of almost Pareto optimal points and thus, the number of possible options for a decision maker.

As a consequence of the inexactness in the function values and the gradients, the approximated set is a superset of the Pareto set. Depending on the errors $\xi_i$ and $\varepsilon_i$, $i = 1, \dots, k$, in the function values and in the gradients, respectively, each point is either contained in the set $\mathscr{P}_{S,\xi}$ (15) or in the set $\mathscr{P}_{S,\varepsilon}$ (13). Based on these considerations, we introduce an inexact dynamical system similar to (14):

$$x_{j+1} = x_j + h_j p_j, \tag{16}$$

where the direction $p_j$ is computed using Algorithm 1 ($p_j = q_u(x_j)$) and the step length $h_j$ is determined by a modified Armijo rule [23] such that $\widetilde{f}_i(x_j + h_j p_j) + \xi_i \leq \widetilde{f}_i(x_{j-1}) + c_1 h_j p_j^\top \nabla \widetilde{f}_i(x_{j-1})$. If the errors $\xi$ and $\varepsilon$ are zero, the computed set is reduced to the set of substationary points. In this situation $\mathscr{P}_{S,\xi} = \mathscr{P}_S$ and $\sum_{i=1}^k \widehat{\alpha}_{min,i} = 0$, hence $p_j = q(x_j)$. Convergence of the dynamical system (16) to an *approximately substationary* point is investigated in the following theorem:

**Theorem 6** *Consider the multiobjective optimization problem* (MOP) *with inexact objective functions and inexact gradients according to* (4) *and* (5), *respectively. Suppose that* $x^*$ *is an accumulation point of the sequence* $(x_j)_{j=0,1,\dots}$ *created by* (16). *Then,*

*(a)* $x^* \in \mathscr{P}_{S,\varepsilon,\xi} = \mathscr{P}_{S,\varepsilon} \cup \mathscr{P}_{S,\xi}$ *where* $\mathscr{P}_{S,\varepsilon}$ *and* $\mathscr{P}_{S,\xi}$ *are defined according to* (13) *and* (15), *respectively.*
*(b) If* $\xi_i = 0$, $\varepsilon_i = 0$, $i = 1, \dots, k$, $x^*$ *is a substationary point of* (MOP).

*Proof* (a) For a point $x_j$ created by the sequence (16) one of the following statements is true:

(i) $x_j \in \mathscr{P}_{S,\varepsilon} \quad \wedge \quad x_j \notin \mathscr{P}_{S,\xi}$ (ii) $x_j \notin \mathscr{P}_{S,\varepsilon} \quad \wedge \quad x_j \in \mathscr{P}_{S,\xi}$

(iii) $x_j \in \mathscr{P}_{S,\varepsilon} \quad \wedge \quad x_j \in \mathscr{P}_{S,\xi}$ (iv) $x_j \notin \mathscr{P}_{S,\varepsilon} \quad \wedge \quad x_j \notin \mathscr{P}_{S,\xi}$

In case (i) $x_j \in \mathscr{P}_{S,\varepsilon}$ which means that the gradients $\nabla \widetilde{f}_i(x_j), i = 1, \ldots, k$, approximately satisfy the KKT conditions. We obtain $\sum_{i=1}^k \widehat{\alpha}_{min,i} = 1$, i.e. the set of valid descent directions is empty ($\mathscr{Q}_u = \emptyset$, cf. Theorem 3). Consequently, $p_j = 0$ and the point $x_j$ is an accumulation point of the sequence (16). In case (ii) the inaccuracies in the function values $\widetilde{f}_i(x_j), i = 1, \ldots, k$ prohibit a guaranteed decrease for all objectives. According to the modified Armijo rule $h_j = 0$ such that $x_j$ is an accumulation point. In case (iii) both $p_j = 0$ and $h_j = 0$. In case (iv) we have $p_j = q(x_j)$. If for any $j \in \{0, 1, \ldots\}$, $x_j \in \mathscr{P}_{S,\varepsilon}$ or $x_j \in \mathscr{P}_{S,\xi}$, we are in one of the cases (i) to (iii) and $x_j$ is an accumulation point. Otherwise, we obtain a descent direction such that the sequence (16) converges to a substationary point $x^* \in \mathscr{P}_{S,sub} \subseteq \mathscr{P}_{S,\varepsilon}$ of (MOP) which is proved in [8].

For part (b), we obtain $\widehat{\alpha}_{min,i} = 0$ by setting the errors $\varepsilon_i, i = 1, \ldots, k$, to zero and hence, the descent direction is $p_j = q(x_j)$ (cf. Algorithm 1). When $\xi_i = 0$, $i = 1, \ldots, k$, the modified Armijo rule becomes the standard Armijo rule for multiple objectives. Consequently, the problem is reduced to the case with exact function and gradient values (case (iv)) in part (a). □

Following along the lines of [8], we can use this result in order to prove convergence of the subdivision algorithm with inexact values:

**Theorem 7** *Suppose that the set* $\mathscr{P}_{S,\varepsilon,\xi} = \mathscr{P}_{S,\varepsilon} \cup \mathscr{P}_{S,\xi}$ *where* $\mathscr{P}_{S,\varepsilon}$ *and* $\mathscr{P}_{S,\xi}$ *are defined according to* (13) *and* (15), *respectively, is bounded and connected. Let $Q$ be a compact neighborhood of* $\mathscr{P}_{S,\varepsilon,\xi}$. *Then, an application of Algorithm 2 to $Q$ with respect to the iteration scheme* (16) *leads to a sequence of coverings* $\mathscr{B}_s$ *which is a subset of* $\mathscr{P}_{S,\varepsilon,\xi}$ *and a superset of the set* $\mathscr{P}_{S,sub}$ *of substationary points* $x \in \mathbb{R}^n$ *of* (MOP), *that is,*

$$\mathscr{P}_{S,sub} \subset \mathscr{B}_s \subset \mathscr{P}_{S,\varepsilon,\xi}.$$

*Consequently, if the errors tend towards zero, we observe*

$$\lim_{\varepsilon_i,\xi_i \to 0,\ i=1,\ldots,k} d_h(\mathscr{B}_s, \mathscr{P}_{S,\varepsilon,\xi}) = d_h(\mathscr{P}_{S,sub}, \mathscr{B}_s) = 0.$$

### 4.2.1 Numerical Realization of the Selection Step

In this section, we briefly describe the numerical realization of Algorithm 2. For details, we refer to [8]. The elements $B \in \mathscr{B}_s$ are $n$-dimensional boxes. In the selection step, each box is represented by a prescribed number of sample points at which the dynamical system (16) is evaluated according to Algorithm 3 (see Fig. 5a for an illustration). Then, we evaluate which boxes the sample points are mapped into

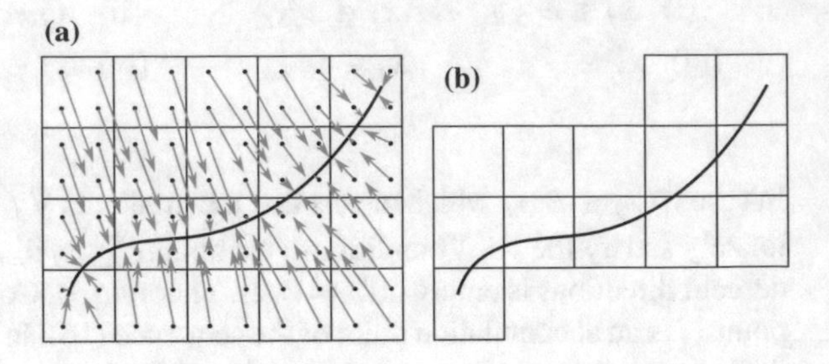


**Fig. 5** Global subdivision algorithm – selection step. **a** Evaluation of the dynamical system (14). **b** All boxes that do not possess a preimage within the collection are discarded

and eliminate all "empty" boxes, i.e. boxes which do not possess a preimage within $\mathscr{B}_s$ (Fig. 5b). The remaining boxes are subdivided and we proceed with the next elimination step until a certain stopping criterion is met (e.g. a prescribed number of subdivision steps).

---

**Algorithm 3** (Descent step under inexactness)

---

**Require:** Initial point $x_0$, error bounds $\xi_i$ and $\varepsilon_i$, $i = 1, \ldots, k$, constant $0 < c_1 < 1$
1: Compute $\widetilde{f}_i(x_0)$ and $\nabla \widetilde{f}_i(x_0)$, $i = 1, \ldots, k$
2: Compute a direction $p$ according to Algorithm 1
3: Compute a step length $h$ that satisfies the modified Armijo rule
$\widetilde{f}_i(x_0 + hp) + \xi_i \leq \widetilde{f}_i(x_0) - \xi_i + c_1 hp^\top \nabla \widetilde{f}_i(x_0)$, e.g. via backtracking [23]
4: Compute $x_{step} = x_0 + hp$

---

## 4.3 Gradient Free Realization

In many applications, gradients are unknown or difficult to compute. In this case, we can use a gradient-free alternative of Algorithm 2 which is called the *Sampling algorithm* in [8]. Algorithm 4 also consists of a subdivision and a selection step with the difference that the selection step is a non-dominance test. Hence, we compare all sample points and eliminate all boxes that contain only dominated points. This way, it is also possible to easily include constraints. In the presence of inequality constraints, for example, we eliminate all boxes for which all sample points violate the constraints and then consequently perform the non-dominance test on the remaining boxes. Equality constrains are simply modeled by introducing two inequality constraints. Finally, a combination of both the gradient-based and the gradient-free algorithm can be applied in order to speed up convergence or to reduce the gradient-based algorithm to the computation of the Pareto set $\mathscr{P}_S$ instead of the set of substationary points $\mathscr{P}_{S,sub}$.

Considering errors in the function values, we can use the sampling algorithm to compute the superset $\mathscr{P}_{S,\xi}$ of the global Pareto set $\mathscr{P}_S$. In the limit of vanishing errors, this is again reduce to the exact Pareto set:

$$\lim_{\xi_i \to 0,\ i=1,\dots,k} d_h(\mathscr{P}_{S,\xi}, \mathscr{P}_S) = 0.$$

---

**Algorithm 4** (Sampling algorithm)

---

Let $\mathscr{B}_0$ be an initial collection of finitely many subsets of the compact set $Q$ such that $\bigcup_{B \in \mathscr{B}_0} B = Q$. Then, $\mathscr{B}_s$ is inductively obtained from $\mathscr{B}_{s-1}$ in two steps:

(i) Subdivision. Construct from $\mathscr{B}_{s-1}$ a new collection of subsets $\widehat{\mathscr{B}}_s$ such that

$$\bigcup_{B \in \widehat{\mathscr{B}}_s} B = \bigcup_{B \in \mathscr{B}_{s-1}} B,$$

$$\operatorname{diam}(\widehat{\mathscr{B}}_s) = \theta_s \operatorname{diam}(\mathscr{B}_{s-1}), \quad 0 < \theta_{min} \leq \theta_s \leq \theta_{max} < 1.$$

(ii) Selection. Define the new collection $\mathscr{B}_s$ by

$$\mathscr{B}_s = \left\{ B \in \widehat{\mathscr{B}}_s \;\middle|\; \nexists \widehat{B} \in \widehat{\mathscr{B}}_s \text{ such that } \widehat{B} \text{ confidently dominates } B \right\}.$$

---

## 5 Results

In this section, we illustrate the results from Sects. 3 and 4 using three examples. To this end, we add random perturbations to the respective model such that (4) and (5) hold. We start with a two dimensional example function $F : \mathbb{R}^2 \to \mathbb{R}^2$ for two paraboloids:

$$\min_{x \in \mathbb{R}^2} F(x) = \min_{x \in \mathbb{R}^2} \begin{pmatrix} (x_1 - 1)^2 + (x_2 - 1)^4 \\ (x_1 + 1)^2 + (x_2 + 1)^2 \end{pmatrix}. \tag{17}$$

In Fig. 6, the box covering of the Pareto set and the corresponding Pareto front obtained with Algorithm 2 are shown without errors and with $\xi = (0, 0)^\top$, $\varepsilon = (0.1, 0.1)^\top$ (Fig. 6a) and $\xi = (0, 0)^\top$, $\varepsilon = (0.0, 0.2)^\top$ (Fig. 6b), respectively. The background in (a) and (b) is colored according to the norm of the optimality condition (1), obtained by solving (QOP), and the white line indicates the upper bound of the error (13). We see that in (a), the box covering is close to the error bound whereas it is less sharp in (b). Consequently, the error estimate is more accurate when the errors are of comparable size in all gradients. The Pareto fronts corresponding to (a) and (b) are shown in (c) and (d). We see that the difference between the Pareto front of the exact solution (red) and the Pareto front of the inexact solution (green) is relatively

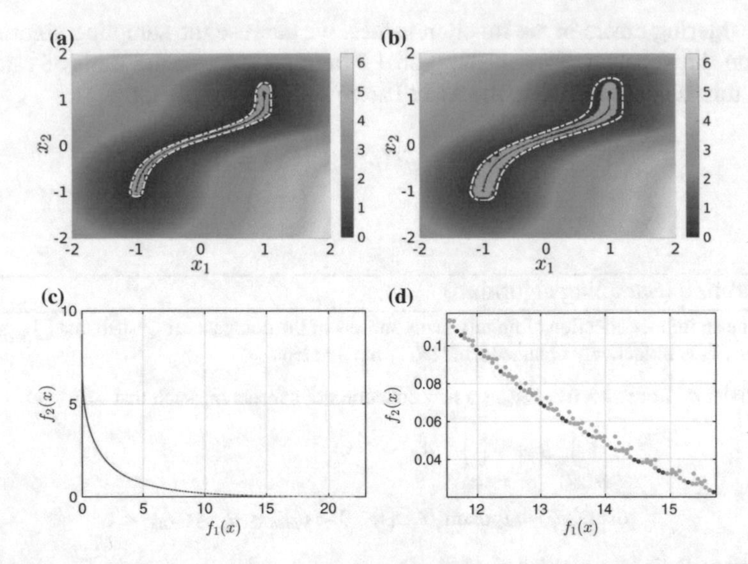


**Fig. 6** **a** Box covering of the Pareto set of problem (17) after 16 subdivision steps ($\text{diam}(\mathscr{B}_0) = 4$, $\text{diam}(\mathscr{B}_{20}) = 1/2^6$). The solution without errors ($\mathscr{P}_{S,sub}$) is shown in *red*, the solution with $\varepsilon = (0.1, 0.1)^\top, \xi = (0, 0)^\top$ ($\mathscr{P}_{S,\varepsilon}$) is shown in *green*. The background color represents the norm of the optimality condition (1) and the white line is the iso-curve $\|q(x)\|_2 = 2\|\varepsilon\|_\infty = 0.2$, i.e. the upper bound of the error. **b** Analog to **a** but with $\varepsilon = (0, 0.2)^\top$ and the iso-curve $\|q(x)\|_2 = 0.4$. **c–d** The Pareto fronts corresponding to **a**. The points are the images of the box centers (color coding as in **a** and **b**)

small but that additional points are computed at the boundary of the front, i.e. close to the individual minima $F_1(x) = 0$ and $F_2(x) = 0$.

In the numerical realization, we approximate each box by an equidistant grid with two points in each direction, i.e. by four sample points in total. This results in a total number of $\approx$50,000 function evaluations for the exact problem. The number of boxes is much higher for the inexact solution, in this case by a factor of $\approx$8. This is not surprising since the equality condition (1) is now replaced by the inequality condition (13). Hence, the approximated set is no longer a $(k-1)$-dimensional object. All sets $B \in \mathscr{B}_s$ that satisfy the inequality condition (13) are not discarded in the selection step. Consequently, at a certain box size, the number of boxes increases exponentially with decreasing diameter $\text{diam}(B)$ (cf. Fig. 10b). The result is an increased computational effort for later iterations. This is visualized in Fig. 7, where the solutions at different stages of the subdivision algorithm are compared. A significant difference between the solutions can only be observed in later stages (Fig. 7e, f). For this reason, an adaptive strategy needs to be developed where boxes satisfying (13) remain within the box collection but are no longer considered in the subdivision algorithm.

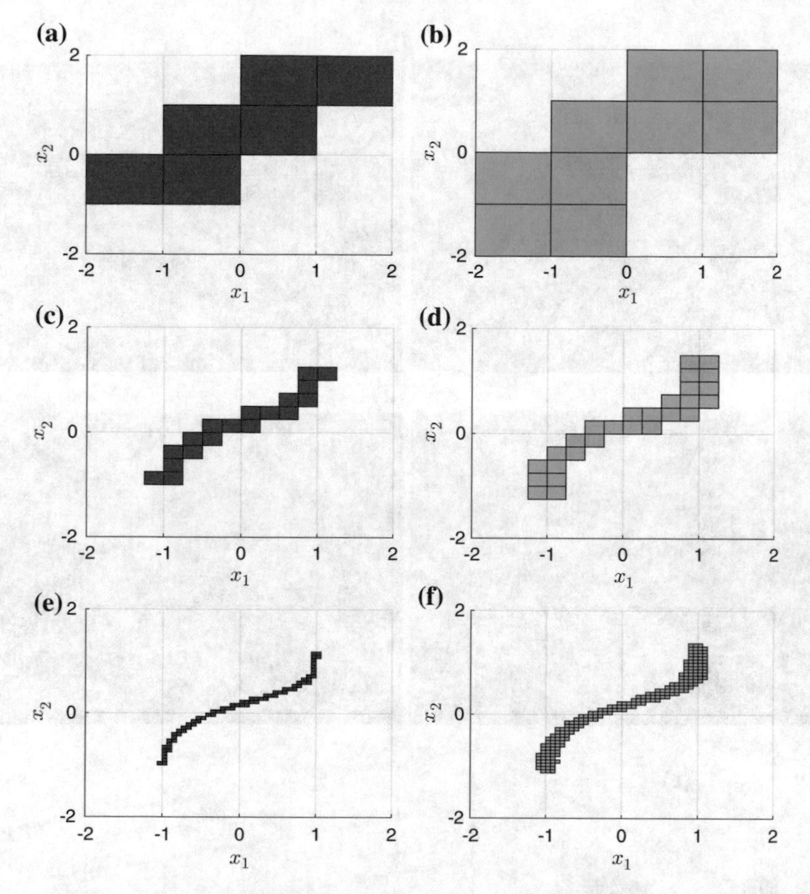


**Fig. 7** Comparison of the exact and the inexact solution of problem (17) after 4, 8 and 12 subdivision steps

As a second example, we consider the function $F : \mathbb{R}^3 \to \mathbb{R}^3$:

$$\min_{x \in \mathbb{R}^3} F(x) = \min_{x \in \mathbb{R}^3} \begin{pmatrix} (x_1-1)^4 + (x_2-1)^2 + (x_3-1)^2 \\ (x_1+1)^2 + (x_2+1)^4 + (x_3+1)^2 \\ (x_1-1)^2 + (x_2+1)^2 + (x_3-1)^4 \end{pmatrix}. \tag{18}$$

The observed behavior is very similar to the two-dimensional case, cf. Fig. 8, where in (b) the box covering for the inexact problem is shown as well as the iso-surface $\|q(x)\|_2 = 2\|\varepsilon\|_\infty$. One can see that the box covering lies inside this iso-surface except for small parts of some boxes. This is due to the finite box size and the fact that at least one sample point is mapped into the box itself. For smaller box radii, this artifact does no longer occur.

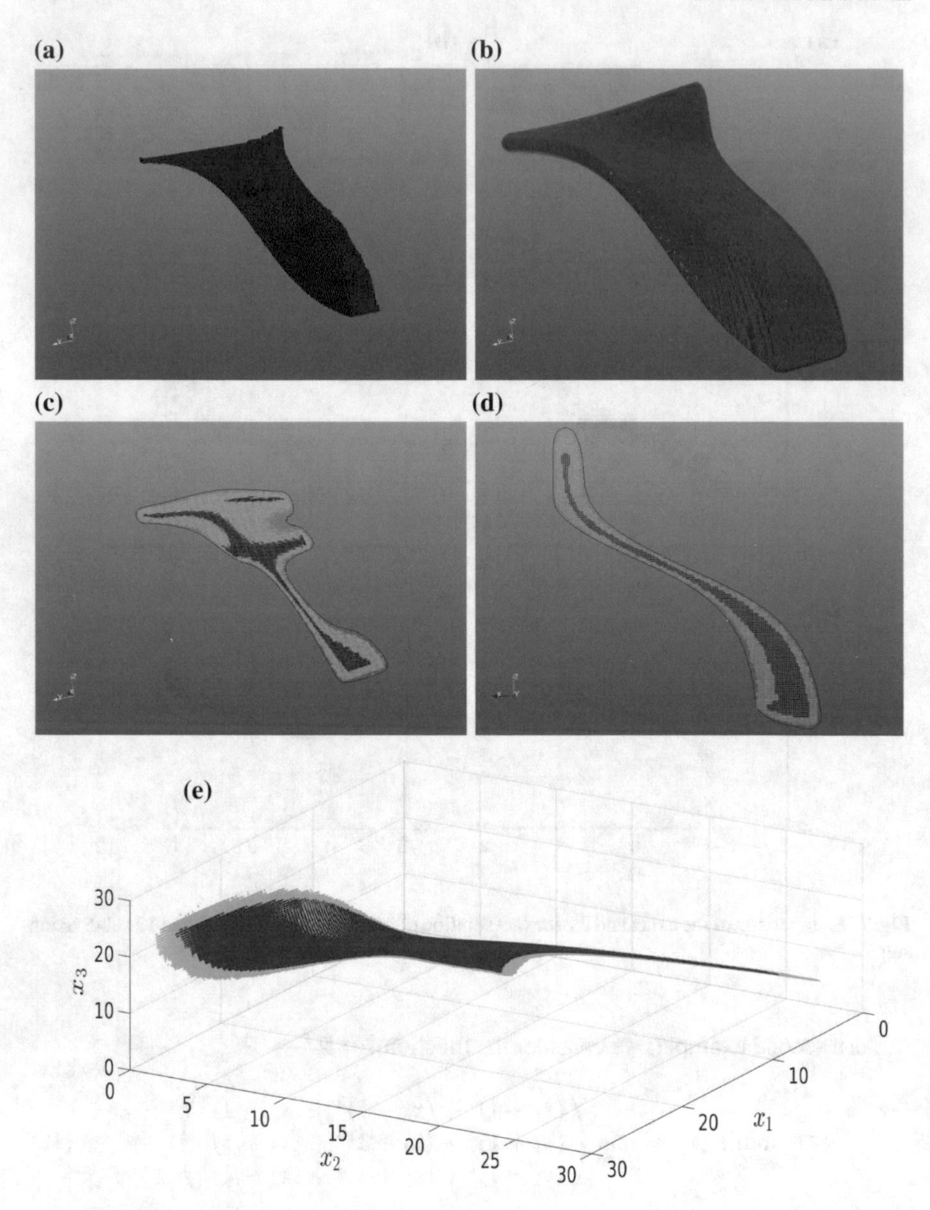


**Fig. 8** Box coverings of the Pareto set (exact solution: *red*, inexact solution: *green*) of problem (18) after 24 subdivision steps ($\text{diam}(\mathscr{B}_0) = 4$, $\text{diam}(\mathscr{B}_{24}) = 1/2^6$). **a** $\mathscr{P}_{S,sub}$. **b** $\mathscr{P}_{S,\varepsilon}$ with $\varepsilon_i = (0.1, 0.1, 0.1)^\top$ ($\xi = (0, 0, 0)^\top$). The iso-surface ($\|q(x)\|_2 = 2\|\varepsilon\|_\infty = 0.2$) is the upper bound of the error. **c–d** Two-dimensional cut planes through the Pareto set. The point of view in **c** is as in **a**, **b** and **d** is a cut plane parallel to the $x_2$ plane at $x_2 = -0.9$. **e** The corresponding Pareto fronts

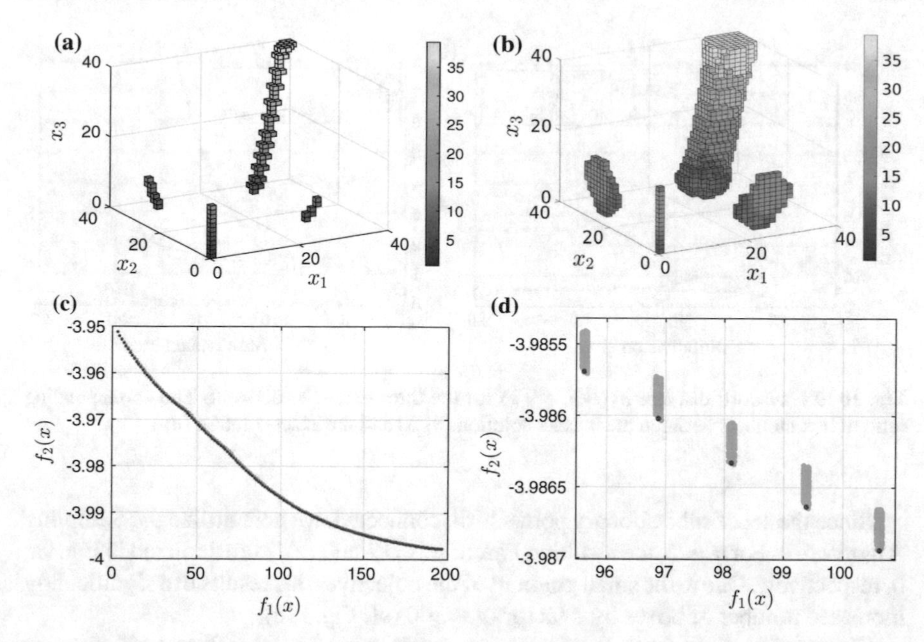


**Fig. 9** **a** Projection of the box covering of $\mathcal{P}_S$ for problem (19) with $n = 5$ after 25 subdivision steps ($\text{diam}(\mathcal{B}_0) = 40$, $\text{diam}(\mathcal{B}_{25}) = 1.25$). The coloring represents the fourth component $x_4$. **b** Box covering of $\mathcal{P}_{S,\xi}$ with inexact data $\widetilde{x}$ with $|\widetilde{x}_i - x_i| < 0.01$ for $i = 1, \ldots, 5$. **c–d** The Pareto fronts corresponding to **a** and **b** in *green* and *red*, respectively. The points are the images of the box centers

Finally, we consider an example where the Pareto set is disconnected. This is an example from production and was introduced in [26]. We want to minimize the failure of a product which consists of $n$ components. The probability of failing is modeled individually for each component and depends on the additional cost $x$:

$$\begin{aligned} p_1(x) &= 0.01 \exp\left(-(x_1/20)^{2.5}\right), \\ p_2(x) &= 0.01 \exp\left(-(x_2/20)^{2.5}\right), \\ p_j(x) &= 0.01 \exp\left(-x_1/15\right), \qquad j = 3, \ldots, n. \end{aligned}$$

The resulting MOP is thus to minimize the failure and the additional cost at the same time:

$$\min_{x \in \mathbb{R}^n} F(x) = \min_{x \in \mathbb{R}^n} \begin{pmatrix} \sum_{j=1}^n x_j \\ 1 - \sum_{j=1}^n \left(1 - p_j(x)\right) \end{pmatrix}. \tag{19}$$

We now assume that the additional cost $x$ is subject so some uncertainty, e.g. due to varying prices, and set $|\widetilde{x}_i - x_i| < 0.01$ for $i = 1, \ldots, n$. Using this, we can estimate the error bounds within the initial box $\mathcal{B}_0 = [0, 40]^n$ and obtain $\xi = (0.05, 2 \cdot 10^{-5})^\top$ and $\varepsilon = (0, 8 \cdot 10^{-7})^\top$.

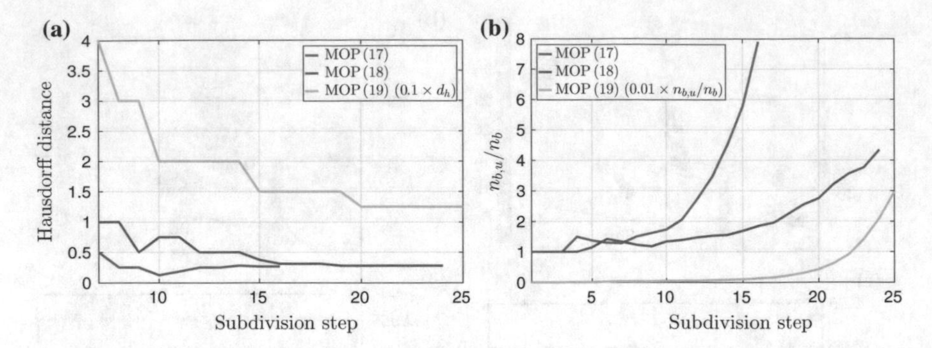


**Fig. 10** **a** Hausdorff distance $d_h(\mathscr{P}_S, \mathscr{P}_{S,\xi})$ for the three examples above. **b** The corresponding ratio of box numbers between the inexact solution ($n_{b,u}$) and the exact solution ($n_b$)

Since the set of substationary points is disconnected, we here utilize the Sampling Algorithm 4. For $n = 5$, the resulting Pareto sets $\mathscr{P}_S$ and $\mathscr{P}_{S,\xi}$ are depicted in Fig. 9a, b, respectively. Due to the small gradient of the objectives, his results in a significantly increased number of boxes by a factor of ≈300 (cf. Fig. 10b).

The quality of the inexact solution can be measured by the Hausdorff distance $d_h(\mathscr{P}_S, \mathscr{P}_{S,\xi})$ [25]. This is depicted in Fig. 10a, where the distance between the exact and the inexact solution is shown for all subdivision steps for the three examples above. We see that the distance reaches an almost constant value in the later stages. This distance is directly influenced by the upper bounds $\varepsilon$ and $\xi$, respectively. However, it cannot simply be controlled by introducing a bound on the error in the objectives or gradients since it obviously depends on the objective functions. Hence, in order to limit the error in the decision space, further assumptions on the objectives have to be made.

# 6 Conclusion

In this article, we present an extension to the subdivision algorithms developed in [8] for MOPs with uncertainties in the form of inexact function and gradient information. An additional condition for a descent direction is derived in order to account for the inaccuracies in the gradients. Convergence of the extended subdivision algorithm to a superset of the Pareto set is proved and an upper bound for the maximal distance to the set of substationary points is given. When taking into account errors, the number of boxes in the covering of the Pareto set increases, especially for later iterations, causing larger computational effort. For this reason, an adaptive strategy needs to be developed where boxes approximately satisfying the KKT conditions (according to (13)) remain within the box collection but are no longer considered in the subdivision algorithm. Furthermore, we intend to extend the approach to constrained MOPs in the future. A comparison to other methods, especially memetic algorithms, would be interesting to investigate the numerical efficiency of the presented method. Finally, we intend to combine the subdivision algorithm with model order reduction techniques in order to solve multiobjective optimal control problems constrained by PDEs (see e.g. [1] for the solution of bicriterial MOPs with the *reference point* approach or [24] for multiobjective optimal control of the Navier–Stokes equations). For the development of reliable and robust multiobjective optimization algorithms, error estimates for reduced order models [34] need to be combined with concepts for inexact function values and gradients. The idea is to set upper bounds for the errors $\xi$ and $\varepsilon$ and thereby define the corresponding accuracy requirements for the reduced order model.

**Acknowledgements** This work is supported by the Priority Programme SPP 1962 "Non-smooth and Complementarity-based Distributed Parameter Systems" of the German Research Foundation (DFG).

## References

1. Banholzer, S., Beermann, D., Volkwein, S.: POD-based bicriterial optimal control by the reference point method. In: 2nd IFAC Workshop on Control of Systems Governed by Partial Differential Equations, pp. 210–215 (2016)
2. Basseur, M., Zitzler, E.: Handling uncertainty in indicator-based multiobjective optimization. Int. J. Comput. Intell. Res. **2**(3), 255–272 (2006)
3. Bosman, P.A.N.: On gradients and hybrid evolutionary algorithms. IEEE Trans. Evol. Comput. **16**(1), 51–69 (2012)
4. Coello Coello, C.A., Van Veldhuizen, D.A., Lamont, G.B.: Evolutionary Algorithms for Solving Multi-objective Problems, vol. 242. Kluwer Academic, New York (2002)
5. Deb, K., Gupta, H.: Searching for robust Pareto-optimal solutions in multi-objective optimization. In: Coello Coello, C.A., Hernández Aguirre, A., Zitzler, E. (eds.) Evolutionary Multi-criterion Optimization, pp. 150–164. Springer, Berlin (2005)
6. Deb, K., Mohan, M., Mishra, S.: Evaluating the $\varepsilon$-domination based multi-objective evolutionary algorithm for a quick computation of Pareto-optimal solutions. Evol. Comput. **13**(4), 501–525 (2005)
7. Dellnitz, M., Hohmann, A.: A subdivision algorithm for the computation of unstable manifolds and global attractors. Numerische Mathematik **75**(3), 293–317 (1997)
8. Dellnitz, M., Schütze, O., Hestermeyer, T.: Covering Pareto sets by multilevel subdivision techniques. J. Optim. Theory Appl. **124**(1), 113–136 (2005)
9. Desideri, J.A.: Mutiple-gradient descent algorithm for multiobjective optimization. In: Berhardsteiner, J. (ed.) European Congress on Computational Methods in Applied Sciences and Engineering (ECCOMAS) (2012)
10. Ehrgott, M.: Multicriteria Optimization. Springer, Berlin (2005)
11. Engau, A., Wiecek, M.M.: Generating $\varepsilon$-efficient solutions in multiobjective programming. Eur. J. Oper. Res. **177**(3), 1566–1579 (2007)
12. Fliege, J., Svaiter, B.F.: Steepest descent methods for multicriteria optimization. Math. Methods Oper. Res. **51**(3), 479–494 (2000)
13. Fliege, J., Drummond, L.M.G., Svaiter, B.F.: Newton's method for multiobjective optimization. SIAM J. Optim. **20**(2), 602–626 (2009)
14. Hernández, C., Sun, J.Q., Schütze, O.: Computing the Set of Approximate Solutions of a Multi-objective Optimization Problem by Means of Cell Mapping Techniques, pp. 171–188. Springer International Publishing, Heidelberg (2013)

15. Hillermeier, C.: Nonlinear Multiobjective Optimization - A Generalized Homotopy Approach. Birkhäuser, Basel (2001)
16. Hughes, E.J.: Evolutionary multi-objective ranking with uncertainty and noise. In: Zitzler, E., Thiele, L., Deb, K., Coello Coello, C.A., Corne, D. (eds.) Evolutionary Multi-criterion Optimization, pp. 329–343. Springer, Berlin (2001)
17. Jahn, J.: Multiobjective search algorithm with subdivision technique. Comput. Optim. Appl. **35**(2), 161–175 (2006)
18. Jin, Y.: Surrogate-assisted evolutionary computation: recent advances and future challenges. Swarm Evol. Comput. **1**(2), 61–70 (2011)
19. Kuhn, H.W., Tucker, A.W.: Nonlinear programming. In: Proceedings of the 2nd Berkeley Symposium on Mathematical and Statsitical Probability, pp. 481–492. University of California Press (1951)
20. Lara, A., Sanchez, G., Coello Coello, C.A., Schütze, O.: HCS: a new local search strategy for memetic multiobjective evolutionary algorithms. IEEE Trans. Evol. Comput. **14**, 112–132. IEEE (2010)
21. Miettinen, K.: Nonlinear Multiobjective Optimization. Kluwer Academic Publishers, Boston (1999)
22. Neri, F., Cotta, C., Moscato, P.: Handbook of Memetic Algorithms, vol. 379. Springer, Berlin (2012)
23. Nocedal, J., Wright, S.J.: Numerical Optimization. Springer Science & Business Media, New York (2006)
24. Peitz, S., Ober-Blöbaum, S., Dellnitz, M.: Multiobjective optimal control methods for fluid flow using model order reduction (2015). arXiv:1510.05819
25. Rockafellar, R.T., Wets, R.J.B.: Variational Analysis, vol. 317. Springer, Berlin (1998)
26. Schäffler, S., Schultz, R., Weinzierl, K.: Stochastic method for the solution of unconstrained vector optimization problems. J. Optim. Theory Appl. **114**(1), 209–222 (2002)
27. Schütze, O., Coello Coello, C.A., Tantar, E., Talbi, E.G.: Computing the set of approximate solutions of an MOP with stochastic search algorithms. In: Proceedings of the 10th Annual Conference on Genetic and Evolutionary Computation, pp. 713–720. ACM (2008)
28. Schütze, O., Vasile, M., Coello Coello, C.A.: Computing the set of epsilon-efficient solutions in multi-objective space mission design. J. Aerosp. Comput. Inf. Commun. **8**(3), 53–70 (2009)
29. Schütze, O., Witting, K., Ober-Blöbaum, S., Dellnitz, M.: Set oriented methods for the numerical treatment of multiobjective optimization problems. In: Tantar, E., Tantar, A.A., Bouvry, P., Del Moral, P., Legrand, P., Coello Coello, C.A., Schütze, O. (eds.) EVOLVE- A Bridge Between Probability, Set Oriented Numerics and Evolutionary Computation. Studies in Computational Intelligence, vol. 447, pp. 187–219. Springer, Berlin (2013)
30. Schütze, O., Alvarado, S., Segura, C., Landa, R.: Gradient subspace approximation: a direct search method for memetic computing. Soft Comput. (2016)
31. Schütze, O., Martín, A., Lara, A., Alvarado, S., Salinas, E., Coello Coello, C.A.: The directed search method for multi-objective memetic algorithms. Comput. Optim. Appl. **63**(2), 305–332 (2016)
32. Singh, A., Minsker, B.S.: Uncertainty-based multiobjective optimization of groundwater remediation design. Water Resour. Res. **44**(2) (2008)
33. Teich, J.: Pareto-front exploration with uncertain objectives. In: Zitzler, E., Thiele, L., Deb, K., Coello Coello, C.A., Corne, D. (eds.) Evolutionary Multi-criterion Optimization, pp. 314–328. Springer, Berlin (2001)
34. Tröltzsch, F., Volkwein, S.: POD a-posteriori error estimates for linear-quadratic optimal control problems. Comput. Optim. Appl. **44**, 83–115 (2009)
35. White, D.J.: Epsilon efficiency. J. Optim. Theory Appl. **49**(2), 319–337 (1986)

# A New Local Search Heuristic for the Multidimensional Assignment Problem

**Sergio Luis Pérez Pérez, Carlos E. Valencia and Francisco Javier Zaragoza Martínez**

**Abstract** The Multidimensional Assignment Problem (MAP) is a natural extension of the well-known assignment problem. The most studied case of the MAP is the 3-dimensional Assignment Problem (3AP), though in recent years some local search heuristics and a memetic algorithm were proposed for the general case. Until now, a memetic algorithm has been proven to be the best-known option to solve MAP instances and it uses some procedures called dimensionwise variation heuristics as part of the improvement of individuals. We propose a new local search heuristic, based on ideas from dimensionwise variation heuristics, which consider a bigger space of neighborhoods, providing higher quality solutions for the MAP. Our main contribution is a generalization of several local search heuristics known from the literature, the conceptualization of a new one, and the application of exact techniques to find local optimum solutions at its neighborhoods. The results of computational evaluation show how our heuristic outperforms the previous local search heuristics and its competitiveness against a state-of-the-art memetic algorithm.

**Keywords** Multidimensional assignment problem · Local search · Exact technique · Heuristic · Neighborhood

---

S.L. Pérez Pérez (✉)
Posgrado en Optimización, Universidad Autónoma Metropolitana Azcapotzalco, Mexico City, Mexico
e-mail: sergio10barca@gmail.com

C.E. Valencia
Departamento de Matemáticas, Centro de Investigación y de Estudios Avanzados del IPN, Mexico City, Mexico
e-mail: cvalencia@math.cinvestav.edu.mx

F.J. Zaragoza Martínez
Departamento de Sistemas, Universidad Autónoma Metropolitana Azcapotzalco, Mexico City, Mexico
e-mail: franz@correo.azc.uam.mx

© Springer International Publishing AG 2018

Y. Maldonado et al. (eds.), *NEO 2016*, Studies in Computational Intelligence 731,
https://doi.org/10.1007/978-3-319-64063-1_8

## 1 Introduction

The Multidimensional Assignment Problem (MAP), abbreviated $s$AP in the case of $s$ dimensions, is a natural extension of the well-known Assignment Problem AP. MAP has been studied since the fifties [23, 30] and was formally described by Pierskalla [28].

In 1972, Karp [22] showed that the problem of deciding whether there exists a 3-dimensional matching of size $k$ is NP-complete and, consequently, the optimization problem of finding the largest 3-dimensional matching is NP-hard. In general, the Multidimensional Matching Problem (MMP), abbreviated $s$MP, is a particular case of the MAP in which is assigned a value of 0 to the present relations and a value of 1 to those non present, the objective is to find a multidimensional assignment of minimum cost, the optimal assignment will contain the relations of the largest s-dimensional matching plus some non present relations which can be discarded. The $s$AP is NP-hard for every $s \geq 3$ as shown in Gary and Johnson [12].

It has been proven that unless $P = NP$, there is no $\varepsilon$-approximate polynomial time algorithm for the multidimensional assignment problem [7]. The special case of 3AP where a distance, verifying the triangle inequalities, is defined on the elements from each dimension, and the cost of the weight is either the sum of the lengths of its distances or the sum of the lengths of its two shortest distances was proven to be $\varepsilon$-approximable [7].

The most studied case of MAP is the 3-dimensional Assignment Problem 3AP [1, 3, 11, 16, 17, 31] though in recent years several algorithms and heuristics were proposed for $s$AP [5, 6, 13, 15, 19–21, 26, 27, 29].

MAP has a wide variety of applications, e.g. for the multisensor data association problem where the objective is to determine which measurements from one or more sensors are related to the same object [2], for the problem of selecting roots of a system of polynomial equations [4], for the school timetabling problem where the goal is to provide an assignment of professors to courses to time slots [10].

The main contribution of our paper is a generalization of several local search heuristics known from the literature, the conceptualization of a new one and the application of exact techniques to find local optimum solutions at its neighborhoods. Furthermore, we evaluate them against state-of-the-art heuristics and meta-heuristics. For the purpose of experimental evaluation we consider several families of instances known from the literature.

In Sect. 2 we state the multidimensional assignment problem. In Sect. 3 we analyze some exact techniques that solve $s$AP and we evaluate its performance by solving some families of instances. In Sect. 4 we describe some previously proposed local search heuristics, then we extend them, and we propose a new one. In Sect. 5 we present the experimental evaluation. Finally, we give our conclusions in Sect. 6.

## 2 The Multidimensional Assignment Problem

The multidimensional assignment problem is also known as the axial multi-index assignment problem. Let $s \geq 2$ be a fixed number of dimensions, a $s$AP is stated as follows. Let $X_1, X_2, \ldots, X_s$ a collection of $s$ disjoint sets. Without loss of generality consider that $n = |X_1| = |X_2| = \cdots = |X_s|$ (otherwise add some dummy elements to equilibrate them). Consider all the members of the Cartesian product $X = X_1 \times X_2 \times \cdots \times X_s$ such that each vector $x \in X$, has assigned a weight $w(x)$ and $x = x_1 x_2 \ldots x_s$ with $x_i \in X_i$ for each $1 \leq i \leq s$. A valid assignment is a collection $A = (x^1, x^2, \ldots, x^n)$ of $n$ vectors if $x_k^i \neq x_k^j$ for each $i \neq j$ and $1 \leq k \leq s$. The weight of an assignment is $w(A) = \sum_{i=1}^{n} w(x^i)$. The objective of $s$AP is to find an assignment of minimal weight. This definition is related to the one that considers a hypergraph $H = (X_1, X_2, \ldots, X_s; E)$ where $H$ is the multipartite hypergraph with vertices in $X_1 \cup X_2 \cup \cdots \cup X_s$ and where the hyperedges of $E$ are the elements of $X$.

Another way to stated the problem is through permutations of sets. Let $\varphi_1, \varphi_2, \ldots, \varphi_s$ be permutations of $X_1, X_2, \ldots, X_s$ respectively. Then $\varphi_1, \varphi_2, \ldots, \varphi_s$ is an assignment with the weight $\sum_{i=1}^{n} w(\varphi_{1i}, \varphi_{2i}, \ldots, \varphi_{si})$. The first permutation can be fixed and without loss of generality the objective is as follows:

$$\min \sum_{i=1}^{n} w(x_i^1, \varphi_{2i}, \ldots, \varphi_{si}) \,. \tag{1}$$

Let $P$ denote the set of all assignments (Cartesian product of permutations) of $n$ items over $s$ disjoint sets. This set has $n!^{s-1}$ elements.

Finally, the following is a 0–1 integer linear programming formulation of this problem. Let $C_{i^1 i^2 \ldots i^s}$ be the cost of assigning simultaneously $x_{i^1}, x_{i^2}, \ldots, x_{i^s}$ with $x_{i^1} \in X_1$, $x_{i^2} \in X_2$, …, $x_{i^s} \in X_s$ the objective is:

$$\begin{aligned}
\min & \sum_{i^1=1}^{n} \sum_{i^2=1}^{n} \cdots \sum_{i^s=1}^{n} C_{i^1 i^2 \ldots i^s} \cdot p_{i^1 i^2 \ldots i^s} \\
\text{subject to} & \sum_{i^2=1}^{n} \sum_{i^3=1}^{n} \cdots \sum_{i^s=1}^{n} p_{i^1 i^2 \ldots i^s} = 1 \quad (1 \leq i^1 \leq n), \\
& \sum_{i^1=1}^{n} \sum_{i^3=1}^{n} \cdots \sum_{i^s=1}^{n} p_{i^1 i^2 \ldots i^s} = 1 \quad (1 \leq i^2 \leq n), \\
& \ldots \\
& \sum_{i^1=1}^{n} \sum_{i^2=1}^{n} \cdots \sum_{i^{s-1}=1}^{n} p_{i^1 i^2 \ldots i^s} = 1 \quad (1 \leq i^s \leq n), \\
& p_{i^1 i^2 \ldots i^s} \in \{0, 1\} \quad (1 \leq i^1, i^2, \ldots, i^s \leq n) \,.
\end{aligned} \tag{2}$$

Alternatively, a maximization optimization problem could be required in which case all the constraints remain the same and the objective function is formulated as a maximization function.

The size of an instance of MAP with $s$ dimensions and $n$ vertices on each is equal to the cardinality of $X$, that is $|X| = n^s$.

## 3 Exact Techniques

In this section we analyze some exact techniques to solve $s$AP. At the end of the section, we provide a very strong implementation that is used as part of the local search heuristic proposed in this work. The next techniques are focused in $s \geq 3$ since it is known that the Assignment Problem (2AP) is solvable in $O(n^3)$ time and $O(n^2)$ space [9, 24].

### *3.1 Brute Force*

A natural solution consists in evaluating all the possible combinations from Eq. 1 to find the optimum. This gives a time complexity of $O(n!^{s-1})$ and a space complexity of $O(n \cdot s)$. The relevance of this technique is because some local search heuristics, as 2-Opt and 3-Opt [19], use this technique as part of its machinery and, it offers a better option for solving very small instances of sAP at least in comparison with more complex techniques, as we show in Sect. 5.

### *3.2 Wise Brute Force*

This approach consists in generating all the first $O(n!^{s-2})$ possible Cartesian product of permutations for the first $s-1$ sets and then apply a 2AP solution for each combination against the last set. This allows a complexity time reduction from $O(n!^{s-1})$ to $O(n!^{s-2}n^3)$, and the space complexity changes from $O(n \cdot s)$ to $O(n \cdot s + n^2)$.

### *3.3 Dynamic Programming Reduction*

This approach consists in reducing the set of all possible candidates from $O(n!^{s-1})$ to $O(2^{(s-1) \cdot n})$ by memorizing the optimal solution for some sub-problems of the original problem in a similar way that is performed in the dynamic programming solution for the Traveling Salesman Problem [8].

Let $S_1, \ldots, S_s$ be sets of $n$ vertices at each. Lets to denote as $S_i^k$ a set of any $k$ vertices of $S_i$, for $1 \leq i \leq n$, and let $X^k = \{k\} \times S_2 \times \cdots \times S_s$ be all members of the Cartesian product between the elements of all the sets. Recall that the set $S_1$ can be fixed. We need to define an appropriate subproblem with an optimal partial solution for the MAP. Suppose we take one vector $x \in X$. We need to delete the corresponding vertices $x_i \in x$ from each set $S_i$, then an optimal partial solution is calculated for the rest of the vertices. The optimal solution consists in choosing the vector $x$ whose sum with its corresponding optimal partial solution is minimum. Then we have an appropriate subproblem.

For a set of subsets of vertices $S_2 \subseteq \{1, \ldots, n\}, \ldots, S_s \subseteq \{1, \ldots, n\}$ with $|S_2| = \cdots = |S_s| = k$ which includes a vector $x^k \in X^k$ (here $X^k = \{k\} \times S_2 \times \cdots \times S_s$) where $x^k = (k, x_2^k, \ldots, x_s^k)$, let $C(S_2, \ldots, S_s)$ be the cost of the minimum assignment that considers $|S_2| = \cdots = |S_s| = k$ vectors with vertices in the corresponding sets $S_2, \ldots, S_s$.

When $k = 0$, we define $C(S_2, \ldots, S_s) = 0$ since the assignment has no vectors.

In order to define $C(S_2, \ldots, S_s)$ in terms of smaller subproblems we need to pick one vector $x^k = (k, x_2^k, \ldots, x_s^k)$ such that $x_i^k \in S_i$ for all $2 \leq i \leq s$. We need to evaluate all the vectors $x^k \in X^k$ in order to get the best $x^k$ such that $C(S_2, \ldots, S_s) = \min_{x^k \in X^k} C(S_2 - x_2^k, \ldots, S_s - x_s^k) + w(x^k)$

Formally, we have the next recursive function:

$$C(S_2, \ldots, S_s) = \begin{cases} 0 & \text{if } |S_2| = \cdots = |S_s| = 0 \\ \min_{x^k \in X^k} (C(S_2 - x_2^k, \ldots, S_s - x_s^k) + w(x^k)) & \text{if } |S_2| = \cdots = |S_s| \geq 1 \end{cases} \quad (3)$$

As we can see in Eq. 3, at each recursive call one vertex is subtracted from each set so that the size of all the sets always remains equilibrated. The process of selecting the vector with the minimum weight among all the available vectors from $X^k$ takes $O(k^{s-1})$ time. The possible state of a vertex $x_i \in S_i$ at the step $k$ is to be present or not, therefore the total number of possible states is $2^{\{s-1\} \cdot n}$. By applying a memoization technique we can avoid repeated recursive calls. The total time complexity for this algorithm is $O(2^{\{s-1\} \cdot n} n^{s-1})$ which is better than previously described algorithms but, still exponential. A disadvantage of this solution is the requirement of exponential space, which is of $O(2^{\{s-1\} \cdot n})$, since it needs to memorize the answers for the derived sub states.

## 3.4 *Branch and Bound Algorithm*

In 1988, Balas and Saltzman presented a branch and bound algorithm for 3AP. They applied a Lagrangian relaxation that incorporated a class of facet inequalities in order to find lower bounds, and a primal heuristic based on the principle of minimizing maximum regret and a variable depth interchange phase for finding upper bounds. They reported results for instances with $n \in \{4, 6, \ldots, 24, 26\}$ vertices and uniformly random generated weights $w(x) \in \{0, \ldots, 100\}$, obtaining running times from some seconds until some minutes. The actual complexity of this algorithm is not provided but, based on the time results, seems to be approximately $O(2^n)$.

In addition, they described an experimental evaluation for the primal heuristic of minimax-regret and for the variable depth interchange heuristic. Both heuristics were tested on instances with $n \in \{20, 25, \ldots, 65, 70\}$ vertices and uniformly random generated weights $w(x) \in \{0, \ldots, 1000\}$, obtaining running times of less than a minute. The time complexity for the minimax-regret heuristic is $O(n^3 \log n)$ and for the variable depth interchange heuristic is $O\left(\binom{n}{2}\right) \sim O(n^2)$.

We are not sure about the generalization of this algorithm to deal with problems in more than 3 dimensions, however, could be a good contribution to determine if it is possible its generalization and, if it were the case, to implement it and evaluate its performance. Such topic is out of the scope of this paper, so we let this line for future work.

### 3.5 MAP-Gurobi

We implemented the 0–1 integer linear programming formulation, provided in Eq. 2, by using the libraries from Gurobi Optimizer 6.0[1] and we called it MAP-Gurobi. Gurobi Optimizer was designed from the ground up to exploit modern architectures and multi-core processors, using the most advanced implementations of the latest algorithms. Even when Gurobi Optimizer is a commercial product, it is easy to get an academic version for purposes of researching [14].

This implementation provide us with a very strong machinery to solve small instances of $s$AP and here is used as part of the proposed heuristics. In order to provide an approach about the time and space complexity for this algorithm, we solved two different sets of small and medium size families of instances. We also tried to solve another type of family of instances of larger size, however, we only had success with the smaller instances of that set, the rest of the instances were not solved, due to the limitation of computer power. When this solver has to solve problems with a very large searching space the memory RAM crashes due to the large amount of required memory to solve this type of instances. We present the results of this experimentation in Sect. 5.

By considering the running times for the family of instances provided by Magos and Mourtos showed in Table 1 we could deduce that complexity of MAP-Gurobi seems to be approximately $O((s-1)^n)$, however this seems to be a big upper bound about its time complexity since in our experimental evaluation Gurobi performed better than this. By the other side, according to the running times for the families of the instances provided by Karapetyan and Gutin [19] we determine that, in addition

[1]The Gurobi Optimizer is the state-of-the-art solver for mathematical programming.

**Table 1** Optimum values and running times for small and medium size problem instances solved by MAP-Gurobi

| Balas and Saltzman dataset | | | | Magos and Mourtos | | | |
|---|---|---|---|---|---|---|---|
| Instance | Optima | Variables | Seconds | Instance | Optima | Variables | Seconds |
| 3_bs_4 | 42.2 | 64 | 0.1 | 4_axial_10 | 480.3 | 10,000 | 1.3 |
| 3_bs_6 | 40.2 | 216 | 0.1 | 4_axial_11 | 603.0 | 14,641 | 2.8 |
| 3_bs_8 | 23.8 | 512 | 0.1 | 4_axial_12 | 711.3 | 20,736 | 3.2 |
| 3_bs_10 | 19.0 | 1000 | 0.1 | 4_axial_13 | 839.7 | 28,561 | 7.5 |
| 3_bs_12 | 15.6 | 1728 | 0.1 | 4_axial_14 | 831.3 | 38,416 | 10.6 |
| 3_bs_14 | 10.0 | 2744 | 0.2 | 4_axial_15 | 822.1 | 50,625 | 24.9 |
| 3_bs_16 | 10.0 | 4096 | 0.3 | 4_axial_16 | 736.9 | 65,536 | 45.0 |
| 3_bs_18 | 6.4 | 5832 | 0.4 | 4_axial_17 | 643.5 | 83,521 | 91.7 |
| 3_bs_20 | 4.8 | 8000 | 0.9 | 4_axial_18 | 608.7 | 1,04,976 | 132.7 |
| 3_bs_22 | 4.0 | 10,648 | 0.8 | 4_axial_19 | 533.5 | 1,30,321 | 147.4 |
| 3_bs_24 | 1.8 | 13,824 | 0.9 | 4_axial_20 | 503.8 | 1,60,000 | 336.3 |
| 3_bs_26 | 1.0 | 17,576 | 1.3 | 5_axial_7 | 407.8 | 16,807 | 1.4 |
| | | | | 5_axial_8 | 587.3 | 32,768 | 4.8 |
| | | | | 5_axial_9 | 471.9 | 59,049 | 14.0 |
| | | | | 5_axial_10 | 341.3 | 1,00,000 | 30.4 |
| | | | | 5_axial_11 | 274.4 | 1,61,051 | 84.0 |
| | | | | 5_axial_12 | 219.7 | 2,48,832 | 300.4 |
| | | | | 5_axial_13 | 177.5 | 3,71,293 | 704.3 |
| | | | | 5_axial_14 | 141.1 | 5,37,824 | 1311.8 |
| | | | | 6_axial_8 | 156.9 | 2,62,144 | 107.9 |

to the size of the input, weights distribution have an impact on the running time, in contrast to the previously described algorithms. Based on that, we can observe that MAP-Gurobi performs better when the weights of the vectors of an instance are set up uniformly distributed random within a range among all the vectors of $X$. This can be observed in Table 2.

Finally, from Table 2 we can observe that more than the number of variables of an instance, what matters is the searching space of an instance, e.g. whereas instances with $s = 5$ dimensions and $n = 18$ vertices were solved, instances with $s = 4$ dimensions and $n = 30$ vertices were not solved which is due to the fact that $18!^4 < 30!^3$.

**Table 2** Optimum values and running times for large size problem instances solved by MAP-Gurobi

| s | n | Variables | Random | | Square root | | Clique | |
|---|---|---|---|---|---|---|---|---|
| | | | Optima | Seconds | Optima | Seconds | Optima | Seconds |
| 3 | 40 | 64,000 | 40.0 | 4.1 | 606.9 | 2.6 | 939.9 | 3.3 |
| 3 | 70 | 3,43,000 | 70.0 | 11.9 | 733.6 | 62.0 | 1157.1 | 50.7 |
| 3 | 100 | 10,00,000 | 100.0 | 50.9 | 838.1 | 1062.1 | 1345.9 | 771.7 |
| 4 | 20 | 1,60,000 | 20.0 | 10.3 | 929.3 | 9.4 | 1901.8 | 8.4 |
| 4 | 30 | 8,10,000 | 30.0 | 56.5 | – | – | – | – |
| 4 | 40 | 25,60,000 | 40.0 | 173.0 | – | – | – | – |
| 5 | 15 | 7,59,375 | 15.0 | 68.0 | 1203.9 | 79.4 | 3110.7 | 43.0 |
| 5 | 18 | 18,89,568 | 18.0 | 186.5 | 1343.8 | 1029.0 | 3458.6 | 196.5 |
| 5 | 25 | 97,65,625 | – | – | – | – | – | – |
| 6 | 12 | 29,85,984 | 12.0 | 493.1 | 1436.8 | 427.3 | 4505.6 | 689.1 |
| 6 | 15 | 1,13,90,625 | – | – | – | – | – | – |
| 6 | 18 | 3,40,12,224 | – | – | – | – | – | – |

# 4 Local Search Heuristics

A local search heuristic is a method that considers some neighborhood of candidate solutions and finds an optimal solution among the solutions of such space. This type of technique moves from a solution to another one in the space of candidate solutions by applying some local changes, until a solution deemed optimal is found or a time bound is elapsed.

In 1991, Balas and Saltzman [3] proposed some of the first local search heuristics for 3AP. We are particularly interested on the $k$-opt heuristic which is slightly different to the classical $k$-opt heuristic used first for the Traveling Salesperson Problem. In 2006, Huang and Lim [16] proposed a local search for 3AP which was later extended by Karapetyan and Gutin [19], they called it Dimensionwise Variation Heuristic. We consider both heuristics and mixed them as one in order to create a new local search heuristic which generalizes both type of heuristics and extend them.

From here, each of the next heuristics requires an initial feasible solution $A = (x^1, x^2, \ldots, x^n)$ somehow generated. A feasible solution can easily generated just by choosing a random permutation for each set of vertices from each dimension but the first, remind we fixed it, and creating the corresponding vectors $x^i$ from combining the i-elements from each dimension as one valid vector $x^i \in A$.

From here, lets consider a toy instance with $s = 4$ dimensions and $n = 5$ vertices. Let $x_i^j$ be the i-th vertex of the dimension $j$ weights are set up as the product of the corresponding indexes of the related vertex from each dimension, so the weights of all the vectors are $w(x_1^1, x_1^2, x_1^3, x_1^4) = 1$, $w(x_1^1, x_1^2, x_1^3, x_2^4) = 2$, ..., $w(x_5^1, x_5^2, x_5^3, x_5^4) = 625$. For simplicity we will write the vectors $(x_a^1, x_b^2, x_c^3, x_d^4)$ as $(a, b, c, d)$ since the dimension which the vertex belongs to can be deducted from the position of the index in the vector.

A feasible solution for an instance looks like:

$$A = \begin{cases} x^1 & : (1, 2, 3, 4) \\ x^2 & : (2, 3, 1, 3) \\ x^3 & : (3, 5, 4, 2) \\ x^4 & : (4, 1, 2, 1) \\ x^5 & : (5, 4, 5, 5) \end{cases}$$

The cost of this solution is $w(A) = w(x^1) + w(x^2) + w(x^3) + w(x^4) + w(x^5) = 24 + 18 + 120 + 8 + 500 = 670$.

We illustrate the structure of a feasible solution because it is considered for the description of the next heuristics. Noticed this structure allow us to have a solution as a matrix with $n$ rows and $s$ columns where the $i$th row is related to the vector $x^i \in A$ and the $j$th column is related to the set of vertices $X_i$.

## 4.1 Dimensionwise Variation Heuristics

Dimensionwise variation heuristics (DVH) are a set of heuristics that performs local improvements over a feasible solution by applying an exact technique. Originally, Huang and Lim [16] applied this technique for 3AP and [19] extended it for more than 3 dimensions, however this technique can be more general.

The original version of DVH works as follows: at one step of the heuristic all the dimensions but a proper subset $F \subset \{1, \ldots, s\}$ are fixed and a matrix $M$ of size $n \times n$ with entries $M_{i,j} = w(v_{i,j})$ is generated. Let $v^d_{i,j}$ denote the dth element of the vector $v_{i,j}$, all the vectors are built according to the next function:

$$v^d_{i,j} = \begin{cases} x^d_i \text{ if } d \in F \\ x^d_j \text{ if } d \notin F \end{cases} \quad \text{for } 1 \leq d \leq s\,. \tag{4}$$

The corresponding 2AP can be solved in $O(n^3)$ time and gives a local optimum for the original instance.

For example, lets consider our randomly generated feasible solution $A$ for our toy instance and lets choose the set $F = \{2, 4\}$, then $M$ is obtained as:

$$\begin{bmatrix} (1,2,3,4) & (2,2,1,4) & \mathbf{(3,2,4,4)} & (4,2,2,4) & (5,2,5,4) \\ (1,3,3,3) & (2,3,1,3) & (3,3,4,3) & \mathbf{(4,3,2,3)} & (5,3,5,3) \\ \mathbf{(1,5,3,2)} & (2,5,1,2) & (3,5,4,2) & (4,5,2,2) & (5,5,5,2) \\ (1,1,3,1) & (2,1,1,1) & (3,1,4,1) & (4,1,2,1) & \mathbf{(5,1,5,1)} \\ (1,4,3,5) & \mathbf{(2,4,1,5)} & (3,4,4,5) & (4,4,2,5) & (5,4,5,5) \end{bmatrix} \rightarrow \begin{bmatrix} 24 & 16 & \mathbf{96} & 64 & 200 \\ 27 & 18 & 108 & \mathbf{72} & 225 \\ \mathbf{30} & 20 & 120 & 80 & 250 \\ 3 & 2 & 12 & 8 & \mathbf{25} \\ 60 & \mathbf{40} & 240 & 160 & 500 \end{bmatrix}.$$

The values marked bold denote the new vectors that will be considered as the new feasible solution after solve the corresponding 2AP. Then, the new feasible solution is:

$$A' = \begin{cases} x^{1'} : (1, 5, 3, 2) \\ x^{2'} : (2, 4, 1, 5) \\ x^{3'} : (3, 2, 4, 4) \\ x^{4'} : (4, 3, 2, 3) \\ x^{5'} : (5, 1, 5, 1) \end{cases}$$

The cost of the new solution $A'$ is $w(A') = w(x^{1'}) + w(x^{2'}) + w(x^{3'}) + w(x^{4'}) + w(x^{5'}) = 30 + 40 + 96 + 72 + 25 = 263$ which is equal to the optimal minimum cost of the 2AP solved.

A complete iteration of this heuristic consists in trying every possible non empty subset $F \subset \{1, \ldots, s\}$. There are $O(2^s)$ possible combinations, then one iteration takes $O(2^s n^3)$ time. This heuristic performs iterations until no improvement is obtained. Even when the theoretical time complexity could be high, Karapetyan and Gutin [19] reported that by at most ten iterations are performed before to converge to a local optimum. We also observed this behavior in our experimental evaluation.

This heuristic considers a searching space of size $O(n!)$ at each step. It is easy to see that at the end of one step of this heuristic the current feasible solution cannot be worst which is thanks to the fact that we are optimizing the derived searching space by solving the corresponding 2AP. Another way to see it is that we are selecting the new vectors among a set of derived vectors that always considers the original vectors plus other combinations such that, if some vectors should be changed in order to get a better the solution, then the 2AP solution obtains the best ones.

## 4.2 Generalized Dimensionwise Variation Heuristics

The simplification of a $s$AP to a 2AP can be applied for any $s$AP with $s \geq 3$. We propose a generalization of this heuristic which consists in reducing a $s$AP to a $t$AP with $2 \leq t < s$.

Our generalized dimensionwise variation heuristic (GDVH) works as follows: let $t$ be an integer value such that $2 \leq t \leq s - 1$ and, based on $t$, suppose to have $F_1, \ldots, F_{t-1}$ non empty proper subsets of $\{1, \ldots, s\}$ such that $F_1 \cap \cdots \cap F_{t-1} = \emptyset$ and $F_1 \cup \cdots \cup F_{t-1} \subset \{1, \ldots, s\}$. At one step of the heuristic all the dimensions but $F_1 \cup \cdots \cup F_{t-1}$ are fixed and a t-dimensional matrix $M^t$ of size $n^1 \times \cdots \times n^t$ (recall the simplification $n^i$ = n for $1 \leq i \leq t$) with entries $M_{i^1,\ldots,i^t} = w(v_{i^1,\ldots,i^t})$ is generated. Let $v^d_{i^1,\ldots,i^t}$ denote the dth element of the vector $v_{i^1,\ldots,i^t}$, all the vectors are built according to the next function:

$$v^d_{i^1,i^2,\ldots,i^{t-1},i^t} = \begin{cases} x^d_{i^1} \text{ if } d \in F_1 \\ x^d_{i^2} \text{ if } d \in F_2 \\ \cdots \\ x^d_{i^{t-1}} \text{ if } d \in F_{t-1} \\ x^d_{i^t} \text{ otherwise} \end{cases} \quad \text{for } 1 \leq d \leq s \; . \tag{5}$$

The corresponding $t$AP instance can be solved by using some exact technique, we suggest our MAP-Gurobi implementation.

For example, lets consider our randomly generated feasible solution $A$ for our toy instance and lets choose $t = 3$ and let $F_1 = \{2, 4\}$, $F_2 = \{3\}$ be the selected proper subsets such that $F_1 \cup F_2 \subset \{1, \ldots, s\}$, then $M$ is obtained as:

$$M_1 \begin{bmatrix} (1,2,3,4) & (2,2,3,4) & (3,2,3,4) & (4,2,3,4) & (5,2,3,4) \\ (1,2,1,4) & (2,2,1,4) & (3,2,1,4) & (4,2,1,4) & (5,2,1,4) \\ (1,2,4,4) & \mathbf{(2,2,4,4)} & (3,2,4,4) & (4,2,4,4) & (5,2,4,4) \\ (1,2,2,4) & (2,2,2,4) & (3,2,2,4) & (4,2,2,4) & (5,2,2,4) \\ (1,2,5,4) & (2,2,5,4) & (3,2,5,4) & (4,2,5,4) & (5,2,5,4) \end{bmatrix} \rightarrow \begin{bmatrix} 24 & 48 & 72 & 96 & 120 \\ 8 & 16 & 24 & 32 & 40 \\ 32 & \mathbf{64} & 96 & 128 & 160 \\ 16 & 32 & 48 & 64 & 80 \\ 40 & 80 & 120 & 160 & 200 \end{bmatrix}.$$

$$M_2 \begin{bmatrix} (1,3,3,3) & (2,3,3,3) & (3,3,3,3) & (4,3,3,3) & (5,3,3,3) \\ (1,3,1,3) & (2,3,1,3) & (3,3,1,3) & (4,3,1,3) & (5,3,1,3) \\ (1,3,4,3) & (2,3,4,3) & (3,3,4,3) & (4,3,4,3) & (5,3,4,3) \\ (1,3,2,3) & (2,3,2,3) & \mathbf{(3,3,2,3)} & (4,3,2,3) & (5,3,2,3) \\ (1,3,5,3) & (2,3,5,3) & (3,3,5,3) & (4,3,5,3) & (5,3,5,3) \end{bmatrix} \rightarrow \begin{bmatrix} 27 & 54 & 81 & 108 & 135 \\ 9 & 18 & 27 & 36 & 45 \\ 36 & 72 & 108 & 144 & 180 \\ 18 & 36 & \mathbf{54} & 72 & 90 \\ 45 & 90 & 135 & 180 & 225 \end{bmatrix}.$$

$$M_3 \begin{bmatrix} (1,5,3,2) & (2,5,3,2) & (3,5,3,2) & (4,5,3,2) & (5,5,3,2) \\ (1,5,1,2) & (2,5,1,2) & (3,5,1,2) & \mathbf{(4,5,1,2)} & (5,5,1,2) \\ (1,5,4,2) & (2,5,4,2) & (3,5,4,2) & (4,5,4,2) & (5,5,4,2) \\ (1,5,2,2) & (2,5,2,2) & (3,5,2,2) & (4,5,2,2) & (5,5,2,2) \\ (1,5,5,2) & (2,5,5,2) & (3,5,5,2) & (4,5,5,2) & (5,5,5,2) \end{bmatrix} \rightarrow \begin{bmatrix} 30 & 60 & 90 & 120 & 150 \\ 10 & 20 & 30 & \mathbf{40} & 50 \\ 40 & 80 & 120 & 160 & 200 \\ 20 & 40 & 60 & 80 & 100 \\ 50 & 100 & 150 & 200 & 250 \end{bmatrix}.$$

$$M_4 \begin{bmatrix} (1,1,3,1) & (2,1,3,1) & (3,1,3,1) & (4,1,3,1) & (5,1,3,1) \\ (1,1,1,1) & (2,1,1,1) & (3,1,1,1) & (4,1,1,1) & (5,1,1,1) \\ (1,1,4,1) & (2,1,4,1) & (3,1,4,1) & (4,1,4,1) & (5,1,4,1) \\ (1,1,2,1) & (2,1,2,1) & (3,1,2,1) & (4,1,2,1) & (5,1,2,1) \\ (1,1,5,1) & (2,1,5,1) & (3,1,5,1) & (4,1,5,1) & \mathbf{(5,1,5,1)} \end{bmatrix} \rightarrow \begin{bmatrix} 3 & 6 & 9 & 12 & 15 \\ 1 & 2 & 3 & 4 & 5 \\ 4 & 8 & 12 & 16 & 20 \\ 2 & 4 & 6 & 8 & 10 \\ 5 & 10 & 15 & 20 & \mathbf{25} \end{bmatrix}.$$

$$M_5 \begin{bmatrix} \mathbf{(1,4,3,5)} & (2,4,3,5) & (3,4,3,5) & (4,4,3,5) & (5,4,3,5) \\ (1,4,1,5) & (2,4,1,5) & (3,4,1,5) & (4,4,1,5) & (5,4,1,5) \\ (1,4,4,5) & (2,4,4,5) & (3,4,4,5) & (4,4,4,5) & (5,4,4,5) \\ (1,4,2,5) & (2,4,2,5) & (3,4,2,5) & (4,4,2,5) & (5,4,2,5) \\ (1,4,5,5) & (2,4,5,5) & (3,4,5,5) & (4,4,5,5) & (5,4,5,5) \end{bmatrix} \rightarrow \begin{bmatrix} \mathbf{60} & 120 & 180 & 240 & 300 \\ 20 & 40 & 60 & 80 & 100 \\ 80 & 160 & 240 & 320 & 400 \\ 40 & 80 & 120 & 160 & 200 \\ 100 & 200 & 300 & 400 & 500 \end{bmatrix}.$$

In this example, can be observed that the vertices of the dimensions in the set $\{1, \ldots, s\} \backslash \{F_1 \cup F_2\}$ are fixed in the corresponding vectors of $M$; the vertices of the dimensions in the set $F_1$ just vary between each matrix $M_i$ and $M_j$ for all $i \neq j$ and $1 \leq i, j \leq 5$; the vertices of the dimensions in the set $F_2$ vary at each matrix $M_i$ for all $1 \leq i \leq 5$. The optimal solution of this 3AP provide us the vectors of the new feasible solution $A'$ which is:

$$A' = \begin{cases} x^{1'} & : (1, 4, 3, 5) \\ x^{2'} & : (2, 2, 4, 4) \\ x^{3'} & : (3, 3, 2, 3) \\ x^{4'} & : (4, 5, 1, 2) \\ x^{5'} & : (5, 1, 5, 1) \end{cases}$$

The cost of the new solution $A'$ is $w(A') = w(x^{1'}) + w(x^{2'}) + w(x^{3'}) + w(x^{4'}) + w(x^{5'}) = 60 + 64 + 54 + 40 + 25 = 243$ which is equal to the optimal minimum cost of the 3AP solved.

Keep in mind that if the reduction to some $t$AP with $3 \leq t \leq s - 1$ still has a big searching space then the resolution of the reduction could take a while or could not be solved due to the computer power, however, the same may occur in reductions to some 2AP with $n$ equal to many thousands of vertices. Table 2 of Sect. 5 provides a good upper bound about the size of possible reductions.

A complete iteration tries every possible combination of $t$ subsets. There are about $O(t^s)$ combinations, therefore one iteration takes $O(t^s \cdot (s-1)^n)$ time where $O((s-1)^n)$ is the complexity of solving an instance of size $O(n^t)$, except for the cases when $t = 2$ because those are the same as in DVH and can be solved in $O(n^3)$. In our experimental evaluation we also found that, as for DVH, less than ten iterations are performed before to converge to a local optimum.

This heuristic considers a searching space of size $O(n!^{t-1})$ at each step and, as for DVH, at the end of one step of this heuristic the current feasible solution cannot be worst.

## 4.3 K-Opt Heuristics

The k-Opt heuristics for 3AP were proposed by Balas and Saltzman [3] and extended for $s$AP by Karapetyan and Gutin [19].

A k-Opt heuristic works as follows: for every possible subset $R$ of $k$ vectors with $R \in A$, solve the corresponding $s$AP subproblem with $k$ vertices on each dimension.

The corresponding $s$AP subproblem with $k < n$ vertices is solved with some exact technique, which may result in the replacement of the selected vectors for a better ones. In particular Balas and Saltzman [3] as well as Karapetyan and Gutin [19] evaluated the 2-Opt and 3-Opt heuristics by using the Brute Force technique because it is faster than other algorithms for instances with 2 or 3 vertices and many dimensions.

For example, lets consider our randomly generated feasible solution $A$ for our toy instance and lets pick $k = 2$ vectors, in particular $x^4$ and $x^5$ then the new two vectors are chosen between the next pairs:

$$\begin{bmatrix} (4,1,2,1) & (4,1,2,5) & (4,1,5,1) & (4,1,5,5) & (4,4,2,1) & (4,4,2,5) & \mathbf{(4,4,5,1)} & (4,4,5,5) \\ (5,4,5,5) & (5,4,5,1) & (5,4,2,5) & (5,4,2,1) & (5,1,5,5) & (5,1,5,1) & \mathbf{(5,1,2,5)} & (5,1,2,1) \end{bmatrix}.$$

The corresponding weights are $\begin{bmatrix} 8 & 40 & 20 & 100 & 32 & 160 & \mathbf{80} & 400 \\ 500 & 100 & 200 & 40 & 125 & 25 & \mathbf{50} & 10 \end{bmatrix}$.

The vectors marked with bold are the new vectors that will be considered as part of the new feasible solution after applying one step of this heuristic. The new feasible solution will be:

$$A = \begin{cases} x^1 & : (1,2,3,4) \\ x^2 & : (2,3,1,3) \\ x^3 & : (3,5,4,2) \\ x^4 & : \mathbf{(4,4,5,1)} \\ x^5 & : \mathbf{(5,1,2,5)} \end{cases}$$

A complete iteration considers each of the $\binom{n}{k}$ possible combinations of vectors. This heuristic repeat iterations until no improvement is performed. By applying the Brute Force algorithm an iteration has a time complexity of $O(\binom{n}{k}k!^{s-1})$. By applying Gurobi-MAP an iteration has a time complexity of $O(\binom{n}{k}(s-1)^k)$.

This heuristic considers a searching space of size $O(k!^{s-1})$ at each step. As in the previously described heuristics, at the end of one step of this heuristic the current feasible solution cannot be worst.

## 4.4 *Generalized Local Search Heuristic*

We introduce a new heuristic called Generalized Local Search Heuristic (GLSH), which extends and combines ideas from GDVH and k-Opt heuristics.

Our GLSH works as follows: let $r$ be an integer value such that $2 \leq r \leq s$ and, based on $r$, suppose to have $F_1, \ldots, F_{r-1}$ non empty proper subsets such that $F_1 \cap \cdots \cap F_{r-1} = \emptyset$ and $F_1 \cup \cdots \cup F_{r-1} \subset \{1, \ldots, s\}$. Notice the difference of 1 between the considered range for $r$ and the integer value $t$ used in GDVH. Let $k$ be an integer value with $2 \leq k \leq n$. The integer values $r$ and $k$ should be chosen such that a reduction of the original problem (this in terms of the searching space) is achieved, which means that the next restrictions should be hold:

$$\begin{array}{c} 2 \leq r \leq s \\ 2 \leq k \leq n \\ r + k < s + n \end{array} . \tag{6}$$

At one step of this heuristic all the dimensions but $F_1 \cup \cdots \cup F_{r-1}$ are fixed and a set $Q$ of $k$ vectors from the feasible solution $A$ are chosen, then a r-dimensional matrix $M^r$ of size $k^1 \times \cdots \times k^r$ (recall the simplification $k^i = k$ for $1 \leq i \leq r$) with entries $M_{i^1,\ldots,i^r} = w(v^{i^1,\ldots,i^r})$ is generated. Let $v^d_{i^1,\ldots,i^r}$ denotes the dth element of the vector $v_{i^1,\ldots,i^r}$, all the vectors are built according to the next function:

$$v^d_{i^1,i^2,\ldots,i^{r-1},i^r} = \begin{cases} x^d_{i^1} \text{ if } d \in F_1 \\ x^d_{i^2} \text{ if } d \in F_2 \\ \ldots \\ x^d_{i^{r-1}} \text{ if } d \in F_{r-1} \\ x^d_{i^r} \text{ otherwise} \end{cases} \quad \text{for } 1 \leq d \leq s . \tag{7}$$

The corresponding $r$AP instance with $k$ vertices at each dimension can be solved by using some exact technique, again, we suggest our MAP-Gurobi implementation. The searching space of this heuristic is $O(k^{r-1}!)$. One of the main differences with the GDVH is that GLSH only considers the vectors in $Q$ instead of the complete list of vectors of $A$ at each time.

This heuristic extends and generalizes GDVH and k-Opt heuristics because by selecting the values $r, k$ as $2 \leq r < s$ and $k = n$ we have the case of a GDVH and by selecting the values $r, k$ as $r = s$ and $2 \leq k < n$ we have a k-Opt heuristic. Finally, by selecting the values $r, k$ as $2 \leq r < s$ and $2 \leq k < n$ we have a particular case of GLSH which is not considered neither in GDVH nor k-Opt heuristics.

The GLSH can works as a GDVH or as a k-Opt, however the parameters $r$ and $k$ could be tunning at each step of this heuristic instead of having them fixed.

Even when one of the main advantages of GLSH is its flexibility to move among different searching spaces due to the possibility of tunning of the parameters $r$ and $k$, in our evaluation experience, we determine that for the particular case when the GLSH is equal to the GDVH the quality solution is comparable to the one of more complex meta-heuristics.

## 5 Experimental Evaluation

In order to test the performance of the proposed heuristics we consider some families of instances known from the literature. Here are briefly described such families of instances.

1. Balas and Saltzman dataset [3] (1991). It includes 60 test instances with the problem size $s = 3$ and $n \in \{4, 6, \ldots, 24, 26\}$. For each $n$, five instances were created with the integer weight coefficients $w(x)$ uniformly random generated in the interval [0, 100]. The names of the instances are denoted as $s$_bs_$n$.

2. Magos and Mourtos [26]. It includes 200 test instances with three different dimension sizes: $s = 4$ and $n \in \{10, 11, \ldots, 19, 20\}$; $s = 5$ and $n \in \{7, 8, \ldots, 13, 14\}$; $s = 6$ and $n = 8$. For each combination of $s$ and $n$, ten instances were created with the integer weight coefficients $w(x)$ uniformly random generated in the interval $[1, n^s]$. The names of the instances are denoted as $s$_axial_$n$.
3. Karapetyan and Gutin [19]. It consists of three different families of instances: Random, Clique and SquareRoot. Each family includes 120 test instances with four different dimension sizes: $s = 3$ and $n \in \{40, 70, 100\}$; $s = 4$ and $n \in \{20, 30, 40\}$; $s = 5$ and $n \in \{15, 18, 25\}$; $s = 6$ and $n \in \{12, 15, 18\}$. For each combination of $s$ and $n$, ten instances were randomly generated.
   - Random. The integer weight coefficients $w(x)$ were uniformly random generated in the interval [1, 100].
   - Clique. This family of instances has weights defined trough s-partite graphs $G = (X_1 \cup \cdots \cup X_s, E)$. The weight $w(e)$ of every edge $e \in E$ was uniformly random generated in the interval [1, 100]. Let $C$ be a clique in $G$ and let $E_C$ be the set of edges induced by this clique, then the weight of a vector, corresponding to the clique $C$, is calculated as follows:

$$w_C(E_C) = \sum_{e \in E_C} w(e) \tag{8}$$

   - SquareRoot. This family of instances is similar to the Clique family but with the difference that it considers the square root of a sum of squares of the involved weights, as is shown in the Eq. 9:

$$w_{SR}(E_C) = \sqrt{\sum_{e \in E_C} w(e)^2} \tag{9}$$

All the algorithms and heuristics were implemented in C++ and its performance was evaluated on a platform with an Intel Core i5-3210M 2.5 GHz processor with 4 GB of RAM under Windows 8. The described testbeds were obtained from the corresponding web pages of the authors Karapetyan [18] and Magos [25].

In order to evaluate the performance of our MAP-Gurobi implementation we consider all the families of instances previously described. The results for the small and medium problem size families of instances are reported in Table 1.

In the left part of Table 1, we illustrate the averaged results for five instances of each type. We can observe that all the instances provided by Balas and Saltzman [3] were solved in about one second. We do not have the Branch and Bound algorithm proposed by Balas and Saltzman [3] so it is difficult to have and idea about if, by considering our computer power, such algorithm is competitive against our Gurobi-MAP implementation when solving 3AP. By the other side, it is not the main goal of this paper, we let this analysis for future work. Recall we are only interested in a competitive exact algorithm aimed to its use as part of our local search heuristics so the development of a faster algorithm to solve small instances of MAP will provide us with a better machinery to increase the competitiveness of our heuristics.

In the right part of Table 1, we illustrate the averaged results for ten instances of each type. We can observe that the instances provided by Magos and Mourtos [26] were solved in higher running times. According to the reported results by Magos and Mourtos [26] they solved the same instances by using a less powerful architecture and a different technique (see [26]) but, in a longer time, e.g. they solved the instance 5_$axial$_14 in approximately seven hundred thousand second whereas our Gurobi-MAP solved it in thirteen hundred seconds. Again, we want to emphasize that this comparison is not the spirit of this paper, it is more in the sense to show our MAP-Gurobi implementation as a good tool to be used as part of the proposed heuristics, however we think that a formal and fair comparison could be exposed in another work in order to determine which is the state-of-the-art algorithm to solve small instances of MAP.

In Table 2, we illustrate the averaged results for ten instances of each combination of $s$ and $n$. We can observe that all the instances with 3 dimensions were solved but, there is a big difference in terms of the solving time between each family of instances. Some of the instances in 4, 5, and 6 dimensions could not be solved. We conclude that the weights distribution have an impact over the computational complexity of our MAP-Gurobi, however, we did not measure such impact, we let such analysis for future work.

Finally, we would like to highlight that our Gurobi-MAP found optimal values for the families of instances that considers $s = 3$ dimensions and $n = 100$ vertices for which other heuristics could not find the optimal value. You can see the previously known values for such instances in the work of Karapetyan and Gutin [21].

From the proposed GLSH, we could test a lot of heuristics however we decided to focused our attention in a particular case of our GLSH, we called such heuristic DVH3, which is the case of a GDVH when we make a reduction of a $s$AP to a 3AP. We compare this heuristic against a state-of-the-art local search heuristic which is in fact the DVH2, i.e. the case of a GDVH when we have the reduction of a $s$AP to a 2AP. We do not show any of the k-Opt heuristics because for values with $2 \leq k \leq 3$ its results are not competitive against DVH2 and for values with $k \geq 4$ its performance is poor compared with its quality solution, which is faraway to DVH2.

The results in the Tables 3, 4 and 5 show the averaged optimal values for ten instances of each type of each family of instances after running each heuristic for 30 iterations and by starting from the same initial random generated feasible solution $A$, except for the case of the memetic heuristic. In the first couple of columns for DVH2 and DVH3 we show the average cost value for 30 iterations of each heuristic and in the next columns is reported the best value found after 30 iterations.

The results for the memetic heuristic are the same reported in Karapetyan and Gutin [21] after consider a running time of 300 s, which corresponds to the best solutions obtained for the corresponding instances. Such memetic algorithm is a state-of-the-art meta-heuristic designed as a general purpose heuristic. This memetic algorithm consists in a genetic algorithm that uses a local search heuristic instead of a mutation function in order to improve the individuals instead of just change them randomly. The memetic algorithm was developed to vary the population size along the remaining running time of the heuristic such that it starts with a wide diversity

**Table 3** Averaged relative solution error for 30 iterations of each heuristic over the family of instances Random

| Instance | Best known | Average solution | | Best found | | |
|---|---|---|---|---|---|---|
| | | DVH2 | DVH3 | DVH2 | DVH3 | Memetic |
| 4r20 | 20.0 | 192.0 | 2.5 | 73.0 | 0.0 | 0.0 |
| 4r30 | 30.0 | 117.3 | 0.0 | 47.3 | 0.0 | 0.0 |
| 4r40 | 40.0 | 74.8 | 0.0 | 27.0 | 0.0 | 0.0 |
| 5r15 | 15.0 | 232.0 | 14.7 | 62.0 | 0.0 | 0.0 |
| 5r18 | 18.0 | 191.1 | 3.9 | 45.6 | 0.0 | 0.0 |
| 5r25 | 25.0 | 124.4 | 0.0 | 32.8 | 0.0 | 0.0 |
| 6r12 | 12.0 | 275.8 | 29.2 | 43.3 | 0.0 | 0.0 |
| 6r15 | 15.0 | 214.0 | 11.3 | 33.3 | 0.0 | 0.0 |
| 6r18 | 18.0 | 175.0 | 2.8 | 28.3 | 0.0 | 0.0 |

**Table 4** Averaged relative solution error for 30 iterations of each heuristic over the family of instances Clique

| Instance | Best known | Average solution | | Best found | | |
|---|---|---|---|---|---|---|
| | | DVH2 | DVH3 | DVH2 | DVH3 | Memetic |
| 4cq20 | 1901.8 | 13.1 | 3.5 | 4.6 | 0.5 | 0.0 |
| 4cq30 | 2281.9 | 19.5 | 6.5 | 9.7 | 2.5 | 0.1 |
| 4cq40 | 2606.3 | 24.0 | 8.7 | 14.5 | 3.8 | 0.5 |
| 5cq15 | 3110.7 | 11.1 | 5.0 | 3.4 | 0.7 | 0.0 |
| 5cq18 | 3458.6 | 13.0 | 6.0 | 5.5 | 1.4 | 0.0 |
| 5cq25 | 4192.7 | 16.3 | 8.2 | 8.7 | 3.6 | 0.1 |
| 6cq12 | 4505.6 | 8.8 | 4.6 | 2.1 | 0.3 | 0.0 |
| 6cq15 | 5133.4 | 10.6 | 5.6 | 3.7 | 1.3 | 0.0 |
| 6cq18 | 5765.5 | 12.2 | 6.8 | 6.1 | 2.4 | 0.1 |

of individuals and, at the end of the heuristic, the combinations (crossings) and improvements (mutations) are focused on the best individuals among the previous iterations. We consider this memetic algorithm because it is the state-of-the-art of meta-heuristics to solve MAP instances.

In Table 3, we can observe that the results of DVH3 for the family of instances Random outperforms the results of DVH2 and are optimal as well as the results obtained by the memetic algorithm.

In Tables 4 and 5, we can observe that the results of DVH3 for the families of instances Clique and SquareRoot outperforms the results of DVH2 and are very close to the results obtained by the memetic algorithm.

In general, we can observe that, for the considered families of instances, the results of the memetic algorithm are always very near to the optimal value, however it is important to say that this memetic algorithm uses DVH2 as part of the local

**Table 5** Averaged relative solution error for 30 iterations of each heuristic over the family of instances SquareRoot

| Instance | Best known | Average solution | | Best found | | |
|---|---|---|---|---|---|---|
| | | DVH2 | DVH3 | DVH2 | DVH3 | Memetic |
| 4sq20 | 929.3 | 16.5 | 4.9 | 5.8 | 0.8 | 0.0 |
| 4sq30 | 1092.4 | 25.6 | 9.9 | 14.7 | 5.0 | 0.1 |
| 4sq40 | 1271.4 | 27.6 | 10.1 | 18.3 | 4.9 | 0.4 |
| 5sq15 | 1203.9 | 13.4 | 6.2 | 4.6 | 0.9 | 0.0 |
| 5sq18 | 1343.8 | 15.4 | 7.1 | 7.3 | 1.9 | 0.0 |
| 5sq25 | 1627.5 | 19.6 | 9.9 | 11.6 | 4.2 | 0.1 |
| 6sq12 | 1436.8 | 9.4 | 5.1 | 2.8 | 0.6 | 0.0 |
| 6sq15 | 1654.6 | 12.1 | 6.5 | 4.6 | 1.7 | 0.0 |
| 6sq18 | 1856.3 | 13.9 | 8.0 | 7.3 | 2.3 | 0.0 |

improvements that are performed for the individuals of the population. We believe that, by using our DVH3 as part of the memetic algorithm better results could even be obtained. We let this analysis for future work.

All the used implementations can be requested to any of the authors via email.

## 6 Conclusions

We presented and evaluated a state-of-the-art implementation, Gurobi-MAP, which solves large size families of instances of MAP. Gurobi-MAP was able to found the optimal solutions for families of instances that other authors solved trough heuristics without found an optimal solution in such cases.

We consider that there is still open a researching line aimed to the development of algorithms that optimally solve larger size instances of MAP, more than those solved in this work. We consider that the Branch and Bound algorithm for 3AP could be improved and generalized aimed to solve $s$AP. We could say that, it is not clear if our MAP-Gurobi outperforms all the previously developed algorithms that solve MAP, however, we highly believe that better algorithms could even be developed.

We presented a GLSH which could provide ideas aimed to the development of more complex heuristics for MAP. We show the results for its particular case DVH3, which improves the quality solution of the previously state-of-the-art local search heuristic DVH2.

Our heuristic is comparable in terms of quality solution to more complex meta-heuristics such as memetic algorithms. Recall that the compared memetic algorithm uses to DVH2 as part of its machinery, then a good assumption is that the use of the proposed heuristics, such as GDVH with $3 \leq t < s$, even could result in solutions of higher quality.

Further experimentation considering other derived heuristics from our GLSH even could result in better heuristics, we let this analysis for future work.

**Acknowledgements** We would like to thank to M. Saltzman who provided us with the instances used for him and E. Balas many years ago and, for show us his interest in the work that we are doing about this problem.
We also want to thank to D. Karapetyan and G. Gutin who provided us with the instances that they used at both of their papers of 2011 about the MAP and, for provided us with the source code that they used for the generation of such families of instances, which was also very useful to accomplish this work.
A special thanks to M. Vargas for provided us with a state-of-the-art implementation that solves 2AP through an auction algorithm.
Finally, we also thank to the Mexican institutions Conacyt and Comecyt who supported to the realization of this researching work.

## References

1. Aiex, R., Resende, M., Pardalos, P., Toraldo, G.: GRASP with path relinking for three-index assignment. J. Comput. **17**, 224–247 (2005). doi:10.1287/ijoc.1030.0059
2. Andrijich, S.M., Caccetta, L.: Solving the multisensor data association problem. Nonlinear Anal.: Theory Methods Appl. **47**, 5525–5536 (2001). doi:10.1016/S0362-546X(01)00656-3
3. Balas, E., Saltzman, M.: An algorithm for the three-index assignment problem. Oper. Res. **39**, 150–161 (1991)
4. Bekker, H., Braad, E., Goldengorin, B.: Using bipartite and multidimensional matching to select the roots of a system of polynomial equations. Comput. Sci. Appl. - ICCSA **2005**, 397–406 (2005). doi:10.1007/11424925_43
5. Bozdogan, A., Efe, M.: Ant colony optimization heuristic for the multidimensional assignment problem in target tracking. In: 2008 IEEE Radar Conference, pp. 1–6 (2008). doi:10.1109/RADAR.2008.4720822
6. Burkard, R., Dell'Amico, M., Martello, S.: Assignment Problems. Society for Industrial and Applied Mathematics, Philadelphia (2009)
7. Crama, Y., Spieksma, F.: Approximation algorithms for three-dimensional assignment problems with triangle inequalities. Eur. J. Oper. Res. **60**, 273–279 (1922). doi:10.1016/0377-2217(92)90078-N
8. Dasgupta, S., Papadimitriou, C.H., Vazirani, U.V.: Algorithms. Mc Graw Hill Higher Education, New York (2008)
9. Edmonds, J., Karp, R.: Theoretical improvements in algorithmic efficiency for network flow problems. J. ACM **19**, 248–264 (1972). doi:10.1145/321694.321699
10. Ferland, J., Roy, S.: Timetabling problem for university as assignment of activities to resources. J. Comput. Oper. Res. **12**, 207–218 (1985). doi:10.1016/0305-0548(85)90045-0
11. Frieze, A.: Complexity of a 3-dimensional assignment problem. Eur. J. Oper. Res. **13**, 161–164 (1983). doi:10.1016/0377-2217(83)90078-4
12. Garey, M., Johnson, D.: Computers and Intractability: A Guide to the Theory of NP-Completeness. W. H. Freeman and Company, New York (1979)
13. Grundel, D., Pardalos, P.: Test problem generator for the multidimensional assignment problem. Comput. Optim. Appl. **30**, 133–146 (2005). doi:10.1007/s10589-005-4558-6
14. Gurobi Optimization, Inc.: Gurobi optimizer reference manual (2015). http://www.gurobi.com

15. Gutin, G., Goldengorin, B., Huang, J.: Worst case analysis of max-regret, greedy and other heuristics for multidimensional assignment and traveling salesman problems. Approximation and Online Algorithms. Lecture Notes in Computer Science, vol. 4368, pp. 214–225 (2007). doi:10.1007/11970125_17
16. Huang, G., Lim, A.: A hybrid genetic algorithm for three-index assignment problem. Eur. J. Oper. Res. **172**, 249–257 (2006)
17. Jiang, H., Xuan, J., Zhang, X.: An approximate muscle guided global optimization algorithm for the three-index assignment problem. Evol. Comput. CEC **2008**, 2404–2410 (2008)
18. Karapetyan, D.: Source codes and extra tables. http://www.cs.nott.ac.uk/~pszdk/?page=publications
19. Karapetyan, D., Gutin, G.: Local search heuristics for the multidimensional assignment problem. J. Heuristics **17**, 201–249 (2011). doi:10.1007/s10732-010-9133-3
20. Karapetyan, D., Gutin, G.: A new approach to population sizing for memetic algorithms: a case study for the multidimensional assignment problem. Evol. Comput. **19**, 345–371 (2011)
21. Karapetyan, D., Gutin, G., Goldengorin, B.: Empirical evaluation of construction heuristics for the multidimensional assignment problem. London Algorithmics 2008: Theory and Practice, pp. 107–122. College Publications, London (2009)
22. Karp, R.: Complexity of Computer Computations: Reducibility Among Combinatorial Problems, pp. 85–103. Springer, Berlin (1972). doi:10.1007/978-1-4684-2001-2_9
23. Koopmans, T., Beckmann, M.: Assignment problems and the location of economic activities. Cowles Found. Res. Econ. **25**, 56–76 (1955)
24. Kuhn, H.: The Hungarian method for the assignment problem. Naval Res. Logist. Q. **2**, 83–97 (1955)
25. Magos, D.: Problem instances for multi-index assignment problems. http://users.teiath.gr/dmagos/MIAinstances
26. Magos, D., Mourtos, I.: Clique facets of the axial and planar assignment polytopes. Discret. Optim. **6**, 394–413 (2009). doi:10.1016/j.disopt.2009.05.001
27. Nguyen, D., Le Thi, H., Pham Dinh, T.: Solving the multidimensional assignment problem by a cross-entropy method. J. Comb. Optim. **27**, 808–823 (2014). doi:10.1007/s10878-012-9554-z
28. Pierskalla, W.: Letter to the editor - the multidimensional assignment problem. Oper. Res. **16**, 422–431 (1968). doi:10.1287/opre.16.2.422
29. Robertson, A.: A set of greedy randomized adaptive local search procedure (GRASP) implementations for the multidimensional assignment problem. Comput. Optim. Appl. **19**, 145–164 (2001). doi:10.1023/A:1011285402433
30. Schell, E.: Distribution of a product by several properties. In: Directorate of Management Analysis, Second Symposium in Linear Programming, vol. 2, pp. 615–642 (1955)
31. Spieksma, F., Woeginger, G.: Geometric three-dimensional assignment problems. Eur. J. Oper. Res. **91**, 611–618 (1996). doi:10.1016/0377-2217(95)00003-8

# Part III
# Electronics and Embedded Systems

# A Multi-objective and Multidisciplinary Optimisation Algorithm for Microelectromechanical Systems

**Michael Farnsworth, Ashutosh Tiwari, Meiling Zhu and Elhadj Benkhelifa**

**Abstract** Microelectromechanical systems (MEMS) are a highly multidisciplinary field and this has large implications on their applications and design. Designers are often faced with the task of balancing the modelling, simulation and optimisation that each discipline brings in order to bring about a complete whole system. In order to aid designers, strategies for navigating this multidisciplinary environment are essential, particularly when it comes to automating design synthesis and optimisation. This paper outlines a new multi-objective and multidisciplinary strategy for the application of engineering design problems. It employs a population-based evolutionary approach that looks to overcome the limitations of past work by using a non-hierarchical architecture that allows for interaction across all disciplines during optimisation. Two case studies are presented, the first focusing on a common speed reducer design problem found throughout the literature used to validate the methodology and a more complex example of design optimisation, that of a MEMS bandpass filter. Results show good agreement in terms of performance with past multi-objective multidisciplinary design optimisation methods with respect to the first speed reducer case study, and improved performance for the design of the MEMS bandpass filter case study.

**Keywords** Microelectromechanical systems · MEMS and multidisciplinary · Multi-objective optimisation · Evolutionary computation

M. Farnsworth (✉) · A. Tiwari
Manufacturing Informatics Centre, Cranfield University, Cranfield, UK
e-mail: m.j.farnsworth@cranfield.ac.uk

A. Tiwari
e-mail: a.tiwari@cranfield.ac.uk

M. Zhu
College of Engineering, Mathematics and Physical Sciences, University of Exeter, Exeter, UK
e-mail: m.zhu@exeter.ac.uk

E. Benkhelifa
School of Computing and Digital Tech, Staffordshire University, Staffordshire, UK
e-mail: e.benkhelifa@staffs.ac.uk

© Springer International Publishing AG 2018

Y. Maldonado et al. (eds.), *NEO 2016*, Studies in Computational Intelligence 731,
https://doi.org/10.1007/978-3-319-64063-1_9

# 1 Introduction

The growth of the application of microelectromechanical systems (MEMS) into an increasing number of disciplines means there is a need to balance the objectives, constraints and functionality of the whole system across these disciplines during the design stage. The design of complex systems found in large engineering environments such as aerospace are often decomposed into a number of disciplines or components and are tackled by specific design teams or departments within an organization [39]. However, this concurrent design approach can lead to sub-optimal trade-offs, as compromises will have to be made when each discipline or component is integrated together to form a whole system. A number of automated methods have been developed and applied over the decades to overcome some of the problems associated with this class of multidisciplinary design optimisation (MDO) problem. The field of MEMS design optimisation mirrors this particular class of design problem due to the nature of the large numbers of interacting components and disciplines in which they act through. Therefore, a design strategy that can aid designers overcome some of the problems associated with this particular problem would be of great benefit. This paper looks to address this by developing and validating a multi-objective and multidisciplinary design optimisation algorithm that allows designers to decompose their multidisciplinary systems and optimise them individually before recombining to the whole system. This is validated using a common speed reducer problem and also a more complex MEMS design problem.

The rest of this paper is organised as follows. Section 2 provides an overview of multidisciplinary optimisation strategies in the literature and the different architectures that are employed. Section 3 presents a novel MDO problem formulation that can handle multi-objective design problems. Section 4 details the case studies and experimental setup used to validate this new MDO approach with results presented in Sect. 5 and finally conclusions follow in Sect. 6.

# 2 Multidisciplinary Optimisation Strategies

## 2.1 Multidisciplinary Optimisation

Complex large scale systems found in many engineering problems today can consist of many components and disciplines coordinating together to form some function or behaviour. In a real design engineering problem, each discipline typically represents a design team concerned with the design of one aspect or component of this complete system. This makes perfect sense as it allows many more people to work upon a particular problem while also allowing specialised designers to focus upon their respective disciplines [2]. There are however drawbacks, with the possibility of each discipline having to interact with others the chances of infeasible/non-viable designs occurring due to conflicts with other engineering teams and their separate disciplines is possible [39]. This is often solved with a post-optimisation trade-off where in order to solve such inconsistencies and obtain a feasible design; changes need to be made which often lead to a sub-optimal solution [39]. Therefore, there is a need to both optimise the individual disciplines and their constituent parts or components all the while maintaining some level of global design optimisation for the system as a whole. MDO is one such class of algorithm which looks to coordinate these individual disciplines and components towards a system design that is optimal as a whole and satisfies all constraints, while maintaining some level of design autonomy [38]. This often involves the decomposition of the original design problem into a set of hierarchical coupled elements often based upon the analysis techniques which are used to analyse the physical or behavioural characteristics of the system, or the possible different physical scales, components within the system. As such the total structural performance of the whole system can be a combination of responses that are evaluated from each level within the hierarchy [11].

Once a hierarchy of decomposed elements is present their coordination and level of autonomy need to be assigned within the optimisation routine. The lowest level of control may be called analysis autonomy where the role of each disciplinary group is limited to the selection and analysis of models [2]. The simplest examples are the single-level methods such as multi-disciplinary feasible (MDF), individual disciplinary feasible (IDF) or an all-at-once approach (AAO) [8] which generally focus upon a centralised decision making process at one level, where analysis can also be undertaken at each discipline or element as shown in Fig. 1a. It is possible to improve these single level methods by utilizing multiple computers or grid systems for distributed analysis, and database management to give improved efficiency and maintainability. However, the reliance on a single optimiser to act as a central decision maker and control all aspects of design for what is often a large scale and complex design problem is still a drawback [25]. The natural progression and next level of autonomy is the inclusion of both analysis models and optimisation algorithms in a distributed multi-level optimisation structure in what can be coined optimisation autonomy. Here each level can contain its own set of analysis and optimisation routines and maintains some element of control over them as shown in Fig. 1b. The coordination

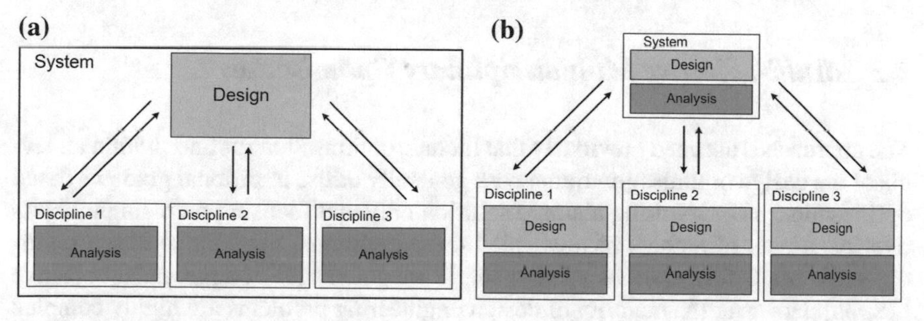


**Fig. 1** Disciplinary autonomy with **a** single-level analysis autonomy and **b** multi-level design autonomy

**Table 1** Chronological review of developments in multidisciplinary optimisation

| Contributors | Date | Description |
|---|---|---|
| [35] | 1985 | OLD, optimisation by linear decomposition |
| [32] | 1988 | CSSO, concurrent subspace optimisation, |
| [8] | 1994 | IDF, individual discipline feasible method, each discipline is solved independently outside of system level |
| [8] | 1994 | MDF, multidisciplinary feasible method, each discipline is directly coupled in some way through input and output analysis, system level controls global/local design variables |
| [3] | 1996 | SAND, simultaneous analysis and design |
| [5] | 1996 | CO, collaborative optimisation, |
| [34] | 1998 | BLISS, bi-level integrated system synthesis, |
| [23] | 2003 | ATC, analytical targeting cascade, |
| [19] | 2005 | QSD, quasi-separable subsystem decomposition, |
| [10] | 2006 | IPD, inexact penalty decomposition, |
| [37] | 2008 | ALC, augmented Lagrangian coordination. |

of a decomposed problem solution such that the overall global solution is found is a challenging task [31] however over the last thirty years a large body of work has been conducted towards this goal [24, 33]. In the literature there are 6 main approaches to MDO which stand out from the rest; these are Optimisation by Linear Decomposition (OLD) [35], Collaborative Optimisation (CO) [5], Concurrent SubSpace Optimisation (CSSO) [32], Bi-Level Integrated System Synthesis (BLISS) [34], Analytical Target Cascading (ATC) [23] and the method of Quasi-separable Subsystem Decomposition (QSD) [19]. Each method differs in the way it coordinates the solution of a decoupled multilevel optimisation problem. Newer methods include the inexact penalty decomposition method [10] and augmented Lagrangian coordination [37]. An overview of the current main approaches to MDO can be found in [11, 12]. A list of the major MDO methods within the literature is shown in Table 1.

## 2.2 Multi-objective Multidisciplinary Optimisation

The approaches outlined previously that include optimisation routines within all levels of the design optimisation framework generally utilise traditional gradient-based optimisation methods using a single solution only and focusing on a single objective. A number of authors have adapted these traditional methods to create multi-objective MDO formulations using a single weighted sum or aggregated objective [28, 36]. However the majority of design engineering problems are highly complex with non-linear responses, discontinuous and multi-modal search spaces and contain both discrete and continuous decision variables. All these factor in a number of pathologies to the search efficiency of the more traditional optimisers, while single solution strategies only provide a single Pareto solution from each run for a designer to choose from. Therefore, looking to incorporate more robust population-based algorithms such as those found within the field of evolutionary computation that focus upon multi-objective design problems could be beneficial. An early example created by [26] featured an immune network system multi-objective genetic algorithm (MOGA) approach (MOGA-INS) for MDO designed to solve hierarchically decomposed multi-objective problems. Each decomposed unit or subsystem contained a MOGA which focused on a specific set of design variables held within the subsystem population representation. Limitations with this approach involved the need for each subsystem to contain the same objectives as all others and being limited to a hierarchical structure. In order to overcome limitations from this previous work [26], the authors in [17] created a multi-objective multidisciplinary optimisation algorithm for hierarchically decomposed problems which allowed for differing objectives within each subsystem. This particular approach used quality metrics as a basis for objective function measurement for individual solutions at the system level. Other multi-objective population based algorithms have been implemented within MDO over the years with varying degrees of implementation and success [1, 15, 21, 30, 42]. A list of multi-objective MDO algorithms found within the literature is shown in Table 2.

## 2.3 Decomposition Methods

One important part of the MDO process is in how the designers go about decomposing the original problem into a set of sub-problems. Decomposition can be seen as identifying weak links between elements that are coupled, and therefore allowing the elements to represent individual though coupled optimisation problems [11]. In general decomposition methodologies can be done in several ways such as object, aspect, sequential and model-based [40]. Model decomposition is a partitioning method based upon functional dependencies between design variables and functions included in the problem [6]. The main approaches to decomposition are the aspect-based and object-based methods, and are discussed further below along with examples of MDO application to design optimisation based upon these decompositions in Tables 3 and 4. Aspect-based decomposition focuses on breaking up the particular problem based upon the actual discipline analysis associated with it. This can be aerodynamics, structural, thermal in the case of aircraft design, or electrical, mechanical, fluidic and structural in the case of a MEMS device. In large-scale design environments the system as a whole can be structured according to the individual components of the system such as turbine engines or wing structures. These often correspond to engineering departments within a company and an object-based decomposition approach mimics this. Decomposing the problem into individual components brings with it a natural mirror to real-world design optimisation along with a simplification and grouping of design variables associated with these components.

**Table 2** Chronological review of developments in multi-objective multidisciplinary optimisation

| Contributors | Date | Applications | Description |
|---|---|---|---|
| [36] | 1997 | Numerous | Multi-objective collaborative MDO, system level contains a weighted sum of subsystem level objectives, subsystems aim to minimise interdisciplinary inconsistencies |
| [26] | 2000 | Speed reducer | MOGA-INS, immune network simulation method integrated with MOGA to give hierarchically decomposed MDO |
| [17] | 2003 | Speed reducer, UAV payload | Hierarchically structured MOGA MDO, requires separable or additively separable objectives |
| [18] | 2004 | Speed reducer | Hierarchical structured MOGA MDO, system level optimiser focuses upon shared design variables/objective while subsystem focus on local variables and objectives |
| [15] | 2004 | Roll stabiliser fin | MORDACE, a MOGA MDO that incorporates robust design with each discipline design solutions able to handle variation from shared data during a compromise at end of routine |
| [28] | 2005 | Race car design | Integrated linear physical programming with collaborative MDO |
| [1] | 2006 | Speed reducer, numerical test problem | Multi-objective collaborative MDO, system level optimiser focuses upon shared design variables/objective while subsystem focus on local variables and objectives |
| [30] | 2007 | Speed reducer, dock design problem | COSMOS, collaborative optimisation strategy for multi-objective systems, optimiser focuses upon shared design variables/objective while subsystem focus on local variables and objectives |
| [21] | 2009 | Container ship | Mixed weighted and multi-objective collaborative MDO utilizing multi-island genetic algorithms on all levels of design |
| [42] | 2010 | Race car design | Particle swarm multi-objective collaborative MDO, a fuzzy decision maker is used to select best design along Pareto front |

**Table 3** Chronological review of developments in aspect-based decomposition

| Contributors | Date | Applications | Description |
|---|---|---|---|
| [25] | 2000 | Supersonic aircraft design | Decomposition of supersonic aircraft into three major disciplines (aerodynamics, structures and mission analysis) |
| [15] | 2004 | Roll stabiliser fin | Sequential optimisation with hydrodynamic optimisation solutions fed into structural subsystem optimiser compared against MORDACE which provided superior performance |
| [39] | 2005 | Structural wing | Decomposition into a system level performance objective and subsystem aerodynamics/structural disciplines |
| [28] | 2005 | Race car design | Consisted of two system level objectives, minimise lap time and maximise normalised weight, with subsystem decomposition into aerodynamic and force disciplines |
| [21] | 2009 | Container ship | Decomposition into static, mode and dynamic disciplinary analysis |
| [42] | 2010 | Race car design | Similar decomposition to [28] with aerodynamic and force disciplinary analysis |

**Table 4** Chronological review of developments in object-based decomposition

| Contributors | Date | Applications | Description |
|---|---|---|---|
| [2] | 2000 | Structural bridge | Decomposition of the main components of bridge structure, the superstructure and deck in a conceptual MDO approach |
| [26] | 2000 | Speed reducer | The design problem objectives and variables are decomposed up into separate subsystems and solved independently before recombining |
| [17] | 2003 | Speed reducer, UAV payload | Payload design with the goal to maximise probability of success, UAV design variables decomposed between subsystem levels |
| [1] | 2006 | Speed reducer, numerical test problem | Decomposition of design problem objectives and variables, similar to [17] |
| [30] | 2007 | Speed reducer, dock design problem | Decomposition of dock structure into separate subsystems containing individual cantilevered beams attached to vertical wall |

## 2.4 MDO Architectures

Microelectromechanical systems often contain a large number of coupled devices or components that provide some form of desired behaviour or function through their collective actions. The system as a whole or the individual components that make it up also often covers a number of disciplinary domains, be they mechanical [14], electrical [13], or more recently fluidic [22] and biological [20]. The increased complexities from designing such multidisciplinary systems can make it harder for designers to build such devices as they often require explicit knowledge in more than one discipline. The application of automated design synthesis and optimisation techniques towards multidisciplinary design problems such as those found in MEMS could greatly speed up the design process and ease the burden of design placed upon the designer. The relationships between the disciplines or components within a design problem often form the basis for the structure the multidisciplinary optimisation routine will take when looking to apply a MDO algorithm. The current state of the art in multi-objective population based MDO employs a multi-level hierarchical structure with an upper and lower level relationship that can be structured to contain the decomposed design problem into a set of discipline or component subsystems.

These two approaches mimic the aspect and object decomposition methodologies described previously and both can be equally applied to the MDO of MEMS. Figure 2 provides an example of how a real world MEMS device, the ADXL150 accelerometer, can be broken up using an aspect based (a) and an object based (b) methodology. Here the aspect based decomposition contains lower level subsystems which undertake specific disciplinary analysis required for design optimisation with design variable, objective and constraints often linked to the individual discipline. The object based decomposition concerns its self with the major constituents of the device or system, with design variables heavily linked to these constituent parts and objectives and constraints often tailored so as to optimise these individual components in such a way as to benefit the global design goals situated at the system level.

The integrated and coupled nature of MEMS and the devices and components within them can mean that it is not always possible to fully decompose a design problem and that there still requires some level of communication between each of the lower level subsystems. In the ADXL150 accelerometer example outlined below, it is conceivable that analysis and design variable information altered within one subsystem is needed by another. The calculation of the electrostatic force is required for the calculation of the mechanics of the device in particular the displacement and stiffness of the suspended springs [38]. The current state of the art in multi-objective population based MDO employs a hierarchical structure with each individual lower level subsystem isolated from all others in a fully decomposed design problem. Such hierarchical structures often require the design problem itself to be hierarchically decomposable with its objectives separable or additively separable which may not always be possible [17]. A non-hierarchical structure however allows communication between the individual subsystems therefore allowing solutions within each subsystem to be provided with the correct disciplinary analyses or subsystem design

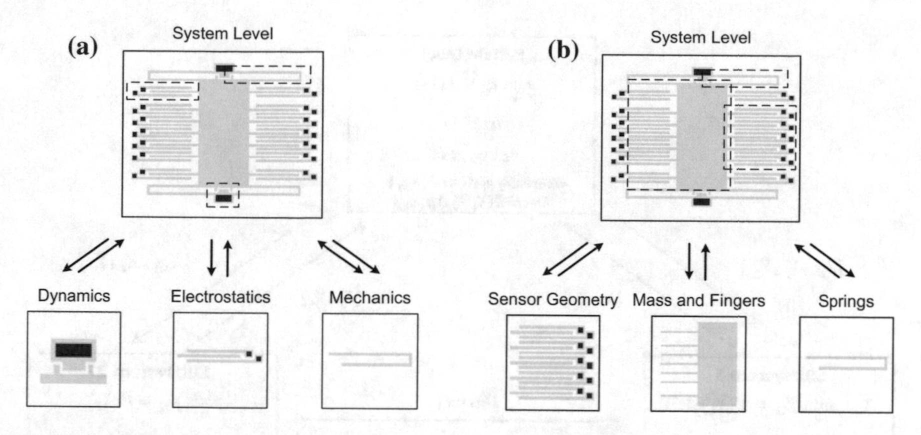


**Fig. 2** Decomposition of ADXL150 accelerometer for MDO using an aspect based (**a**) or object based (**b**) methodology [38]

variables. A number of ways have been presented on how to transfer coupled variables in order to reconcile each of the subsystems into the formation of a complete solution. The cooperative co-evolutionary algorithm set out in [29] looks to choose the current best solution from each sub species and recombine them with the chosen solution in the current subsystem to be evaluated. In [30] a different approach looks to pass approximations of coupled variables from the system level to each subsystem. The difference between the real, but inaccessible, value and the approximate values decreases during the optimisation process. Updated coupling values from each subsystem are sent at every system level invocation and then passed on to all other subsystem levels later on, however they soon become approximations again as each subsystems optimisation routine evolves. The next section outlines the multi-objective and multidisciplinary optimisation algorithm designed to handle non-hierarchical communication between subsystems and used throughout this paper.

# 3 Methodology

## *3.1 MDO Problem Formulation*

In applying the MDO algorithm we first begin with the decomposition of the design problem into a number of subsystems each with their own decision variables, local objectives and constraints. The decision on how this decomposition is undertaken is up to the user and within the MDO literature there are a number of methodologies two of which, aspect and object have been discussed. There are also similar methods for identifying the important functions, analysis and objectives in a design problem for

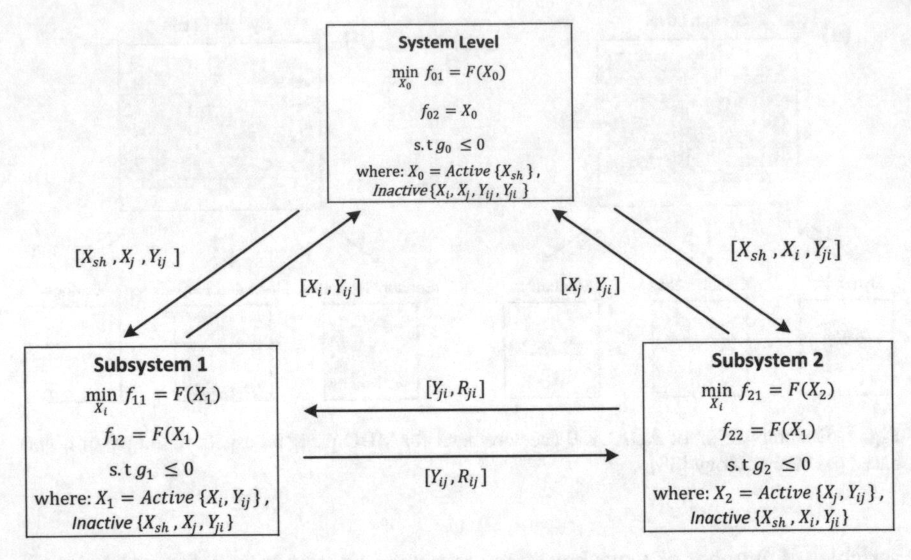


**Fig. 3** Multidisciplinary optimisation non-hierarchical structures for decomposed problem

example axiomatic design [7] which can also lend their support. The decomposition of a multi-objective problem into a number of subsystems is shown in Fig. 3. Here the default design problem is held and optimised within the system level with the original objectives $f_{01}/f_{02}$ and constraints $g_0$ active and a chosen set of decision variables $X_{sh}$ open for variation. The decision on what variables are included within the $X_{sh}$ set are up to the designer however they are often decision variables that are common to more than one subsystem [18] and often hard to separate so are shared throughout all subsystems. All other decision variables are closed to the system level and remain fixed.

The subsystems are constructed as a non-hierarchical design with communication both from the system to subsystem or parent to child level and from subsystem to subsystem occurring. Each of the subsystems contains its own local objectives $f_{11}/f_{12}$ and these can be unique, additively separable from one of the system level objectives or one of the system level objectives in its own right as shown in Fig. 3. In a similar vein the constraints $g_1/g_2$ held within each subsystem can also be unique or taken from the system level design problem. The active decision variables within each subsystem consist of local disciplinary design variables $X_i/X_j$ and the coupled disciplinary design variables $Y_i/Y_j$. Where the local disciplinary design variables are fixed to each subsystem the coupled design variables are not and as a result they are transferred from their local subsystem to all other subsystems within the structure every cycle. Finally, not all problems can be fully decomposable in respect of their disciplinary analysis and as a result more than one system may rely on information garnered from another. Therefore, when applicable coupled analysis response variables can also be passed between the child subsystems, with the origin of the subsystem analysis passing on these variables to any other subsystem that requires them. A default

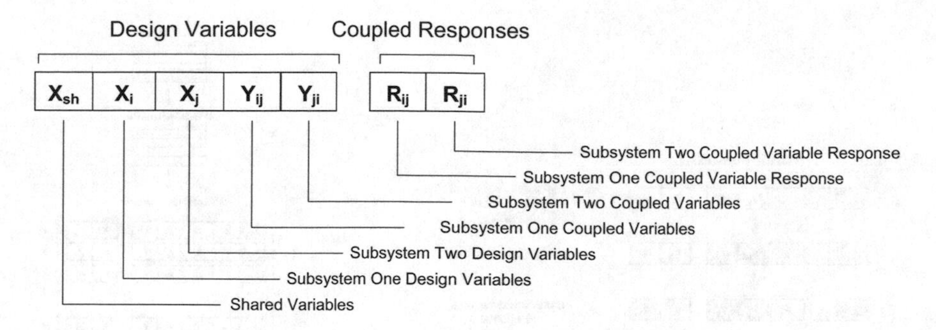


**Fig. 4** Multidisciplinary optimisation design and response variable sets for a chromosomal representation

chromosomal representation of the various design and response variable sets for a single solution is shown in Fig. 4.

The overall process of the multidisciplinary optimisation algorithm can be broken down into a number of key steps, in this instance linked to a multi-objective population based optimiser and they are described below.

**First** The first step begins with the initialisation of the system level population; in particular the various variable sets as shown in Fig. 5. Any coupled variable response values are set to null to be filled later after functional evaluation. The filled system level population set popCurrent is than ready for variation. Functional evaluation of each individual is based upon system level objectives $f_{01}$, $f_{02}$.

**Second** A selection set is chosen from the current system level population ready for variation and the creation of an offspring population set. Only the system levels shared variables $X_{sh}$ are varied based upon the chosen optimisers operators. At the system level this offspring set is then used depending upon the chosen the optimisers replacement operators as the basis for the next popCurrent set. However as shown in Fig. 5 the newly created system level offspring population set is also passed on to the each of the subsystems within the multidisciplinary optimisation structure.

**Third** The next step shown in Fig. 5 moves on to the subsystem level of the design process, upon receiving the offspring sets the individual solutions are used to fill the local subsystem populations. Subsystem populations with a lower number of solutions than the supplied offspring set are filled using a truncation operator. For each subsystem the local population sets now need to be evaluated using local objectives $f_{11}$, $f_{12}/f_{21}$, $f_{22}$ and constraints before then undertaking a standard routine of selection, variation and replacement. Each of the subsystems variation operators are restrained to only alter their local disciplinary design variables $X_i/X_j$ and the local coupled disciplinary design variables $Y_i/Y_j$. After variation has occurred, any coupled variable within each subsystem offspring solution is passed on to all other subsystems, as a result all subsystem offspring set sizes are fixed to the same size. Finally functional and constraint evaluation of each

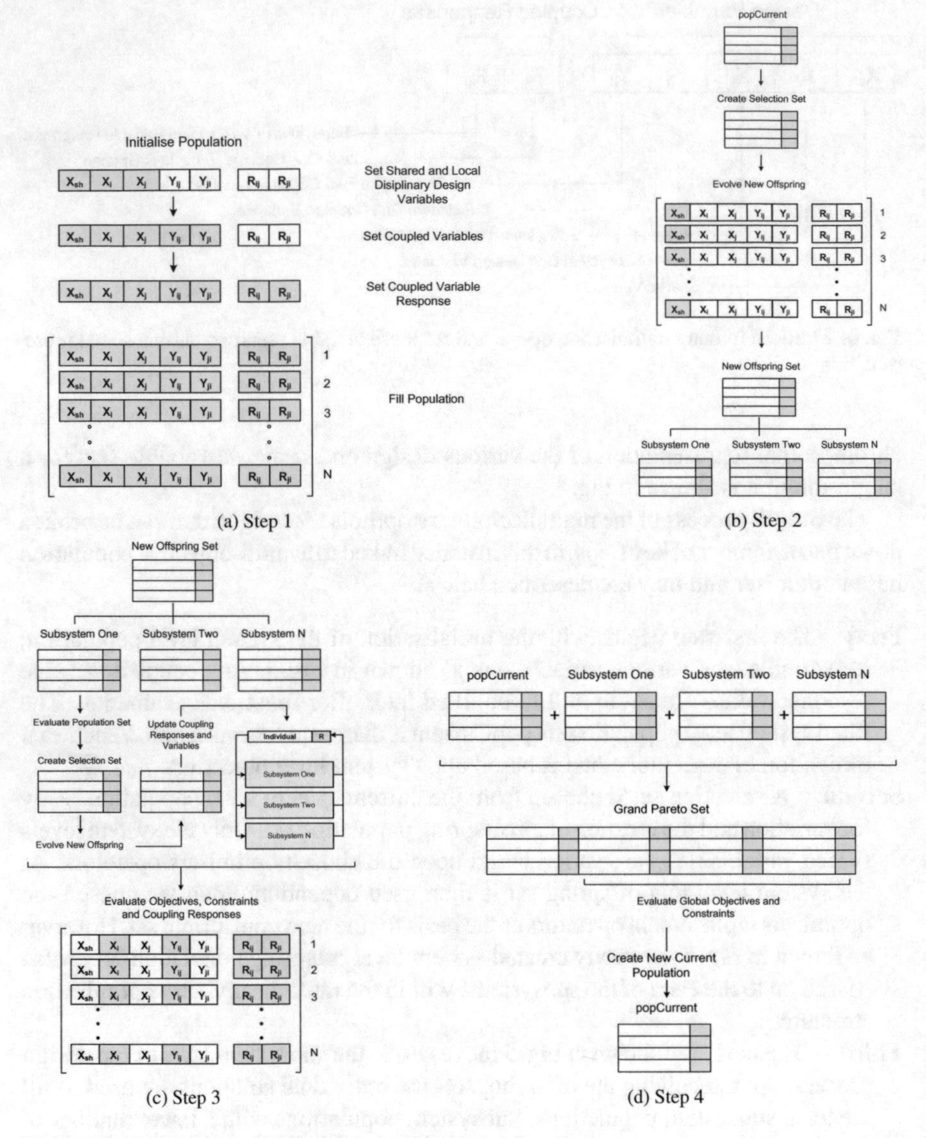


**Fig. 5** Multi-objective multidisciplinary optimisation process

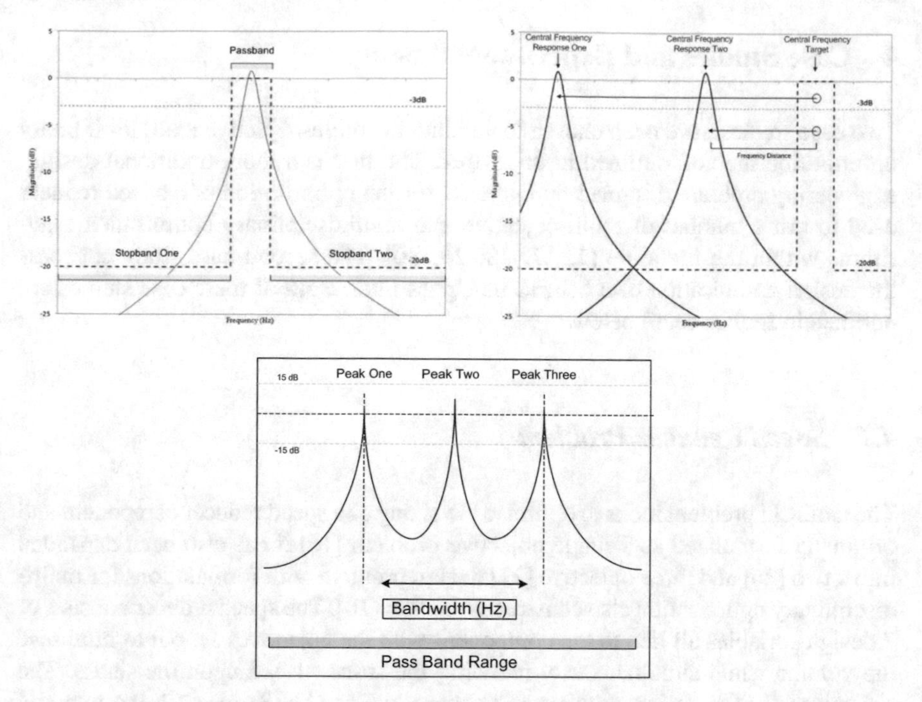


**Fig. 6** MEMS bandpass filter synthesis breakdown for filter objective

subsystem offspring population set is undertaken and where necessary coupled disciplinary analysis variable values are also transferred to any subsystem solutions that may require them for functional or constraint evaluation. The local subsystem offspring sets are then combined with their local population sets before replacement operators update each subsystem with a new population set. This iterative process then continues for a fixed number of cycles before ending and moving on to the next step.

**Fourth** The final step in Fig. 5 looks to take the evolved subsystem population sets and combine them into a Total Population set for evaluation of objectives and constraints at the system level $f_{01}$, $f_{02}$. The size of the total population set is fixed to the total sum of all subsystem population sets. This total population set is then combined with the system level popCurrent set to form a unified Grand Pareto set which is then used to create a new popCurrent set using the optimisers replacement operators. Upon completion of this step the process begins again at step two and the whole process is repeated until a chosen criterion is used to determine whether it should be stopped.

## 4 Case Studies and Experimental Setup

Two case studies have been chosen to validate the multi-objective multidisciplinary optimisation method outlined in this paper. The first is a more traditional design-engineering problem designed by Golinski for the optimisation of a speed reducer used to test a number of multi-objective and multidisciplinary optimisation algorithms within the literature [1, 17, 18, 26, 30]. The second case study concerns the design optimisation of a MEMS bandpass filter. Both of these case studies are outlined in further detail below.

### *4.1 Speed Reducer Problem*

The Golinski problem looks to optimise the sizing of a speed reducer component and originally formulated as a single objective problem [16] it has also been expanded into a two [26] and three objective [17] design problem with formulations for multidisciplinary optimisation also constructed within [18]. The speed reducer consists of 7 design variables all tied to the component with the objectives set out to minimise the volume while simultaneously reducing the stress placed upon the shafts. The objectives for the design problem are shown in Eqs. (1)–(3) for both the two and three objective problem and 4 and 5 for the decomposed objectives. In the case of the two objective design problem only objectives $f_1$ and $f_2$ are used, while for the multidisciplinary optimisation approach two subsystems are used, the first focusing on objectives $f_{1,1}$ and $f_2$ and the other $f_{1,2}$ and $f_2$ for the two objective design problem and $f_{1,2}$ and $f_3$ for subsystem two in the three objective problem. Also associated with the speed reducer problem are 11 inequality constraints outlined in Eqs. (6)–(16). Finally Table 5 holds the decision variables for the speed reducer problem along with their type and upper/lower bounds.

**Table 5** Speed reducer variable information

| Variable tag | Sub tree type | Lower bound | Upper bound |
|---|---|---|---|
| Variable 1 | Real-valued | 2.6 | 3.6 |
| Variable 2 | Real-valued | 0.7 | 0.8 |
| Variable 3 | Integer | 17 | 28 |
| Variable 4 | Real-valued | 7.3 | 8.3 |
| Variable 5 | Real-valued | 7.3 | 8.3 |
| Variable 6 | Real-valued | 2.9 | 3.9 |
| Variable 7 | Real-valued | 5.0 | 5.5 |

$$f_1 = 0.7854x_1x_2^2\left(\frac{10x_3^2}{3}14.933x_3 - 43.0934\right) - 1.508x_1(x_6^2 + x_7^2) + 7.477(x_6^3 + x_7^3) + 0.7854(x_4x_6^2 + x_5x_7^2) \tag{1}$$

$$f_2 = \frac{\sqrt{\left(\frac{745x_4}{x_2x_3}\right)^2 + 1.69 \times 10^7}}{0.1x_6^3} \tag{2}$$

$$f_3 = \frac{\sqrt{\left(\frac{745x_5}{x_2x_3}\right)^2 + 1.575 \times 10^7}}{0.1x_7^3} \tag{3}$$

$$f_{1,1} = 0.7854x_1x_2^2\left(\frac{10x_3^2}{3}14.933x_3 - 43.0934\right) - 1.508x_1x_6^2 + 7.477x_6^3 + 0.7854x_4x_6^2 \tag{4}$$

$$f_{1,2} = -1.508x_1x_7^2 + 7.477x_7^3 + 0.7854x_5x_7^2 \tag{5}$$

$$g_1 \equiv \frac{1}{x_1x_2^2x_3} - \frac{1}{27} \leq 0 \tag{6}$$

$$g_2 \equiv \frac{1}{x_1x_2^2x_3^2} - \frac{1}{397.5} \leq 0 \tag{7}$$

$$g_3 \equiv \frac{x_4^3}{x_2x_3x_6^4} - \frac{1}{1.93} \leq 0 \tag{8}$$

$$g_4 \equiv \frac{x_5^3}{x_2x_3x_7^4} - \frac{1}{1.93} \leq 0 \tag{9}$$

$$g_5 \equiv x_2x_3 - 40 \leq 0 \tag{10}$$

$$g_6 \equiv \frac{x_1}{x_2} - 12 \leq 0 \tag{11}$$

$$g_7 \equiv 5 - \frac{x_1}{x_2} \leq 0 \tag{12}$$

$$g_8 \equiv 1.9 - x_4 + 1.5x_6 \leq 0 \tag{13}$$

$$g_9 \equiv 1.9 - x_5 + 1.1x_7 \leq 0 \tag{14}$$

$$g_10 \equiv f_2 - 1300 \leq 0 \tag{15}$$

$$g_10 \equiv f_3 - 1100 \leq 0 \tag{16}$$

### 4.2 MEMS Bandpass Filter

Large engineering design problems for example those found within the aeroplane industry can be difficult or impossible to undertake as a whole due to the large number of design variables, constraints and disciplinary analyses of the problem. In reality the design problem is often decomposed and each individual component solved or optimised separately by a design team, often focusing on specific variables, constraints and objectives. MEMS are inherently multidisciplinary through the interaction of the mechanical and electronic components of the device. The application of MEMS into fields such as biology or chemistry through lab on-chip devices increases the number of disciplines a designer or design team must understand and integrate into the design process. A MEMS bandpass filter forms the basis of the second case study. It consists of an array of coupled folded flexure resonator tanks that collectively function as the filter itself, examples can be found in [27, 41].

The design of a bandpass filter in this instance consists of a single discipline in the form of electrical circuit simulation. Modelled as an electrical equivalent circuit, it contains equivalent elements for the mechanical resonator tanks and coupling springs that make up the bandpass device. Each of these components plays an important role in how the frequency transmission of the bandpass filter is shaped. A detailed breakdown of the modelling, simulation and optimisation approach can be found in [13]. The aim of this design problem is to create a solution whose bandpass characteristics match the targets outlined by the designer.

A number of design objectives have been created to solve this particular problem and are outlined in Fig. 6. A frequency transmission from a single micromechanical resonator consists of a number of frequency data points plotted against the magnitude in units of dB as seen top left of Fig. 6. The quality and performance of the filter transmission can be measured by simply calculating where each data point lies within the pass band and stop band ranges outlined and measured against their target magnitude, in this case 0 dB for points within the passband and −20 dB within the stop band regions. The overall frequency performance can then be quantified as a sum of the total deviation from each of these ranges for the data points within the frequency transmission. Ideally all data points that lie within the pass band will have 0 insertion loss and no gain giving a magnitude of 0 dB, while all points within the stop band will be −20 dB or less and therefore have a deviation of 0 for both regions.

**Table 6** Bandpass filter problem information

| Variable tag | Sub tree type | Lower bound | Upper bound |
|---|---|---|---|
| Voltage | Real-valued | 1 | 200 |
| Tank number | Integer | 1 | 9 |
| Finger number | Integer | 1 | 200 |
| Thickness (m) | Real-valued | 2 | 30 |
| Capacitance (F) | Real-valued | 3E-15 | 8E-15 |
| Inductance (H) | Real-valued | 40000 | 80000 |
| Coupling spring capacitance (F) | Real-valued | 3E-15 | 8E-15 |
| Tank | Branch | N/A | N/A |
| Objectives | | Constraints | |
| Bandpass filter response error | Minimise | N/A | |
| Bandpass central frequency error | Minimise | | |

**Table 7** Bandpass filter parameter ranges

| | Bandpass filter characteristics |
|---|---|
| Passband | 9.5–10.5 kHz |
| Stopband 1 | 1 Hz–9.5 kHz |
| Stopband 2 | 10.5–15 kHz |
| Central frequency | 10 kHz |

Central frequency of the bandpass filter is important when wanting to design a frequency transmission for a targeted portion of the spectrum. The central frequency of a transmission is simply calculated as the distance of the peak frequency data point to the desired central frequency outlined by the designer. The objective shown on the top right in Fig. 6 is both a targeted design goal and a guide to the optimiser, allowing individual or coupled resonator transmission responses to move closer to the targeted region of interest. The design variables for this problem are shown in Table 6 and represent the values attributed to each resonator tank of the electrical circuit equivalent model. The representation is a varied length chromosome dependent on the number of tanks present. The bandpass filter characteristics are shown in Table 7 for this particular design target.

The object based decomposition of the bandpass filter begins with classifying the customer requirements at the highest level, in this instance the characteristics of a bandpass filter with low insertion loss and high bandwidth over a target frequency range. The global objectives to try and meet these targets have already been outlined previously in the filter response and central frequency objectives and in order to undertake a MDO approach new objectives are required as the system is decomposed. The decomposition of these objectives and the bandpass filter device into separate

**Table 8** Bandpass multidisciplinary optimisation objectives

| | System level | | Subsystem 1 | | Subsystem 2 | |
|---|---|---|---|---|---|---|
| | Objective 1 | Objective 2 | Objective 1 | Objective 2 | Objective 1 | Objective 2 |
| Objective type | Minimise | Minimise | Minimise | Minimise | Minimise | Maximise |
| Objective description | Filter response | Central frequency | Pass band error | Central frequency | Stop band error | Bandwidth |
| Constraint type | N/A | | Inequality | | N/A | |
| Constraint description | N/A | | Stop band error $\leq$ 1000 | | N/A | |

system and subsystem elements can then occur to try and aid the overall design process. The literature [4, 41] points to the effect each individual resonator has on central frequency of the bandpass filter, both individual and coupled resonators and their mass, stiffness and damping values reflect their central frequency peak within the whole transmission shape. The pass band ripple and insertion loss are also heavily influenced by the constituent resonators that make up the bandpass filter [41]. Subsystem one is tasked with solving this particular functional requirement, with specific objectives and constraints as shown in Table 8. The design of a filter which has a flat pass band characteristic within the target frequency range means the addition of a pass band error objective. This new objective is calculated exactly as in the filter response however, only the pass band is taken into consideration.

The bandwidth of the filter device should be sufficient enough to cover the target pass band range, however the frequency transmission than needs to possess a sufficient roll off either side of the pass band into the stop band region to be effective. Both the bandwidth and stop band functional requirements are heavily influenced by the number of resonator tanks within the filter and the coupling spring stiffness that couple them [27, 41]. Subsystem two contains objectives designed to focus on these particular functional requirements, the construction of a bandwidth objective, the goal of which is to maximum the bandwidth of the first and last peaks of the filter transmission as shown at the bottom of Fig. 6 is included. The bandwidth is calculated as the distance in Hz between the first and last peaks of the bandpass filter divided by the average gain of the two peaks. Each peak is calculated simply as a point where either side shows a decline in the magnitude dB, and it must lie within the pass band range and have a magnitude between 15 and −15 dB. In the case where only one peak is present, then the bandwidth is set to a value of 1. The final objective for subsystem two is for stop band error and like the pass band error objective is calculated from the filter response.

In addition to the objectives a new constraint is added to the overall design process. Subsystem one contains a constraint to the total stop band error of the frequency transmission; this is to stop certain frequency transmissions from dominating at a detriment to the overall design optimisation, these transmissions characteristically have a frequency response of 0 dB from start to finish of the bandpass target range giving them 0 pass band error, but large stop band error.

## 4.3 Experimental Setup

The improved design synthesis and optimization of MEMS devices is the targeted outcome of this approach through the application of automated optimization heuristics in conjunction with available MEMS modeling and simulation tools.

From the field of evolutionary computation two of the current state-of-the-art multi-objective algorithms have been chosen to undertake design synthesis, firstly NSGAII [9] and finally SPEA2 [43]. Both algorithms have been explored in terms of performance and applied successfully over a number of areas and problems outside and within MEMS design synthesis. These multi-objective algorithms are used as the base optimisation to compare the outlined MDO method for both case studies outlined. The default parameters for both algorithms across both case studies are shown in Table 9.

The speed reducer case study uses a mixed integer and real-valued tree-based chromosome structure, while the bandpass filter case study uses a varied length tree-based structure to account for the decrease or increase in resonator tank numbers. In order to account for this varied length approach new crossover and mutation operators have been introduced or adapted from those present within both NSGAII and SPEA2. The details of these can be found in [13] but they essentially allow for the insertion and removal of resonator tanks and their associated variables within the chromosome during variation. The standard single level representation of the bandpass filter design problem is shown in Fig. 7. This also includes an overview of related structural tags and node markers. Structural tags relate to specific branch nodes within the representation and the nodes that control the count or number present. Node markers are used to provide additional information for a particular

**Table 9** Default algorithm parameters NSGAII and SPEA2

| Algorithm parameter | Default value |
|---|---|
| Population size | 100 |
| Offspring size | 100 |
| Selection size | 100 |
| Replacement size | 200 |
| SBX distribution index | 20 |
| Polynomial mutation distribution Index | 20 |
| Probability of SBX crossover | 0.8 |
| Probability of mutation | 0.142857 |
| Generations | 100 |
| Tests | 5 |

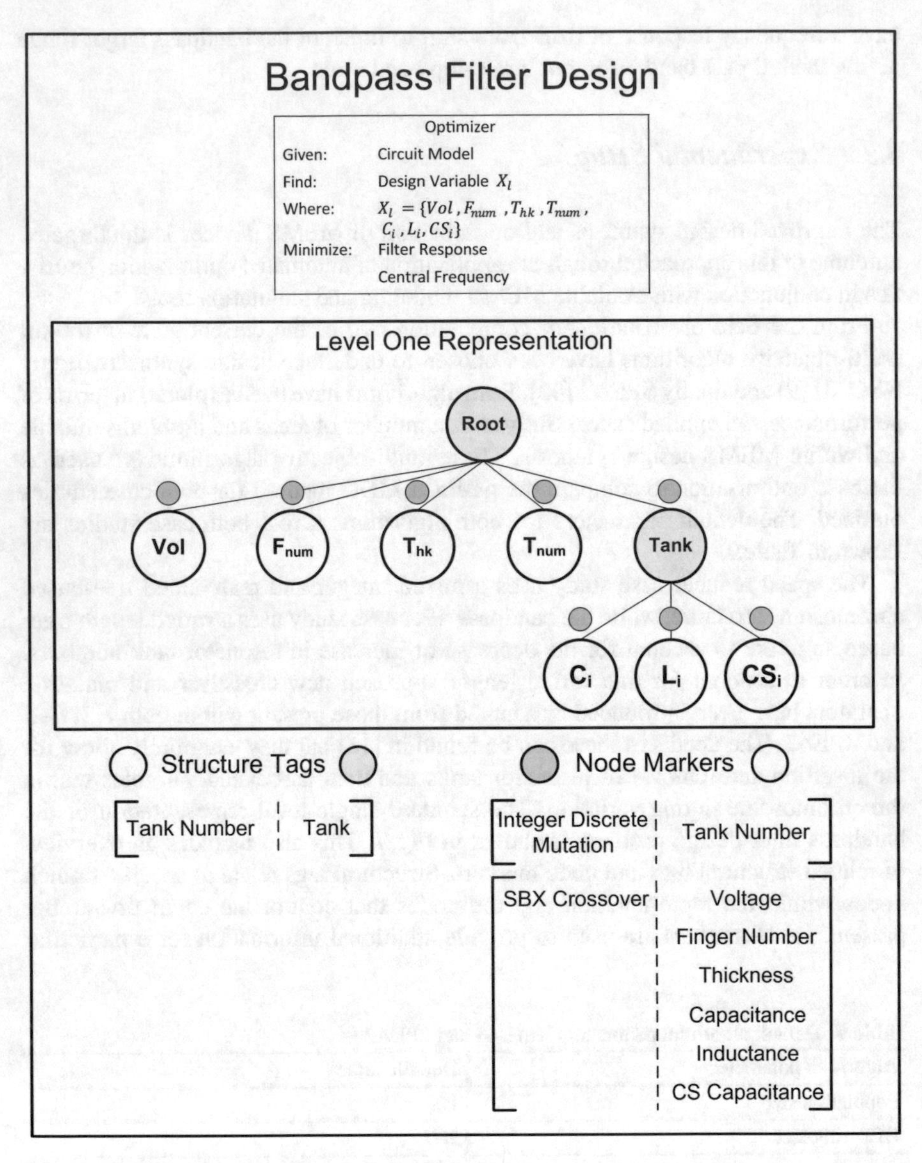


**Fig. 7** Bandpass filter design template, with overview of problem, default representation, associated structure tags, node markers and global variables

operator such as SBX crossover within NSGAII, marking certain nodes that can have crossover performed or not.

As discussed in order to undertake the MDO approach each case study has been decomposed into a number of subsystems. The speed reducer follows past examples [17, 18] and the representation used throughout this experiment is shown in Fig. 8 for both the two and three objective problems. The MEMS bandpass filter case study is detailed in Fig. 9 and contains the system, subsystem and coupled variables

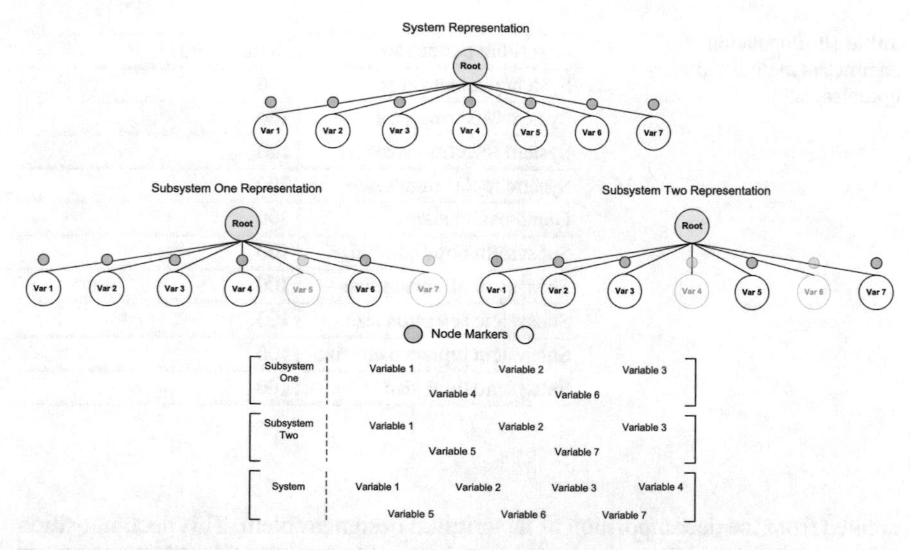


**Fig. 8** Speed reducer multidisciplinary optimisation representation

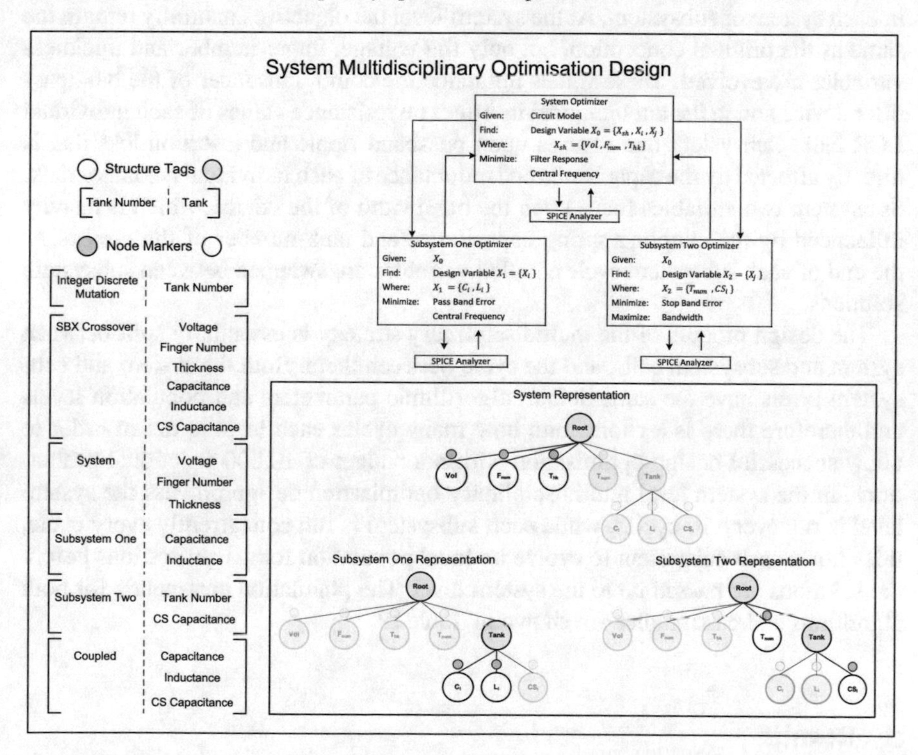


**Fig. 9** Bandpass filter multidisciplinary optimisation design template, with overview of problem, default representation, associated structure tags, node markers and global variables

**Table 10** Population parameters multidisciplinary optimisation

| Algorithm parameter | Default value |
|---|---|
| System population size | 100 |
| System offspring size | 100 |
| System selection size | 100 |
| System replacement size | 200 |
| Grand pareto size | 300 |
| Subsystem population size | 100 |
| Subsystem offspring size | 100 |
| Subsystem selection size | 100 |
| Subsystem replacement size | 100 |
| Subsystem total size | 200 |

created from the decomposition of the original design problem. This decomposition is based upon the relation certain variables have with the particular objectives set out in each system or subsystem. At the system level the objectives naturally remain the same as the original conception, but only the voltage, finger number and thickness variables are evolved. These values influence the comb transducer of the bandpass filter device and in the circuit model its effect on resistance values of each individual LCR tank. Subsystem one focuses upon passband ripple and insertion loss that is directly affected by the capacitance and inductance of each individual resonator tank. Subsystem two variables focus upon the bandwidth of the device, which is heavily influenced by the coupling spring capacitance and tank number of the device. At the end of each subsystem cycle coupled variables are swapped between subsystem solutions.

The design process of the multidisciplinary strategy is essentially split between system and subsystem calls, and the cycle between them. Both the system and subsystem levels have the same default algorithmic parameters and population levels and therefore there is a choice into how many cycles each level is run in order to allow successful design optimisation within a budget of 10,000 functional evaluations. In the system level multidisciplinary optimisation design process the system level is run every 10 cycles, while each subsystem is run concurrently every cycle, this allows each subsystem to evolve its local population for 10 generations before the solutions are passed up to the system level. The population parameters for both algorithms and case studies are shown in Table 10.

# 5 Results

The results for both case studies are presented below and consist of a number of final population sets, hypervolume values, solution characteristics and additional analyses. In order to assess the performance of each of the algorithms on the two case studies the hypervolume metric is used to both evaluate the Pareto spread and dominance of the objective space each of the tests final populations has produced. The mean and bound hypervolume results for each algorithm are shown for each example, with the chosen nadir point for the objectives indicated below the specific table.

## 5.1 *Speed Reducer Problem*

Looking at the results in the two and three objective speed reducer problems there is similar performance in terms of population spread across both algorithms and strategies as shown in Fig. 10. The hypervolume results shown in Table 11 indicate some difference in performance across the five runs with the single level NSGAII outperforming all others on the two objective problem and the MDO algorithm outperforming the others on the three objective problem. If we are to compare these results with past examples found in [26] for the two objective and [17] for the three objective problem we see similar performance, with less functional evaluations required and more solutions generated. As a validation of the outlined MDO algorithm the speed reducer case study has proven a successful demonstration of its capabilities.

## 5.2 *MEMS Bandpass Filter*

The results presented are the individual final population sets for each of the five tests performed by each algorithm of the MDO strategy, shown in Fig. 11. Also shown are the best frequency transmissions found by the MDO strategy for each algorithm, ranked by the filter frequency objective in Fig. 12. The objective values of the best result ranked by the filter frequency objective for every run are held in Table 12. Finally the hypervolume values for both algorithms are shown in Table 13, with the best results shaded.

Undertaken by both NSGAII and SPEA2 the characteristic performance of both shows the single level SPEA2 outperforming the single level NSGAII in this example. Though the SPEA2 MDO implementation performed similar to the single level strategy its results seem to indicate that this is possibly not due to convergence to a suboptimal front as was prevalent in the single level strategy. Instead the Pareto population sets for the SPEA2 MDO results are more fragmented, with both poor and good fitness response fronts in the objective space. The structured nature of the MDO strategy, with separate subsystems and objectives can result in an inefficient use of resources or in this case functional evaluation cost as populations migrating from the system to subsystem levels need to be re-evaluated. This cost equates to about 1600 functional evaluations of lost search, a significant amount, which could have lead to the algorithm refining those transmissions from a 1400 filter response error to a lower one. NSGAII MDO on the other hand shows a robust performance

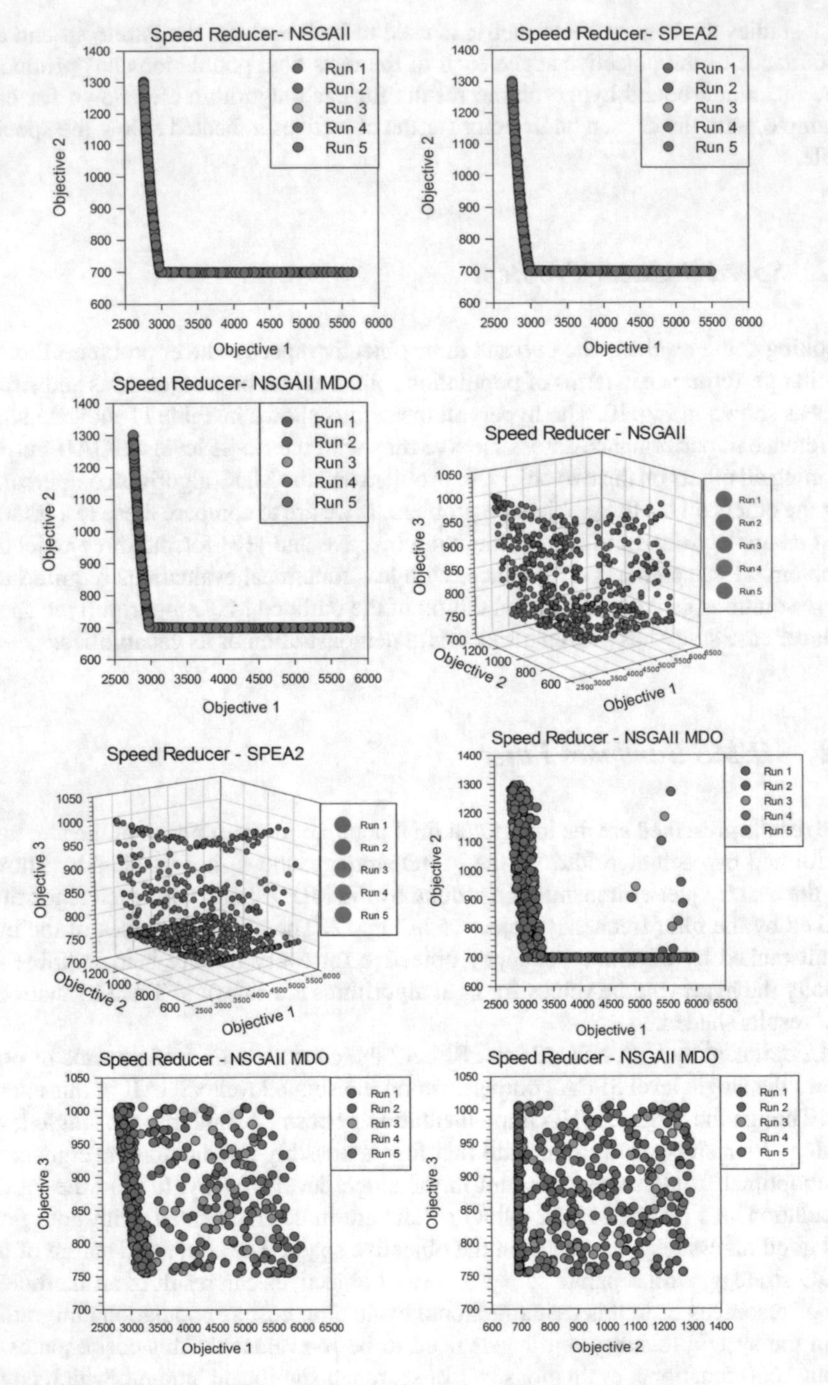


**Fig. 10** Final population sets two and three objective speed reducer problem

**Table 11** Speed reducer two and three objective design problem hypervolume metric values

| Speed reducer 2 objectives | | | |
|---|---|---|---|
| | NSGAII | SPEA2 | MDO NSGAII |
| $S^U$ | 1927425.898 | 1926371.720 | 1926094.169 |
| $S^M$ | 1927096.298 | 1925150.499 | 1925560.005 |
| $S^L$ | 1926533.553 | 1921631.936 | 1924012.468 |
| **Speed reducer 3 objectives** | | | |
| | NSGAII | SPEA2 | MDO NSGAII |
| $S^U$ | 669348872.86 | 674722574.432 | 672953351.4 |
| $S^M$ | 667099790.88 | 667656079.534 | 670425237.7 |
| $S^L$ | 664741675.34 | 660545560.310 | 666497637.0 |

$*(S^U S^M S^L)^1$ [5800, 1350] $*(S^U S^M S^L)^2$ [6000, 1350, 1100]

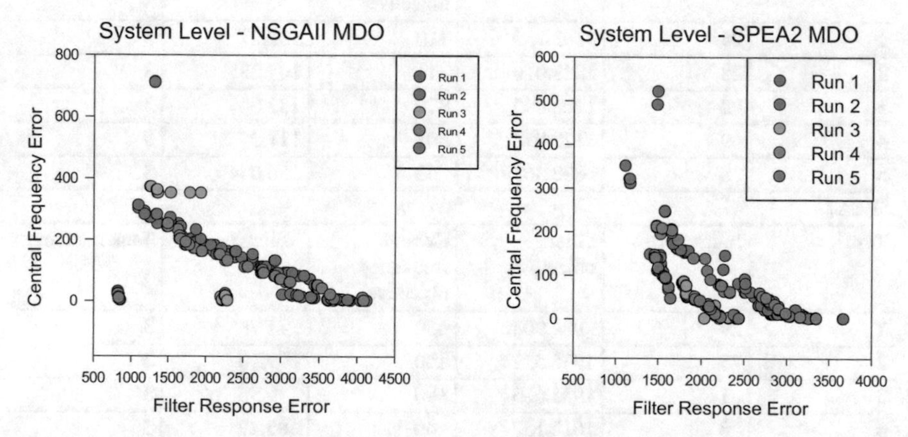


**Fig. 11** MEMS bandpass filter run 1–5 final population sets for (*left*) NSGAII multidisciplinary optimisation and (*right*) SPEA2 multidisciplinary optimisation

over the 5 runs, significantly performing better than its single level counterpart as shown in Table 12. An interesting characteristic of this approach is that a number of population fronts are continuous over a large range of the frequency fitness error objective. Whether the internal workings of the NSGAII algorithm and its crowding operator has had a positive effect on the solution spread which is filtered through to the MDO strategy in some way is a possibility though not certain and for brevity is left unexplored. The overall performance of both algorithms when compared against the single level strategies is shown in Fig. 13, with NSGAII outperforming the single level significantly and SPEA2 showing similar performance.

Two examples of a generational system population set for SPEA2 MDO are shown in Fig. 14 showing each population set progressively moving towards the optimal in separated fronts indicating how each cycle contributes to Pareto objective fitness. A major component of the MDO strategy is its structured system and subsystem hierarchy and the exchange of genetic material between each subsystem every gen-

**Table 12** Bandpass filter results

NSGAII

| Test | Index | Filter objective | Central frequency objective | Voltage | Tank number |
|---|---|---|---|---|---|
| 1 | 0 | 1750.493 | 101 | 61.34 | 3 |
| 2 | 0 | 1680.625 | 235 | 105.75 | 3 |
| 3 | 31 | 1248.642 | 32 | 190.34 | 3 |
| 4 | 0 | 3315.054 | 910 | 30.64 | 2 |
| 5 | 0 | 2148.439 | 206 | 144.01 | 3 |

Multidisciplinary optimization NSGAII

| Test | Index | Filter objective | Central frequency objective | Voltage | Tank number |
|---|---|---|---|---|---|
| 1 | 25 | 1103.233 | 310 | 183.27 | 3 |
| 2 | 28 | 1323.016 | 711 | 105.25 | 3 |
| 3 | 2 | 1268.895 | 370 | 133.04 | 3 |
| 4 | 9 | 1128.922 | 308 | 111.34 | 3 |
| 5 | 8 | 832.363 | 23 | 56.044 | 3 |

SPEA2

| Test | Index | Filter objective | Central frequency objective | Voltage | Tank number |
|---|---|---|---|---|---|
| 1 | 29 | 984.904 | 430 | 11.785 | 3 |
| 2 | 15 | 1936.521 | 160 | 199.10 | 3 |
| 3 | 1 | 1925.665 | 180 | 139.56 | 3 |
| 4 | 8 | 1012.157 | 40 | 85.27 | 3 |
| 5 | 80 | 1643.993 | 175 | 48.73 | 3 |

Multidisciplinary optimization SPEA2

| Test | Index | Filter objective | Central frequency objective | Voltage | Tank number |
|---|---|---|---|---|---|
| 1 | 4 | 1117.323 | 350 | 147.73 | 3 |
| 2 | 15 | 1493.894 | 490 | 177.23 | 3 |
| 3 | 4 | 1789.087 | 82 | 134.85 | 3 |
| 4 | 60 | 1489.723 | 212 | 13.81 | 3 |
| 5 | 15 | 1412.127 | 147 | 70.87 | 3 |

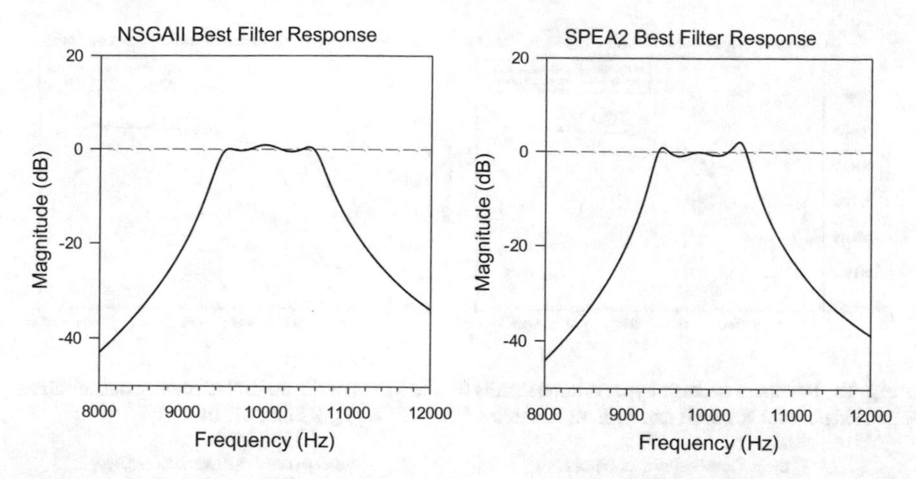


**Fig. 12** System level best filter response ranked by filter frequency objective for (*left*) NSGAII multidisciplinary optimisation and (*right*) SPEA2 multidisciplinary optimisation

**Table 13** Hypervolume results for NSGAII and SPEA2

| NSGAII | | |
|---|---|---|
| Hypervolume | Single level | Multidisciplinary optimisation |
| $S^U$ | 43755994.223 | 45837863.064 |
| $S^M$ | 39634066.167 | 44060899.260 |
| $S^L$ | 32507262.763 | 43035839.206 |
| SPEA2 | | |
| Hypervolume | Single level | Multidisciplinary optimisation |
| $S^U$ | 44945558.089 | 44084774.987 |
| $S^M$ | 42433896.104 | 42559736.279 |
| $S^L$ | 40276362.685 | 41041427.413 |

*($S^U$ $S^M$ $S^L$) [10000, 5000]

eration. The partitioning of the design process into system and subsystem cycle events breaks up the design process as population sets are transferred from system to subsystem and vice versa. Figure 15 highlights the generational change to each subsystems population set through the individual hypervolume values of each run. The effect on subsystem one is negligible with steady improvement throughout the design process and typical of the system MDO level hypervolume results. However subsystem two shows a marked decrease in hypervolume or performance of the Pareto front after every system level update, where a new offspring population set is passed to each subsystem. This is in part due to the possible loss of good solutions through the passing of the offspring set, or the disparity of the genotype/phenotype of these solutions from the system level objective space, which uses the standard central frequency and filter response objectives, to subsystem two objectives of stop

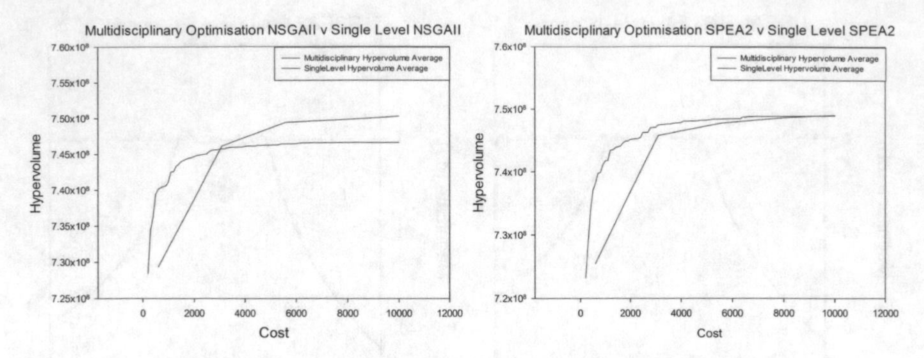


**Fig. 13** Bandpass average hypervolume results for the 5 runs of the multidisciplinary optimisation and single level NSGAII and SPEA2 strategies * ($S^U$ $S^M$ $S^L$) [182000, 4150]

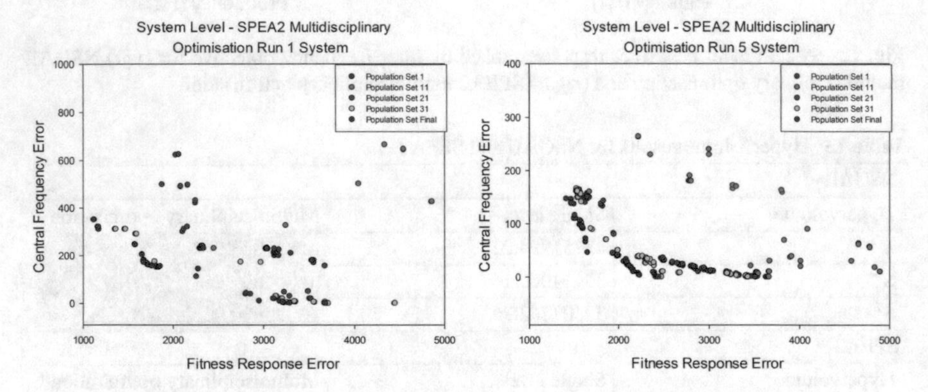


**Fig. 14** Generational system population plots for SPEA2 multidisciplinary optimisation runs 1 (*left*) and 5 (*right*) - each plot contains 5 equally distant generational plots

band and bandwidth. There is a slight exaggeration of the hypervolume dip because solutions with a stop band of 0 are not found at the system level, these are evolved afterwards. The loss in the hypervolume performance of the population set is often linked to the bandwidth objective of subsystem two, where solutions at the system level often have smoother pass bands with peaks within the pass band range giving smaller bandwidths then those evolved locally before. The subsystem then has to search and evolve past solutions with an equivocal bandwidth.

Exploring the effect of each subsystem and the decision variables under their control and how the genes and their alleles evolve over the design process are presented next. Shown in Figs. 16 and 17 are a series of generational filter frequency transmissions for the best solution found in the system, subsystem one and two population sets for NSGAII MDO run 1 and SPEA2 run 1 respectively. Each solution chosen was ranked by the filter frequency error, passband error and bandwidth objective respectively and the objective values and tank values for each filter are shown below the response.

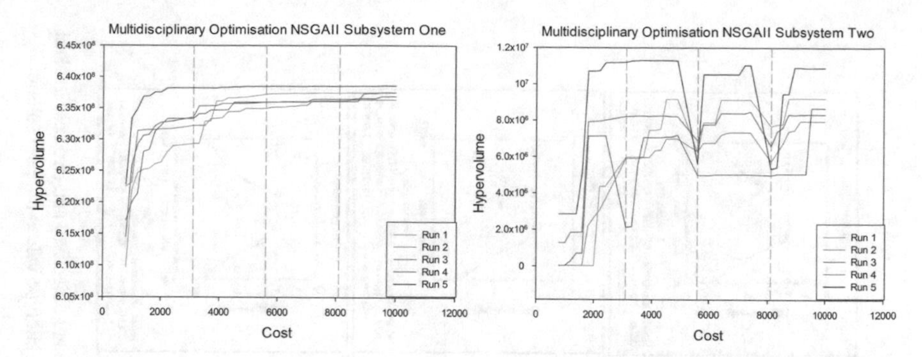


**Fig. 15** Hypervolume results for the 5 runs of the NSGAII multidisciplinary optimisation strategy for subsystem one (*Left*) * ($S^U$ $S^M$ $S^L$) [4000, 160000] and subsystem two (*right*) * ($S^U$ $S^M$ $S^L$) [12000, 2.0]

The frequency transmissions for both examples NSGAII run 1 and SPEA2 run 1 show an incremental improvement over the three separate partitions, cycle 1, 11 and 21, and are followed with phenotypes that show minimal change over the following examples in cycles 31 and 41. The only deviation from this is in Fig. 17 and subsystem two results that switch to a phenotype that is locally better than that of the best global system level solution. Looking at the specific subsystems, each one begins with the evolution of tailored frequency transmission characteristics, with subsystem one containing solutions focusing on the pass band region predominately with little regard for the stop band if only to remain unconstrained. The overall characteristic of the subsystem one solution in Figs. 16 and 17 takes on the shape of a typical bandpass filter, though with an unrefined pass band, at cycle 11 until convergence to the final phenotype at cycle 21. Subsystem two focuses upon both bandwidth and the stopband region of the frequency transmission. The bandwidth of the frequency transmission between two or more peaks is established around cycle 11 in both NSGAII and SPEA2 examples and this is evolved to give wider bandwidths further on in the design process. The next question is to what effect each subsystem and the solutions they evolve have on the global system level where the designer wishes to evolve the solutions they want to match the target filter characteristics. Interestingly in both examples the phenotypes of one subsystem match more closely the phenotype of the system level solution, here NSGAII subsystem two and SPEA2 subsystem one show closer affinity. The genotypic values in Figs. 16 and 17 of each of these solutions also show a close correlation, with individual capacitance and inductance values for each resonator tank closely resembling their system level counterparts, in particular SPEA2 subsystem one being identical. The frequency transmission of the system level solution begins to match more closely with the best subsystem one solution from cycle 21 onwards, probably in part due to the pass band objective playing a more dominant part in the system level frequency error objective at this stage of the design process. However subsystem two frequency transmission results on a number of examples do not mirror the example found or retained at the system level. The

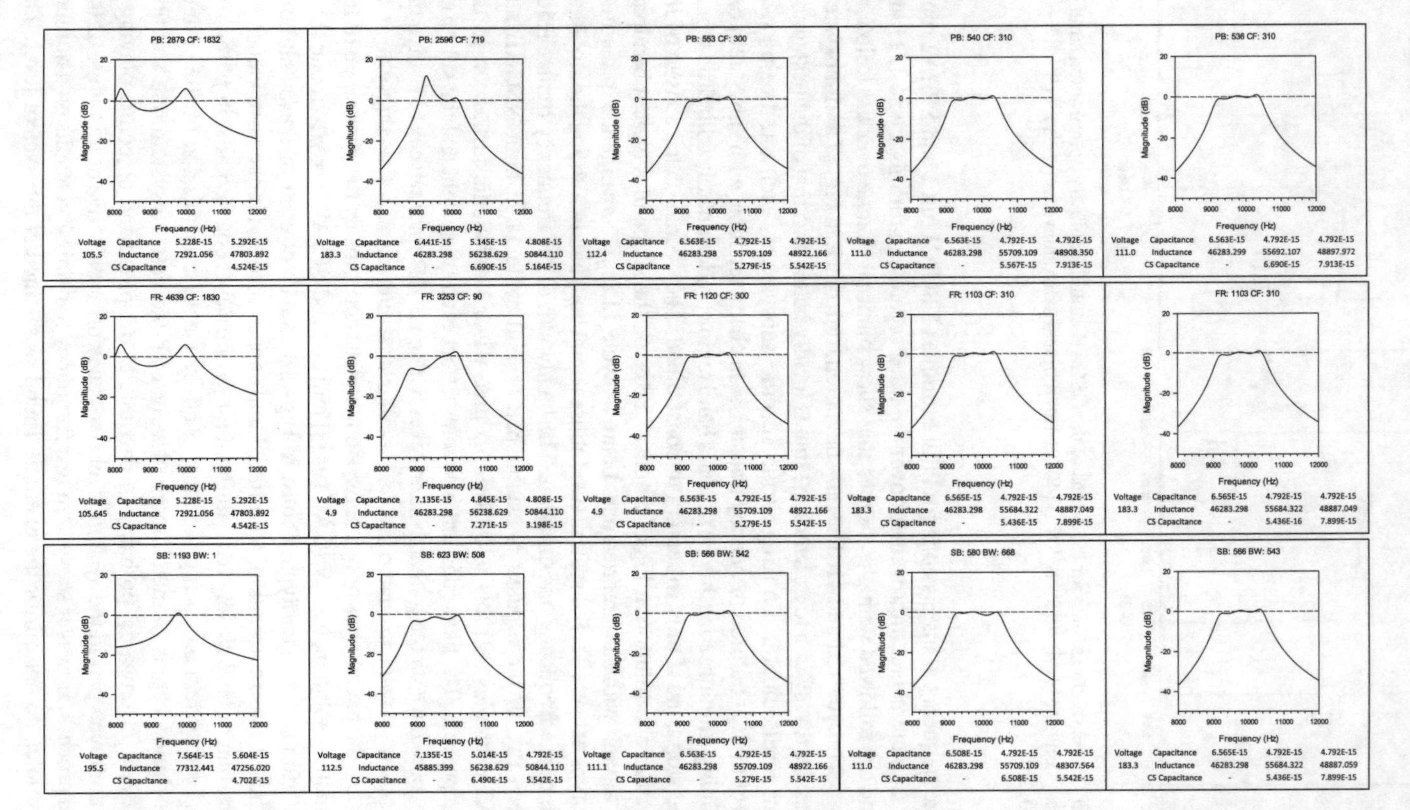


**Fig. 16** Best filter transmission plots for population (ranked by filter response objective), subsystem one (ranked by passband objective) and two best (ranked by bandwidth objective) over 5 generations (1, 11, 21, 31, 41) for NSGAII run 1. Each plot includes objective values and genotype values for each solution

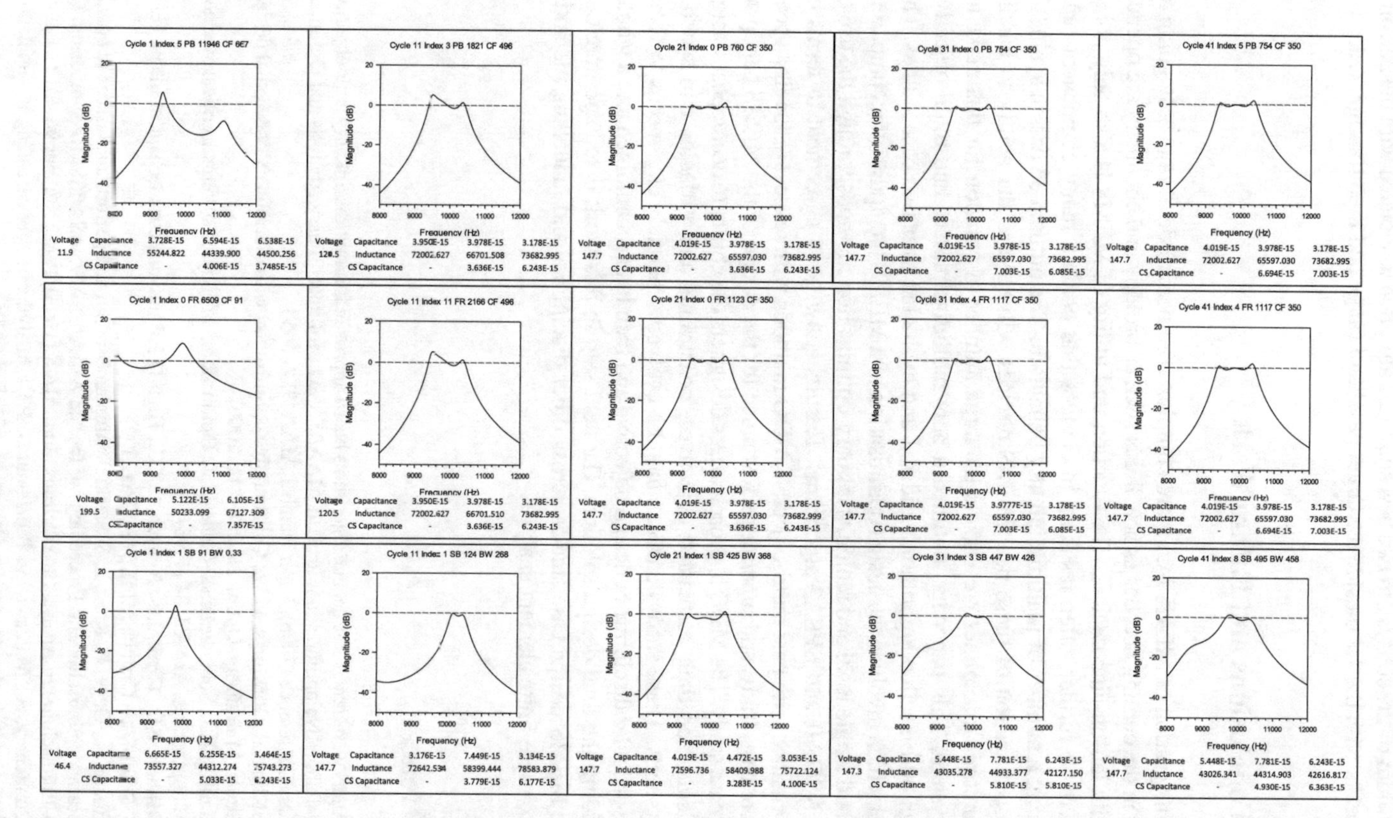


**Fig. 17** Best filter transmission plots for system population (ranked by filter response objective), subsystem one (ranked by passband objective) and two (ranked by bandwidth objective) over 5 generations (1, 11, 21, 31, 41) for SPEA2 run 1. Each plot includes objective values and genotype values for each solution

bandwidth of the subsystem two best solutions are often associated with the 2nd and 3rd peaks with the 1st outside the target passband range and therefore ignored.

## 6 Conclusions and Future Work

The integration of MEMS into more complex commercial devices is only going to grow in the coming decades as new fields such as biology and chemistry are opened up with lab-on-chip devices. The function and utility of MEMS is also only going to increase further, often resulting in devices that contain many components and covering a number of multidisciplinary behaviour. The automated design synthesis and optimisation of these new MEMS devices will require the ability to handle the multiple disciplines present and the large number of components that make up the system. This paper has outlined a new multidisciplinary and multi-objective design optimisation algorithm, validated and evaluated over two case studies. The last of which involved the design optimisation of a MEMS bandpass filter comparing standard single level and multidisciplinary optimisation methods through the use of both NSGAII and SPEA2 algorithms. Results show good agreement in terms of performance with past multi-objective MDO methods with respect to the first speed reducer case study, and superior performance for the design of the MEMS bandpass filter case study. The MDO approach offers designers the ability to decompose design problems, and their associated objectives, constraints and variables into specific subsystems. These subsystems can then be evolved separately as a means to focus on a particular discipline or design objective and then later recombined into a whole, providing the final design solution. The next step in this work is to experiment on more MEMS design case studies across other disciplines and modelling methods, for example finite element analysis.

## References

1. Aute, V., Azarm, S.: A genetic algorithms based approach for multidisciplinary multiobjective collaborative optimization. In: 11th AIAA/ISSMO Multidisciplinary Analysis and Optimization Conference (2006), Paper Number AIAA 2006-6953
2. Balling, R., Rawlings, M.W.: Collaborative optimization with disciplinary conceptual design. Struct. Multidiscip. Optim. **20**, 232–241 (2000)
3. Balling, R.J., Sobieszczanski-Sobieski, J.: Optimization of coupled systems: a critical overview of approaches. AIAA J. **34** (1996)
4. Bannon, F.D., Clark, J.R., Nguyen, C.T.C.: High-Q HF microelectromechanical filters. IEEE J. Solid-State Circuits **35**(4), 512–526 (2000)
5. Braun, R., Gage, P., Kroo, I., Sobieski, I.: Implementation and performance issues in collaborative optimisation. In: Proceedings of the 6th AIAA/USAF/NASA/ISSMO Symposium On Multidisciplinary Analysis and Optimisation. AIAA (1996), Paper Number 96-4017
6. Choudhary, R., Malkawi, A., Papalambros, P.Y.: Analytical target cascading in simulation-based building design. Autom. Constr. **14**, 551–568 (2005)

7. Conceicao Antonio, C.A.: Optimization of geometrically nonlinear composite structures based on load-displacement control. Compos. Struct. **46**, 345–356 (1999)
8. Cramer, E.J., Dennis, J.E., Frank, P.D., Lewis, R.M., Shubin, G.R.: Problem formulation for multidisciplinary optimization. SIAM J. Optim. **4**, 754–776 (1994)
9. Deb, K., Agarwal, S., Pratap, A., Meyarivan, T.: A fast elitist non-dominated sorting genetic algorithm for multi-objective optimization: NSGA-II. In: Proceedings of the 6-th International Conference Parallel Problem Solving from Nature (PPSN-VI), pp. 849–858 (2000)
10. DeMiguel, A.V., Murray, W.: A local convergence analysis of bilevel decomposition algorithms. Optim. Eng. **7**, 99–133 (2006)
11. de Wit, A.J., van Keulen, E.: Overview of methods for multi-level and/or multi-disciplinary optimization. In: 6th AIAA Multidisciplinary Design Optimization Specialist Conference. AIAA (2010), Paper Number 2010-2914
12. de Wit, A.J., Lipka, A., Ramm, E., van Keulen, E.: Multilevel optimization of material and structural layout. In: 3rd European Conference on Computational Mechanics, Solids, Structures and Coupled Problems in Engineering (2006)
13. Farnsworth, M., Benkhelifa, E., Tiwari, A., Zhu, M.: A novel approach to multi-level evolutionary design optimization of a mems device. Evolvable Systems: From Biology to Hardware. LNCS, vol. 6274, pp. 322–334. Springer, Berlin (2010)
14. Fedder, G.K., Mukherjee, T.: Physical design for surface-micromachined mems. In: Proceedings of the 5th ACM/SIGDA Physical Design Workshop 15–17 April, Reston, VA, USA, pp. 53–60 (1996)
15. Giassi, A., Bennis, F., Maisonneuve, J.J.: Multidisciplinary design optimisation and robust design approaches applied to concurrent design. Struct. Multidiscip. Design **28**, 356–371 (2004)
16. Golinski, J.: Optimal synthesis problems solved by means of nonlinear programming and random methods. J. Mech. **5**, 287–309 (1970)
17. Gunawan, S., Azarm, S., Wu, J., Boyars, A.: Quality-assisted multi-objective multidisciplinary genetic algorithms. AIAA J. **41**(9), 1752–1762 (2003)
18. Gunawan, S., Farhang-Mehr, A., Azarm, S.: On maximizing solution diversity in a multiobjective multidisciplinary genetic algorithm for design optimization. Mech. Based Design Struct. Mach. **32**(4), 491–514 (2004)
19. Haftka, R.T., Watson, L.T.: Multidisciplinary design optimisation with quasiseparable subsystems. Optim. Eng. **6**, 9–20 (2005)
20. Hostis, F.l., Green, N.G., Morgan, H., Akaisi, M.: Solid state AC electroosmosis micro pump on a chip. In: International Conference on Nanoscience and Nanotechnology, ICONN, Brisbane, Qld, July (2006)
21. Huang, H.Y., Wang, D.Y.: Static and dynamic collaborative optimization of ship hull structure. J. Mar. Sci. Appl. **8**, 77–82 (2009)
22. Isoda, T., Ishida, Y.: Seperation of cells using fluidic mems device and a quantitative analysis of cell movement. Trans. Inst. Electr. Eng. Jpn **126**(11), 583–589 (2006)
23. Kim, H.M., Nestor, F.M., Panos, P.Y., Tao, J.: Target cascading in optimal system design. J. Mech. Design **125**, 474–480 (2003)
24. Kodiyalam, S., Sobieszczanski-Sobieski, J.: Multidisciplinary design optimization some formal methods, framework requirements, and application to vehicle design. Int. J. Veh. Design **23**, 3–22 (2001)
25. Kroo, I., Manning, V.: Collaborative optimization: Status and directions. In: 8th AIAA/NASA/ISSMO Symposium on Multidisciplinary Analysis and Optimization (2000), Paper Number AIAA 2000-4721
26. Kurapati, A., Azarm, S.: Immune network simulation with multiobjective genetic algorithms for multidisciplinary optimization. Eng. Optim. **33**, 245–260 (2000)
27. Lin, L., Nguyen, C.T.C., Howe, R.T., Pisano, A.P.: Micro electromechanical filters for signal processing. Microelectromech. Syst. (1992)
28. McAllister, C.D., Simpson, T.W., Hacker, K., Lewis, K., Messac, A.: Integrating linear physical programming within collaborative optimization for multiobjective multidisciplinary design optimization. Struct. Multidiscip. Optim. **29**, 178–189 (2005)

29. Potter, M.A., De Jong, K.A.: A cooperative coevolutionary approach to function optimization. In: Davidor, Y., Schwefel, H.P., Manner, R. (eds.) Proceedings of the Third International Conference on Parallel Problem Solving in Nature (PPSN-III), Jerusalem, Israel (1994)
30. Rabeau, S., Depince, P., Bennis, F.: Collaborative optimization of complex systems: a multidisciplinary approach. Int. J. Interact. Design Manuf. **1**, 209–218 (2007)
31. Schoeffler, J.D.: Static multilevel systems. Optimization Methods for Large-Scale Systems: With Applications. Mcgraw-Hill Book Company, New York (1971)
32. Sobieszczanski-Sobieski, J.: Optimization by decomposition: a step from hierarchic to nonhierarchic systems. In: Recent Advances in Multidisciplinary Analysis and Optimization, NASA/Air Force Symposium, NASA (1988), cP-3031
33. Sobieszczanski-Sobieski, J.: Sensitivity analysis and multidisciplinary optimization for aircraft design: recent advances and results. J. Aircr. **27**, 993–1001 (1990)
34. Sobieszczanski-Sobieski, J., Argte, J.S., Sandusky, R.R.J.: Bi-level integrated system synthesis (BLISS). In: Proceedings of the 7th AIAA/USAF/NASA/ISSMO Symposium On Multidisciplinary Analysis and Optimization. AIAA (1998), Paper Number 98-4916
35. Sobieszczanski-Sobieski, J., James, B.B., Dovi, A.R.: Structural optimization by multi-level decomposition. AIAA J. **23**, 124–142 (1985)
36. Tapetta, R.V., Renaud, J.E.: Multiobjective collaborative optimization. J. Mech. Design **119**, 403–411 (1997)
37. Tosserams, S., Etman, L.F.P., Rooda, J.E.: Augmented Lagrangian coordination for distributed optimal design in MDO. Int. J. Numer. Methods Eng. **73**, 1885–1910 (2008)
38. Tosserams, S., Etman, L.F.P., Rooda, J.E.: A micro-accelerometer MDO benchmark problem. Struct. Multidiscip. Optim. **41**, 255–275 (2010)
39. Tribes, C., Dube, J.F., Trepanier, J.Y.: Decomposition of multidisciplinary optimization problems: formulations and applications to a simplified wing design. Eng. Optim. **37**(8), 775–796 (2005)
40. Wagner, T.: A general decomposition methodology for optimal design. Ph.D. thesis, The University of Michigan (1993)
41. Wang, K., Nguyen, C.T.C.: High-order medium frequency micromechanical electronic filters. J. Microelectromech. Syst. **8**(4), 534–556 (1999)
42. Zadeh, P.M., Roshanian, J., Farmani, M.R.: Particle swarm optimization for multiobjective collaborative multidisciplinary design optimization. In: 51st AIAA/ASMEA/ASCE/AHS/ASC Structures, Structural Dynamics, and Materials Conference. AIAA (2010), AIAA Paper Number 2010-3081
43. Zitzler, E., Laumanns, M., Thiele, L.: SPEA2: improving the strength pareto evolutionary algorithm for multiobjective optimization. In: Evolutionary Methods for Design Optimization and Control with Applications to Industrial Problems, Athens, Greece, pp. 95–100. International Center for Numerical Methods in Engineering (2001)

# Coefficients Estimation of MPM Through LSE, ORLS and SLS for RF-PA Modeling and DPD

**E. Allende-Chávez, S.A. Juárez-Cázares, J.R. Cárdenas-Valdez, Y. Sandoval-Ibarra, J.A. Galaviz-Aguilar, Leonardo Trujillo and J.C. Nuñez-Pérez**

**Abstract** This paper shows and compares three techniques based on the least squared error for the estimation of the constant coefficients of the memory polynomial model used for the modeling of power amplifiers for radio-frequency and for the construction of a pre-distorter. The first technique is the conventional linear regression using the least square error method. The second technique is the order recursive least squares which can be used for exploring the most adequate nonlinearity order and memory depth of the memory polynomial model by comparing subsequent errors. The sequential least squares method is useful when the measurements of a system are coming sample by sample and the parameters of the model should be adjusted on-line. The mathematical background of the three methods is shown; as an experimental validation of this methods they were simulated in Matlab for the measurements of a 10W NPX Power Amplifier based on the transistor CLF1G0060 GaN HEMTs. An NMSE of −19.83 dB was reached for the best model.

E. Allende-Chávez · J.R. Cárdenas-Valdez · Y. Sandoval-Ibarra · L. Trujillo
Tecnológico Nacional de México, Instituto Tecnológico de Tijuana,
Tijuana, B.C., Mexico
e-mail: edgar.allende@tectijuana.edu.mx

J.R. Cárdenas-Valdez
e-mail: jose.cardenas@tectijuana.edu.mx

Y. Sandoval-Ibarra
e-mail: ysandoval@citedi.mx

L. Trujillo
e-mail: leonardo.trujillo@tectijuana.edu.mx

S.A. Juárez-Cázares · J.A. Galaviz-Aguilar · J.C. Nuñez-Pérez (✉)
Instituto Politécnico Nacional, CITEDI, Tijuana, B.C., Mexico
e-mail: nunez@citedi.mx

S.A. Juárez-Cázares
e-mail: sjuarez@citedi.mx

J.A. Galaviz-Aguilar
e-mail: jgalaviz@citedi.mx

© Springer International Publishing AG 2018

Y. Maldonado et al. (eds.), *NEO 2016*, Studies in Computational Intelligence 731,
https://doi.org/10.1007/978-3-319-64063-1_10

Also in order to linearize the power amplifier a pre-distorter was constructed through indirect learning architecture achieving a 50 dBm spurious free dynamic range and a 25 dBc reduction in the adjacent power ratio.

**Keywords** ILA · LSE · MPM · ORLS · Power amplifier · SLS

## 1 Introduction

Code Division Multiple Access (WCDMA) and Orthogonal Frequency Division Multiplexing (OFDM) are examples of the complex systems of digital modulation used nowadays in most of the Radio Frequency (RF) communication systems, these were developed to face the problem of an increasing demand in the data transmission and its structure have the purpose of optimizing the use of available bandwidth. The resulting spectral signals from the previous modulations outcome in a non-constant envelope signal with huge Peak to Average Power Ratio (PAPR), usually around 10 dB which means in order to accomplish fine communication it is necessary to have a high linear Power Amplifier (RF-PA) but also it should be very efficient [1–3].

RF-PA is one of the most important elements in the transmission chain of an RF communication system and this is due that the signal to transmit cannot go directly to the antenna, it needs to be raised to a certain level to achieve good signal to noise ratio which is an important factor in the receiver stage. Nevertheless the RF-PA is an element inherently nonlinear and this causes distortions in the output signal. The non-linear amplification results in certain number of problems which derive in bandwidth interferences some of these issues are memory effects in short term, inter-modulation products, etc. They also generate spectral regrowth in the adjacent channels, these should be avoided because this can lead to legal issues since it is strictly regulated [4, 5]. RF-PA linearization is an important issue in telecommunication in order to increase its performance, some techniques to accomplish it are: feed-forward, feed-back and the most extended Digital Pre-Distortion (DPD) [1, 3, 6–8].

DPD techniques in literature include DPD through indirect learning and DPD through direct learning and frequency selective pre-distortion [2, 3, 6, 9]. The linearization of a RF-PA requires having a mathematical model which describes the device behavior; available behavioral models include models with and without memory [4, 10, 11], artificial neural networks [12], Volterra series based models and most recently modeling based on genetic programming [13–15]. The memory polynomial model is used in this paper to describe the RF-PA behavior since it has shown to have good accuracy and an acceptable computational complexity.

Beside the common application of linear regression based on Least Square Error (LSE) there are statistical recursive techniques which using the same principles of LSE calculate the optimal values of the constant coefficients in a model but are more suitable in different contexts. In this paper we discuss about Order Recursive Least Squares (ORLS) and Sequential Least Squares (SLS). ORLS is used to explore the most adequate nonlinearity order and memory depth in a model since it can compare the error of sequential model orders or memory depths using recursion and avoiding matrix inversions in the process. It has been applied for estimation in OFDM systems and the development of adaptive filters [16, 17]. SLS is pretty useful for the estimation of parameters when the data is coming sample by sample, some of its applications are design of Infinite Impulse Response filters (IIR) and Finite Impulse Response (FIR) filters, implementation of adaptive filters and error estimation systems in aeronautics [18–20]. Characteristics of SLS make it suitable for the automatic adjust of parameters of a behavioral model when the measurements are being performed.

The Indirect Learning Architecture (ILA) has been used successfully for the development of digital pre-distorters, one of its main advantages is that the user can perform it by only using measurements taken from the Device Under Test (DUT), this makes possible the development of the pre-distorter without having the DUT physically [21, 22].

This work shows the mathematical background to apply the ORLS and SLS techniques to extract the coefficients of memory polynomial model used for RF-PA modeling, the experimental validation of these techniques is performed through a simulation of the algorithms in Matlab, also SLS is used for the development of a complete pre-distorter.

The present paper is organized as follows: in Sect. 2 some mathematical models are given which are used to model PA, in this section the Memory Polynomial Model(MPM) is introduced, this paper is focused on it. Section 3 is used to show the three methods to extract the coefficients of the MPM based on least square error approach. The development of a pre-distorter through LSE and SLS using ILA is shown in Sect. 4. The results discussion are given in Sect. 5. Finally conclusions are given in Sect. 6.

## 2 Mathematical Modeling of a PA's Behavior

Volterra series are a nonlinear behavioral model which has the ability to capture the memory effects of a system, so it is suitable for the modeling of a wide range of electronic devices. Because the inherent nonlinearity of the RF-PA and its condition in which the output not only depends on the present input but in previous inputs too, Volterra series constitute a feasible method to model the behavior of an RF-PA [10, 13, 14]. Volterra series for RF-PA modeling are described by the Eq. (1)

$$y(n) = \sum_{k=1}^{K} \sum_{i_1}^{M} \cdots \sum_{i_{2k-1}}^{M} \left[ h_{2k-1}(i_1, i_2, \ldots, i_{2k-1}) * \prod_{j-1}^{k+1} x(n - i_j) \prod_{\lambda+1}^{2k+1} x^*(n - i_j) \right], \quad (1)$$

where
$()^*$ - denotes the complex conjugate
$x(n)$ - is the discrete input at time $n$
$y(n)$ - is the discrete output at time $n$
$K$ - is the nonlinearity order of the model
$M$ - is the memory depth of the model
$h_{2k-1}(i_1, i_2, \ldots, i_{2k-1})$ - is the Volterra kernel of order $k$.

If the Eq. (1) is inspected it can be noticed that the number of coefficients in the equation grows exponentially when the order of nonlinearity and the memory depth are increased, this problem impacts directly in the number of logic resources when the model is implemented in hardware for an application like DPD, because of the previous fact the complete Volterra model results no practical. There are models based on Volterra series which reduce its computation complexity and try to maintain its accuracy, these models are Winner model, Hammerstein and MPM.

## 2.1 Memory Polynomial Model

The MPM consists in the combination of several delay taps and nonlinear functions. MPM represents a truncation of the general Volterra series which only considered the terms of the main diagonal in Volterra kernels thus the number of constant coefficients is significantly reduced compared with the complete Volterra series. The block diagram of the MPM is given in Fig. 1. Equation (2) depicts MPM model.

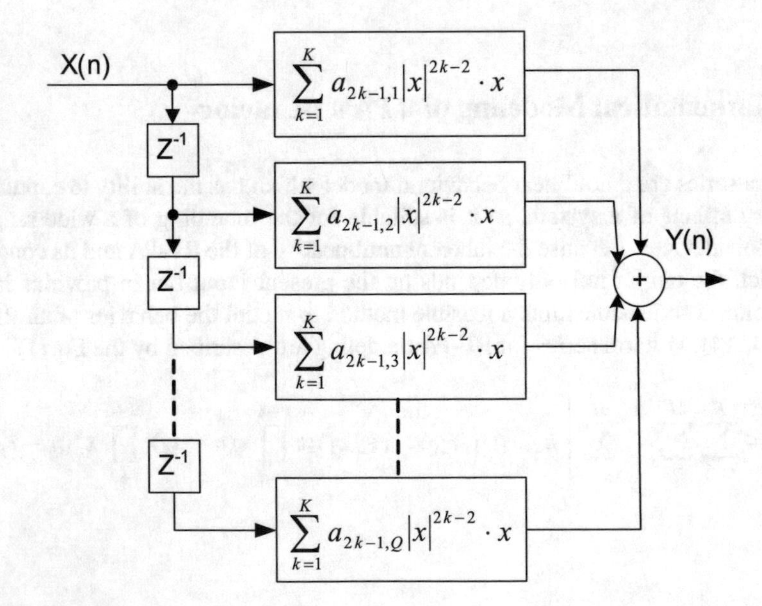


**Fig. 1** MPM block diagram

$$y(n) = \sum_{q=0}^{Q} \sum_{k=1}^{K} a_{2k-1,q} |x(n-q)|^{2(k-1)} x(n-q), \tag{2}$$

where

$x(n)$ - is the discrete input at time $n$
$y(n)$ - is the discrete output at time $n$
$K$ - is the nonlinearity order of the model
$Q$ - is the memory depth of the model

# 3 Techniques for Constant Coefficients Estimation of the MPM

Estimation of the constant coefficients of the MPM is very important, because based on them the generated model will be more or less accurate to the real behavior of the RF-PA, it should be considered that relying on the value used for coefficients in the model, the performance of it doing predictions of the system will vary.

The classical way to estimate the constant coefficients of the MPM is using the linear regression method through Least Square Error (LSE) [14], this method has the characteristic of a direct application and has been used successfully in previous research such as [11, 14]. However this common method does not give any information about the most adequate nonlinearity order or memory depth of the model being adjusted. The LSE method also needs a big number of measurements to make the model more accurate, if an on-line identification is needed, a system with big store capacity will be required.

## 3.1 LSE

With the purpose of estimate the coefficients of the MPM, Eq. (2) should be represented in a matrix form [14], so for that the output data can be represented as Eq. (3) shows.

$$\mathbf{Y} = [y(0), y(1), \ldots, y(N-1)], \tag{3}$$

where

$\mathbf{Y}$ - represent the vector of outputs
$y(n)$ - is the discrete output at time $n$
$N$ - represents the number of observations

A matrix **H** of observation can be defined as:

$$\mathbf{H} = \left[\mathbf{H}_0, \mathbf{H}_1, \ldots, \mathbf{H}_Q\right], \tag{4}$$

where:

$$\mathbf{H_q} = \begin{bmatrix} h_{1,q}(0), & h_{3,q}(0), & \cdots & h_{2k-1,q}(0) \\ h_{1,q}(1), & h_{3,q}(1), & \cdots & h_{2k-1,q}(1) \\ \vdots, & \vdots, & \ddots, & \vdots \\ h_{1,q}(N-1), & h_{3,q}(N-1), & \cdots & h_{2k-1,q}(N-1) \end{bmatrix}, \tag{5}$$

and

$$h_{2k-1} = |x(n-q)|^{2*(k-1)} * x(n-q). \tag{6}$$

The complex coefficients can be expressed as

$$\mathbf{a} = \left[\mathbf{a}_0, \ldots, \mathbf{a}_q, \ldots, \mathbf{a}_Q\right]^T, \tag{7}$$

where $a_q$ express

$$\mathbf{a}_q = \left[a_{1,q}, a_{3,q}, \ldots, a_{2k-1,q}\right], \tag{8}$$

thus Eq. (1) can be represented in matrix notation as

$$\mathbf{Y} = \mathbf{H} * \mathbf{a}. \tag{9}$$

The coefficients which minimize the error between the output of the model and **Y** can be estimated using LSE through Eq. (10)

$$\hat{\mathbf{a}} = \left[\hat{\mathbf{a}}_0, \ldots, \hat{\mathbf{a}}_q, \ldots, \hat{\mathbf{a}}_Q\right] = \mathbf{H}^+ * \mathbf{Y}, \tag{10}$$

where $\mathbf{H}^+$ represents the pseudo-inverse of the matrix **H** which can be calculated by (11)

$$\mathbf{H}^+ = (\mathbf{H}^T * \mathbf{H}^{-1}) * \mathbf{H}^T. \tag{11}$$

And the least square error is given by:

$$J_{min} = (\mathbf{Y} - \mathbf{H} * \hat{\mathbf{a}})^T * (\mathbf{Y} - \mathbf{H} * \hat{\mathbf{a}}). \tag{12}$$

## 3.2 ORLS

Most of the time there is not any information about which will be the most adequate order of nonlinearity and memory depth, when this occurs ORLS can help to decide how to adjust these parameters in the model since ORLS was thought to explore subsequent orders of a model by comparing the least squared error of a previous order model with the one of a current model being evaluated.

If MPM is started with order $K = 1$ and memory $Q = 0$, the matrix $\mathbf{H}$ is obtained

$$\mathbf{H} = \begin{bmatrix} x(0) \\ x(1) \\ \vdots \\ x(N-1) \end{bmatrix}. \tag{13}$$

To increment the order of nonlinearity of the model matrix $\mathbf{H}$ can be expressed as

$$\mathbf{H}_{k+1} = [\mathbf{H}_k, \mathbf{h}_{k+1}], \tag{14}$$

with

$$\mathbf{h}_{k+1} = [|x_0|^{2(k-1)}x_0, |x_1|^{2(k-1)}x_1, \ldots, |x_{N-1}|^{2(k-1)}x_{N-1}]. \tag{15}$$

The new coefficients for the increased order can be calculated by

$$\hat{\mathbf{a}}_{k+1} = \begin{bmatrix} \hat{\mathbf{a}}_k - \dfrac{(\mathbf{H}_k^T\mathbf{H}_k)^{-1}\mathbf{H}_k^T\mathbf{h}_{k+1}\mathbf{h}_{k+1}^T\mathbf{P}_k\mathbf{Y}}{\mathbf{h}_{k+1}^T\mathbf{P}_k\mathbf{h}_{k+1}} \\ \dfrac{\mathbf{h}_{k+1}^T\mathbf{P}_k\mathbf{Y}}{\mathbf{h}_{k+1}^T\mathbf{P}_k\mathbf{h}_{k+1}} \end{bmatrix}, \tag{16}$$

where

$$\mathbf{P}_k = \mathbf{I} - \mathbf{H}_k(\mathbf{H}_k^T\mathbf{H}_k)^{-1}\mathbf{H}_k^T. \tag{17}$$

In order to reduce calculations the inversion in Eq. (17) can be eliminated by defining Eq. (18)

$$\mathbf{D}_k = (\mathbf{H}_k^T\mathbf{H}_k)^{-1}, \tag{18}$$

and using the recursive formula shown in Eq. (19)

$$\mathbf{D}_{k+1} = \begin{bmatrix} \mathbf{D}_k + \dfrac{\mathbf{D}_k\mathbf{H}_k^T\mathbf{h}_{k+1}\mathbf{h}_{k+1}^T\mathbf{H}_k\mathbf{D}_k}{\mathbf{h}_{k+1}^T\mathbf{P}_k\mathbf{h}_{k+1}} & -\dfrac{\mathbf{D}_k\mathbf{H}_k^T\mathbf{h}_{k+1}}{\mathbf{h}_{k+1}^T\mathbf{P}_k\mathbf{h}_{k+1}} \\ -\dfrac{\mathbf{h}_{k+1}^T\mathbf{H}_k\mathbf{D}_k}{\mathbf{h}_{k+1}^T\mathbf{P}_k\mathbf{h}_{k+1}} & \dfrac{1}{\mathbf{h}_{k+1}^T\mathbf{P}_k\mathbf{h}_{k+1}} \end{bmatrix}. \tag{19}$$

Equation (20) defines $\mathbf{P}_k$ which represents the projection matrix

$$\mathbf{P}_k = \mathbf{I} - \mathbf{H}_k \mathbf{D}_k \mathbf{H}_k^T . \tag{20}$$

The least square error can be computed through

$$J_{mink+1} = J_{mink} - \frac{(\mathbf{h}_{k+1}^T)\mathbf{P}_k \mathbf{Y}}{\mathbf{h}_{k+1}^T \mathbf{P}_k \mathbf{h}_{k+1}} . \tag{21}$$

If an increase in memory depth is needed the matrix $\mathbf{H}$ should be increased by $k$ columns delaying them $q - 1$ times, matrix $\mathbf{H}$ should be increased one column by time in order to preserve the concept, so $\mathbf{H}$ will be increased $k$ times in each memory depth increment.

Performing a comparison of the squared error between two consecutive nonlinear orders or memory depths, it can be determined if the increment in complexity of the model is justified by the improvement in accuracy. The last fact is useful when a correct tradeoff between these two characteristics is searched in the generated model to implement it in hardware.

### 3.3 SLS

The use of SLS for the development of adaptive filters [18, 20] inspires the idea of using them for an adaptive estimation of the constant coefficients in MPM. When a RF-PA is being measured and a MPM with defined parameters of nonlinearity order and memory depth is being used for modeling the behavior, the constant coefficients of the model can be updated with each new sample through SLS. Last scenario brings also the idea of adaptive pre-distorters. Figure 2 shows the SLS scheme used in this work. In [20] the derivations of the SLS formulas are showed. The estimators of the constant coefficients of the MPM are given by:

$$\hat{\mathbf{a}} = \mathbf{S}_D(n)\mathbf{S}_D(n), \tag{22}$$

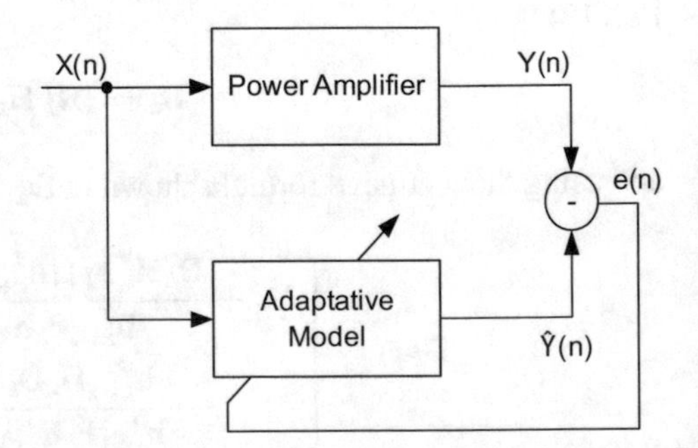


**Fig. 2** SLS scheme

where

$$\mathbf{S}_D(n) = \frac{1}{\lambda}\left[\mathbf{S}_D(n-1) - \frac{\mathbf{S}_D(n-1)\mathbf{X}(n)\mathbf{X}^T(n)\mathbf{S}_D(n-1)}{\lambda + \mathbf{X}^T(n)\mathbf{S}_D(n-1)\mathbf{X}(n)}\right], \tag{23}$$

and

$$\mathbf{P}_D = \lambda\mathbf{P}_D(N-1) + y(n)\mathbf{X}(\mathbf{n}), \tag{24}$$

where

$\mathbf{X}(n)$ - is the observation for the data $x(n)$
$\lambda$ - is the forgetting factor which helps to give more weight to the new samples than the old ones and it takes values in the range $\lambda \leq 1$.

At the beginning of the algorithm values for the parameters $\mathbf{S}_D(-1)$ and $\mathbf{P}_D(-1)$ should be assigned, these values are

$$\mathbf{S}_D = \delta\mathbf{I}, \tag{25}$$

where

$\delta$ - is a big value, a value of $10^5$ was used for this application.
$\mathbf{I}$ - is an identity matrix with size $k(q+1)$.

And

$$\mathbf{P}_D = \mathbf{x}(-1) = [0, 0, \ldots, 0]^T. \tag{26}$$

The size of vector $\mathbf{P}_D$ is also determined by $k(q+1)$. It can be noticed that the estimation of coefficients through SLS is more convenient and flexible when an on-line identification is necessary, it also represents a reduction in the computation and storage requirements since not any matrix inversion is performed and the biggest charge in memory is a square matrix of size $k(q+1)$.

## 4 DPD

As it was discussed in Sect. 1 modern modulation techniques have the characteristic to bring the work of RF-PA near to the saturation level due to the no-linearity of it, many methods for the RF-PA linearization have been developed, the most extended is Digital Pre-Distortion (DPD) which is cost effective and appropriate for new communications systems [23].

DPD technique consists of adding an extra block before the input of the RF-PA, this block has the property of being the inverse mathematical model of the RF-PA, resulting system acts like a linear system in the theory, but due to some issues in

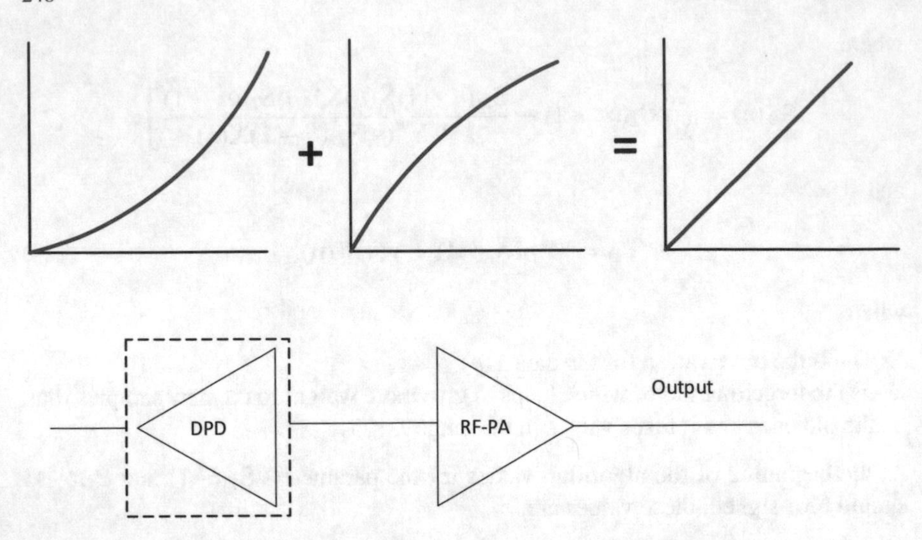


**Fig. 3** Linearization of RF-PA through DPD

the identification of the model of the RF-PA and in the inverse model development residual distortion can be observed [24].

Figure 3 depicts the idea of DPD. Since DPD needs the development of an inverse mathematical model of the RF-PA, it should be consider which model was adopted for the RF-PA modeling in order to optimize the linearization, for example for an MPM RF-PA modeling based, the inverse of the model can be another MPM, in this case coefficients for the inverse model should be calculated in a similar way as for RF-PA modeling, but we need to adopt a methodology to perform this estimation, there are two principal architectures for this, Indirect Learning Architecture (ILA) and Direct Learning Architecture (DLA) [25].

In this work for the development of a DPD through an MPM inverse model ILA was employed, ILA is adopted given its ease of application and is more adequate to prove the feasibility of the proposed techniques for coefficients extraction. Also, ILA is used since in previous research [26] it has shown better suppression of the spectral regrowth and has a faster convergence.

## *4.1 ILA*

The ILA block diagram is shown in Fig. 4. ILA uses two similar models for the pre-distorter and the post-distorter and works as follows. The input $x(n)$ goes to the pre-distorter and the output of it $z(n)$ goes to the RF-PA model, the output $y(n)$ divided by $g$ is the input of the post-distorter, $g$ is the intended gain, the output of the post-distorter $\hat{z}$ is compared with $z(n)$ and the work here is to minimize the error between this two signals, this could be achieve by collecting a certain number of

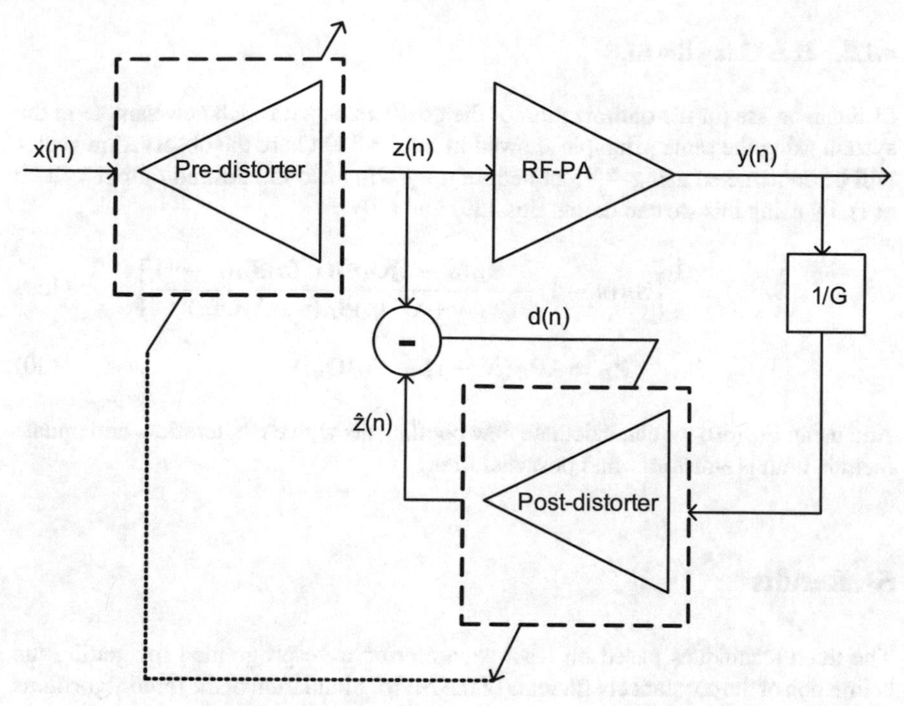


**Fig. 4** ILA block diagram

samples and minimize the error in a single step by using LSE method or doing it sample by sample using a recursive technique such as SLS [23].

### 4.1.1 ILA Using the LSE

The ILA can be implemented to converge in a single step collecting a considerable number of samples of the system and implementing the next methodology.

An observation matrix **U** can be constructed using the inputs $\frac{\mathbf{Y}}{g}$ in a similar way as in the Sect. 3.1 was done with $x(n)$, the matrix equation for the system would be

$$\mathbf{Z} = \mathbf{U}\mathbf{a}, \tag{27}$$

and the coefficients which minimize the error between **Z** and $\hat{\mathbf{Z}}$ will be

$$\hat{\mathbf{a}} = (\mathbf{U}^T * \mathbf{U}^{-1}) * \mathbf{U}^T * \mathbf{Z}. \tag{28}$$

### 4.1.2 ILA Using the SLS

SLS can be use for the optimization of the coefficients with each new sample in the system using the same principle showed in Sect. 3.3 but here the observation vector will be constructed using $\frac{y(n)}{g}$, let denote it by $O(n)$ and the desired output will be $z(n)$, by using this we can define Eqs. (29) and (30)

$$\mathbf{S}_D(n) = \frac{1}{\lambda}\left[\mathbf{S}_D(n-1) - \frac{\mathbf{S}_D(n-1)\mathbf{O}(n)\mathbf{O}^T(n)\mathbf{S}_D(n-1)}{\lambda + \mathbf{O}^T(n)\mathbf{S}_D(n-1)\mathbf{O}(n)}\right], \tag{29}$$

$$\mathbf{P}_D = \lambda\mathbf{P}_D(N-1) + z(n)\mathbf{O}(n). \tag{30}$$

And using Eq. (30) we can calculate new coefficients with each iteration, and update them in both pre-distorter and post-distorter.

## 5 Results

The three techniques based on least square error were programed for getting the estimation of the constant coefficients of the MPM. Simulation of the three algorithms was made in Matlab where a Graphical User Interface (GUI) was developed and on it the user can select a set of data to be analyzed and the way in which the constant coefficients of the MPM will be estimated, user also can select the nonlinear order, memory depth and if the data is complex or real. After the estimation Normalized Mean Square Error (NMSE) of the generated model is showed and the calculated coefficients are sent to the workspace of Matlab, Fig. 5 shows the main window of the developed GUI.

Figures 6, 7, 8, 9, 10 and 11 show the AM-AM and AM-PM plots generated by the GUI for the data of PA model NXP 10W with 65,536 samples. The PA electric characteristics are detailed in Table 1. Plots were generated with the three techniques and with a nonlinearity order of 5 and memory depth of 2, Table 2 indicates the coefficients estimated by each technique and the respective NMSE for the generated model.

As it can be confirmed in Figs. 6, 7, 8, 9, 10 and 11 the three techniques of estimation have comparable results since all of them are based on a least square error approach. These plots demonstrate that the estimation of output of the RF-PA through MPM adjusted by the implemented techniques has a good accuracy and models the behavior of the RF-PA behavior correctly, so the generated models by our system can be applied for any linearization method, the estimation of RF-PA output was made using complex numbers to calculate the amplitude and phase with only one estimation.

Both LSE and SLS with forgetting factor of 1 gave almost the same coefficients, more clear variation is observed in the case of ORLS this is because only 10,000 samples were used for coefficients estimation since the limitation of memory

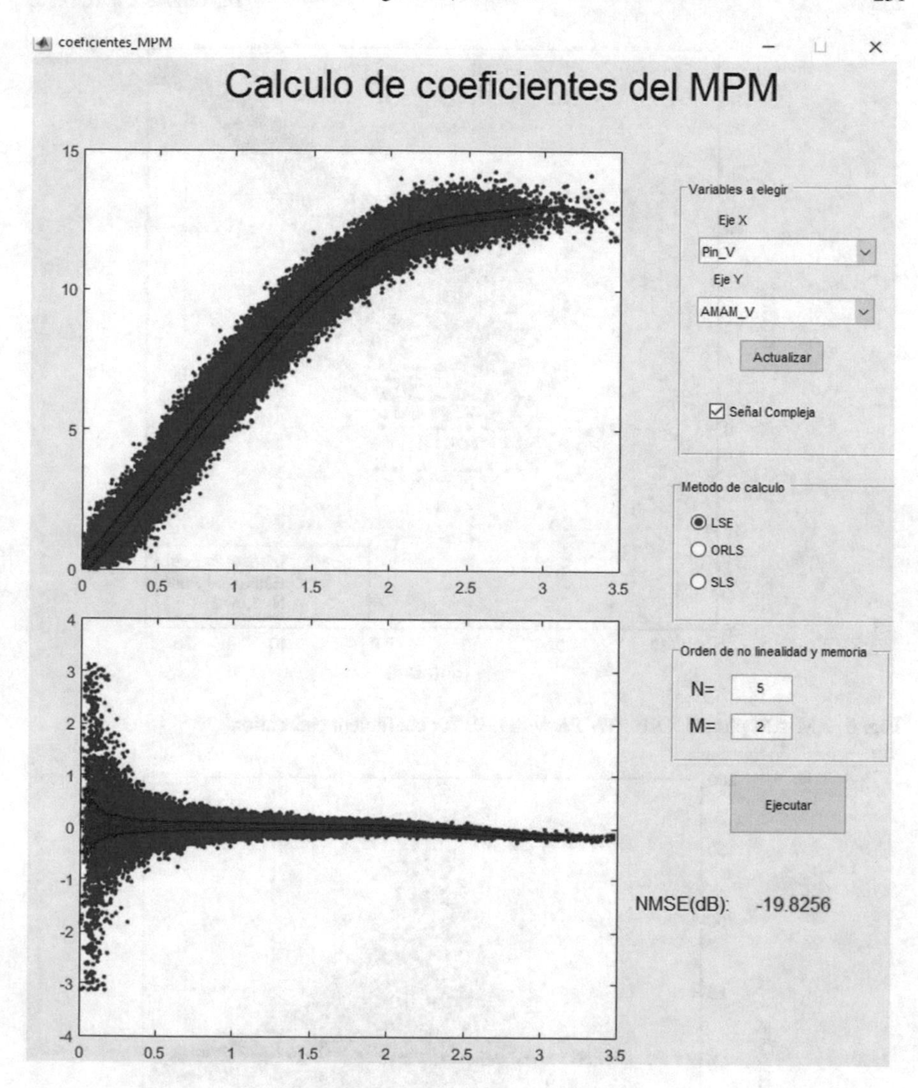


**Fig. 5** GUI developed for the simulation of LSE, ORLS and SLS

available in our system however coefficients given by ORLS model correctly the behavior of the RF-PA.

In Table 3 the Halstead parameters calculated for our implementations of the algorithms and the mean of the execution time of 100 runs of each algorithm with the same parameters are shown. Halstead parameters are used to measure computational complexity of an algorithm [27], as a first approach in this work they were used to give an idea of the volume of logic resources required by each method since they consider the number of evaluated functions in the algorithm. It is important to mention that for our purposes the most relevant Halstead parameters are the calculated in length

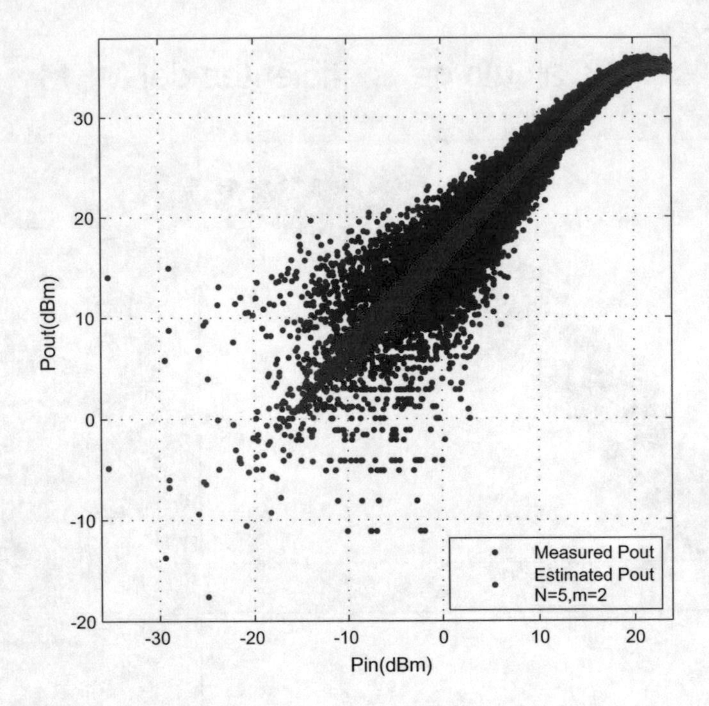


**Fig. 6** AM-AM plot of NXP 10W PA with LSE for coefficient estimation

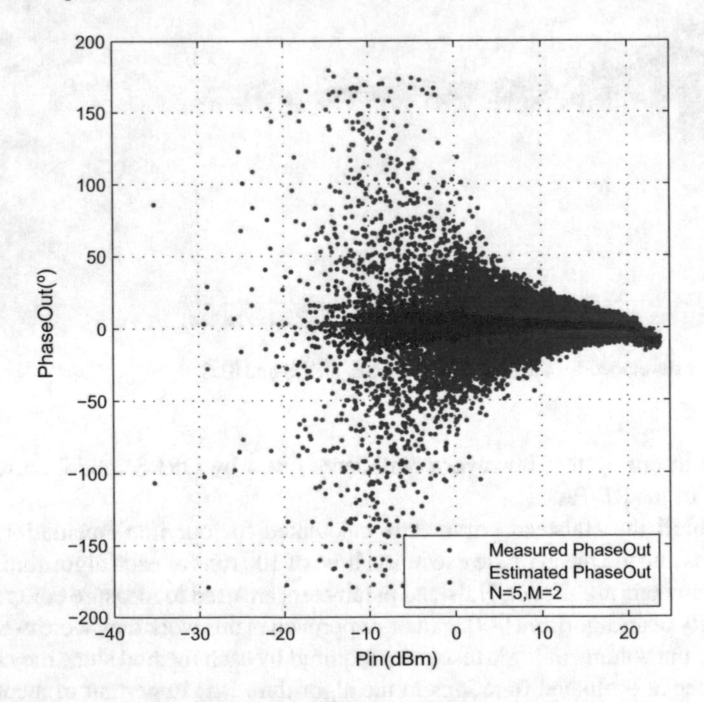


**Fig. 7** AM-PM plot of NXP 10W PA with LSE for coefficient estimation

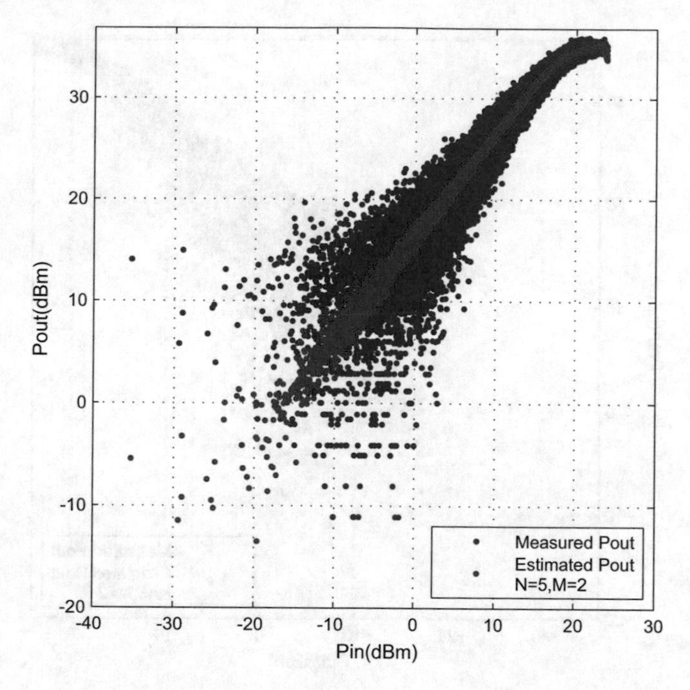


**Fig. 8** AM-AM plot of NXP 10W PA with ORLS for coefficient estimation

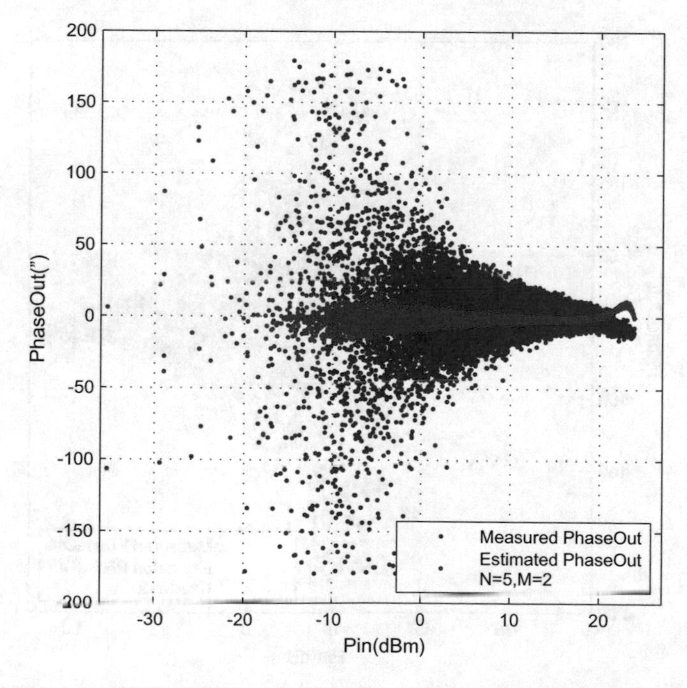


**Fig. 9** AM-PM plot of NXP 10W PA with ORLS for coefficient estimation

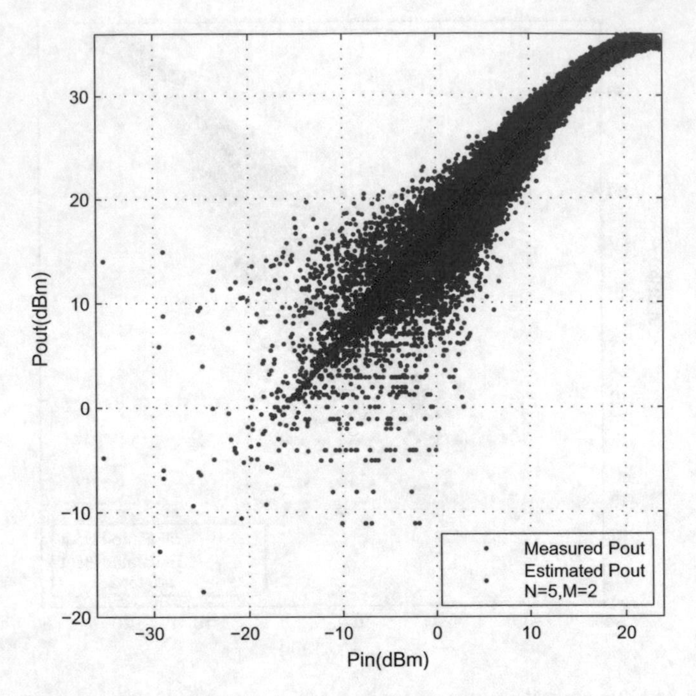


**Fig. 10** AM-AM plot of NXP 10W PA with SLS for coefficient estimation

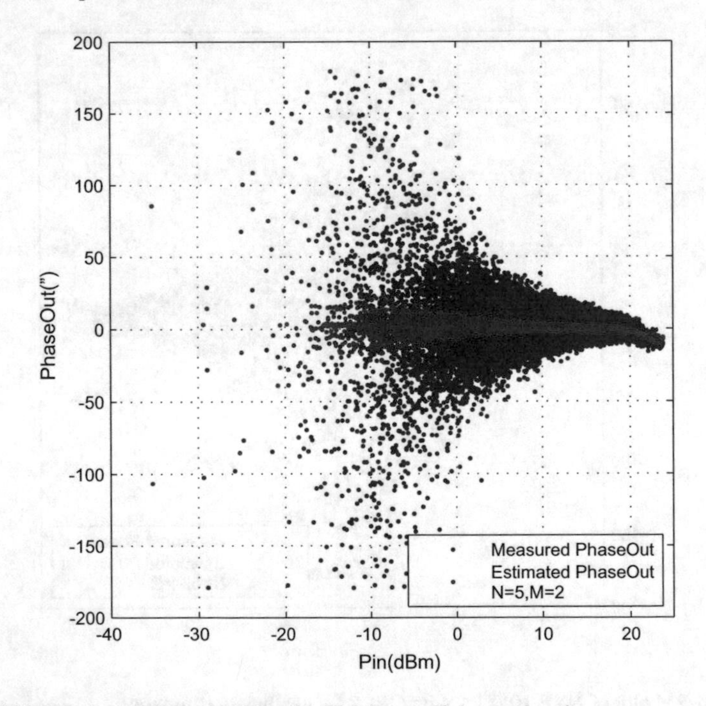


**Fig. 11** AM-PM plot of NXP 10W PA with SLS for coefficient estimation

**Table 1** NXP10W PA specifications

| Parameter | NXP 10W PA @ 2 GHz |
|---|---|
| Gain (1 dBm) | 36 dBm |
| Class | AB ($V_{ds} = 50$ V, $I_{ds} = 54$ mA) |
| Operation frequency | 500–2500 MHz |
| Drain efficiency ($\eta_d$) | 21% |

**Table 2** Coefficients estimated by each technique and NMSE of the generated model

| Method | Coefficients | NMSE |
|---|---|---|
| | 0.2581+4.1497i, 3.7880−7.2138i, −2.4810+3.2497i<br>0.5067−0.4630i, −0.3401+0.8840i, 0.1313−0.4585i<br>−0.3183+0.0459i, 0.2040−0.1000i, −0.0753+0.0677i<br>0.0370−0.0091i, −0.0260+0.0116i, 0.0102−0.0078i | |
| LSE | −0.0014+0.0006i, 0.0011−0.0007i, −0.0005+0.0004i<br>6.4327−0.7727i, 1.2244+0.1144i, −1.0410+0.4351i<br>−0.2058+0.1801i, 1.1960−0.0940i, −0.7330+0.0431i<br>0.0078−0.0706i, −0.4630+0.0867i, 0.2810−0.0758i<br>−0.0087+0.0159i, 0.0669−0.0160i, −0.0392+0.0131i | −19.8256 |
| ORLS | 0.0005−0.0012i, −0.0028+0.0013i, 0.0016 − 0.0009i<br>5.2582+4.1495i, 3.7879−7.2132i, −2.4809+3.2494i<br>0.5067−0.4629i, −0.3401+0.8839i, 0.1313−0.4585i<br>−0.3183+0.0458i, 0.2039−0.0999i, −0.0753+0.0677i<br>0.0370 − 0.0091i −0.0260 + 0.0116i 0.0102 − 0.0078i | −18.4474 |
| SLS | −0.0014 + 0.0006i 0.0011 − 0.0007i −0.0005 + 0.0004i | −19.8256 |

**Table 3** Halstead complexity and execution time

| Method | Halstead parameters | Execution time (s) |
|---|---|---|
| LSE | h1 = 3, h2 = 12, N1 = 11, N2 = 22, vocabulary = 15, length = 33, calculated_length = 47.7744, volume = 128.9273, difficulty = 2.75, effort = 354.5503, time = 19.6972, bugs = 0.04297 | 0.03077 |
| ORLS | h1 = 4, h2 = 55, N1 = 36, N2 = 72, vocabulary = 59, length = 108, calculated_length = 325.9747, volume = 635.3254, difficulty = 2.6181, effort = 1663.3975, time = 92.4109, bugs = 0.2117 | 74.7581 |
| SLS | h1 = 5, h2 = 31, N1 = 25, N2 = 50, vocabulary = 36, length = 75, calculated_length = 165.1897, volume = 387.7443, difficulty = 4.0322, effort = 1563.4853, time = 86.8602, bugs = 0.1292 | 3.4735 |

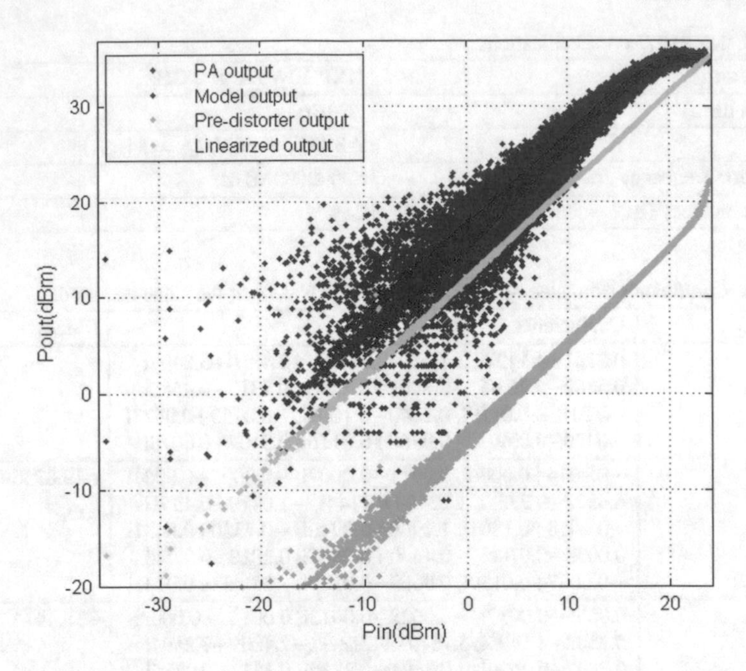


**Fig. 12** AM-AM linearization through DPD using ILA and LSE

and volume which give a direct idea of the number of operations performed by the program and this will influence directly in the logic resources needed for hardware implementation.

A pre-distorter was built for this RF-PA using ILA and both techniques LSE and SLS to obtain the coefficients for inverse model using the full dataset. Figures 12 and 13 show the real RF-PA behavior, MPM of the RF-PA adjusted by LSE, the LSE pre-distorter output and LSE linearized output for AM-AM and AM-PM curves.

Figures 14 and 15 show the real RF-PA behavior, MPM of the RF-PA adjusted by SLS the SLS pre-distorter output and SLS linearized output for AM-AM and AM-PM curves. As it can be seen in the plots, both pre-distorters give a linearized output in the AM-AM plot and maintain the phase near to zero in the AM-PM plot.

As it can be seen in the plots, both pre-distorters give a linearized output in the AM-AM plot and maintain the phase near to zero in the AM-PM plot.

Figure 16 depicts the spectral analysis of the used data, the input signal is clearly in the limits of used bandwidth since the power in near bands is negligible but when it is amplified by the RF-PA an increment of power in the near bands is clearly observed due to the nonlinearities of RF-PA, when the pre-distorter is added it can be seen that this band interferences are significantly attenuated.

The last step in this chain of modeling and pre-distortion is to implement the generated method in a Field Programmable Gate Array (FPGA) development board. An FPGA is a device which contains logic blocks that can be configured and interconnected to implement different functionalities with the main advantage that non-

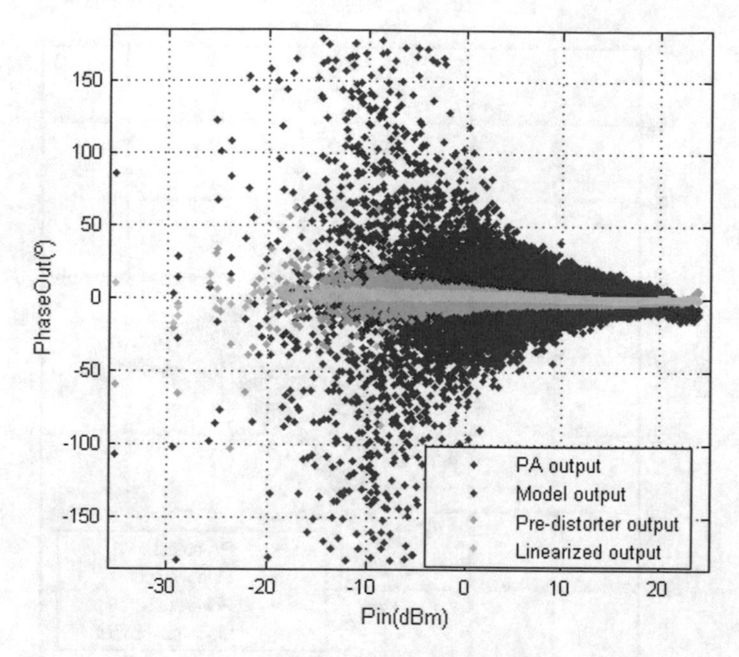


**Fig. 13** AM-PM linearization through DPD using ILA and LSE

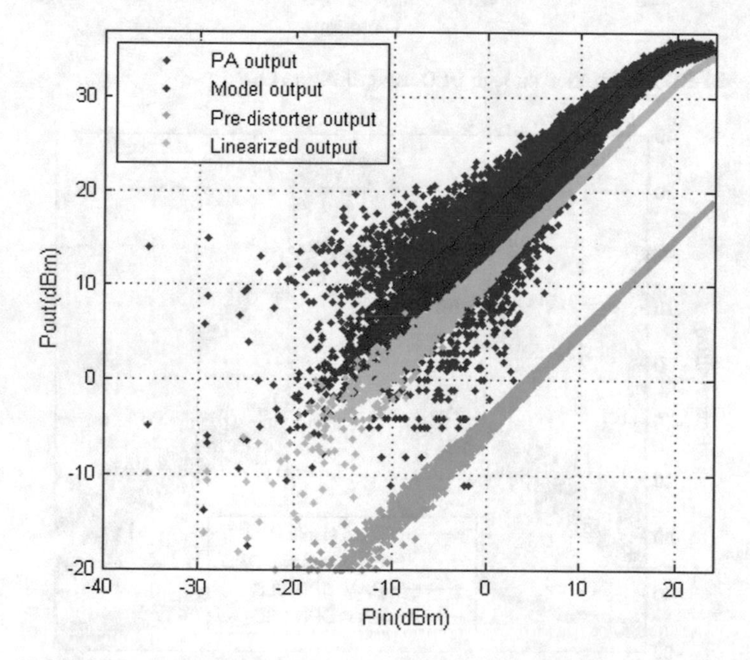


**Fig. 14** AM-AM linearization through DPD using ILA and SLS

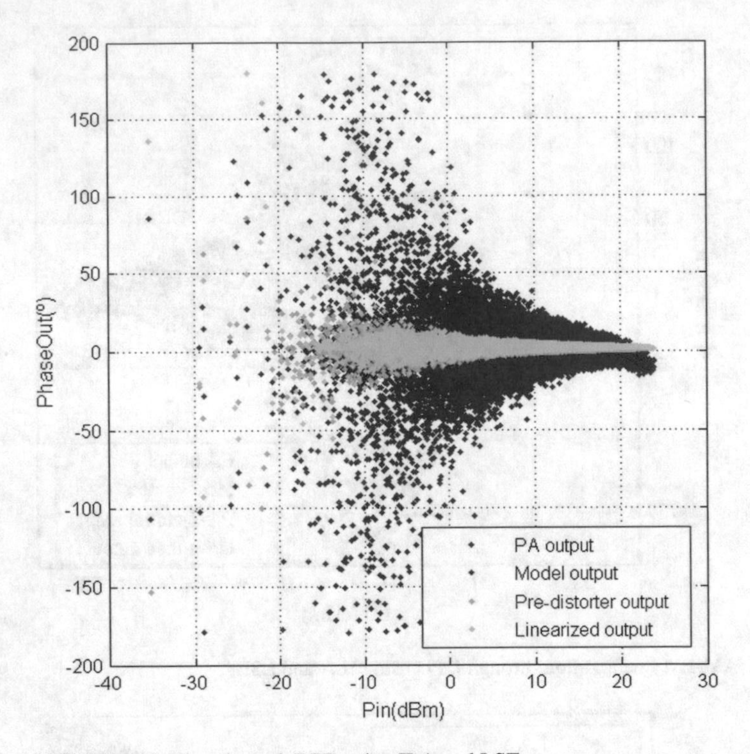


**Fig. 15** AM-PM linearization through DPD using ILA and LSE

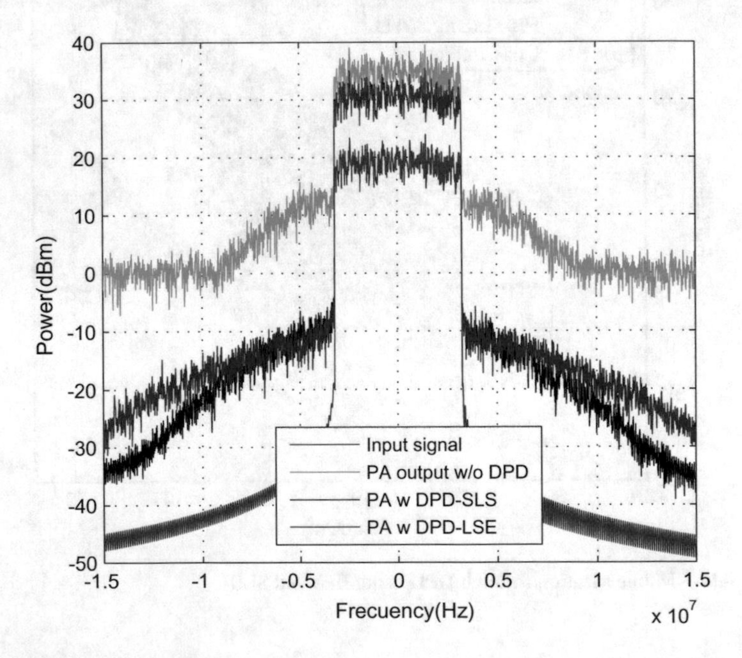


**Fig. 16** Power spectral plot

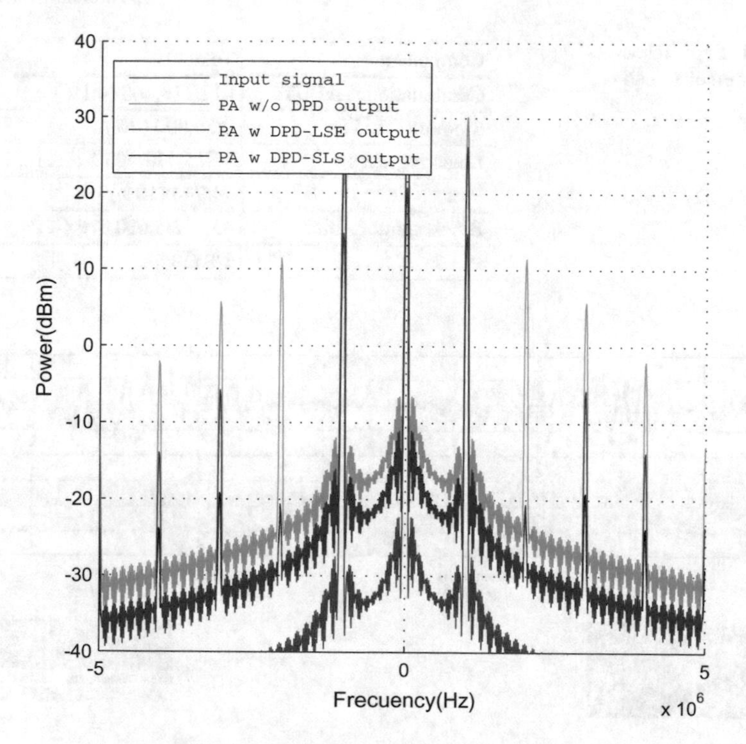


**Fig. 17** Amplitude modulated signal power spectral plot

predefined architecture is defined such a Digital Signal Processor. The FPGA platform was chosen because the development of pre-distorters in chip is planned in a future, so the FPGA represents an adequate platform given its flexibility for prototyping and debugging.

The emulation of generated models was made using the DSP Stratix III developed board and the tool DSP Builder by Altera which can generate Hardware description language from Simulink models. In order to perform the emulation, an amplification of an amplitude modulated signal was simulated with the generated models and then using the Digital Analog Converter (DAC) it can be analyzed in the oscilloscope, the power spectral of the implemented wave is illustrated in Fig. 17. The implementation was developed using Look Up Tables (LUT's), in Table 4 details about the resources used are given and the results are showed in Fig. 18 which shows the amplification without DPD and with SLS DPD.

**Table 4** Logic resources used for emulate DPD

| Component | Total |
|---|---|
| Combinational ALUTs | 412/113,600 (<1%) |
| Memory ALUTs | 0/56,800 (0%) |
| Logic registers | 1,215/113,600 (<1%) |
| Pins | 74/744 (10%) |
| Block memory bits | 43,212/5,630,976 (<1%) |
| PLLs | 1/8 (13%) |

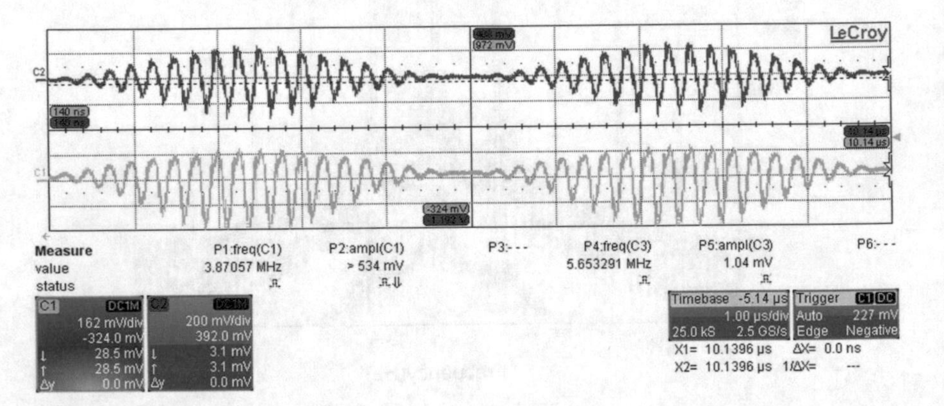


**Fig. 18** FPGA implementation signal w/o DPD (*down*) and signal w DPD-SLS (*UP*)

## 6 Conclusions

In this work three methods for the estimation of coefficients of the MPM were showed and implemented, first LSE which is the common linear regression, it is fast for calculating coefficients nevertheless for its application it is necessary to have a previous data measurement and an idea of what nonlinear and memory depth to use. We propose ORLS and SLS as alternatives of estimation when LSE is not pretty applicable. The ORLS technique does not need matrix inversions but it requires enough memory to storage a square matrix of the same size of the number of measurements to use in the calculations, the advantage of ORLS is that it can give an idea of how to parameterize our MPM and it will give the opportunity to weight our model between calculation complexity and accuracy. SLS method results more usable when the data are coming sample by sample, it also does not require big storage capacity because the biggest charge in memory is the matrix $\mathbf{S}_D$ its implementation does not use matrix inversions which reduce the computational cost, SLS is the most feasible alternative if an online identification is going to be performed or if and adaptive model is required. The viability of the two proposed methods has been proved through the simulation of them and the results where those similar coefficients were found. Finally it can be said that with the same data and configuration, the three methods will give the same coefficients since they derive from the common least square approach, making them different only by their advantages in particular scenarios.

Two of the techniques described, LSE and SLS, were used to develop DPD predistorters both of them showed good performance doing linearization in the RF-PA model, a better work in large bandwidth span was observed in the SLS predistorter which attenuates better the inter-modulation products. The two linearization outputs had a spurious free dynamic range of 50 dBm and a reduction of the adjacent channel power ratio of 25 dBc could be achieved. In a real scenario where an on-line linearization must be performed, SLS technique can represent a big reduction in the memory capacity required by the system making it suitable for its implementation in FPGA boards. FPGA based systems represent a flexible and feasible platform for DPD implementation.

**Acknowledgements** The authors wish to thank PhD. Patrick Roblin, Professor at Ohio State University, for its support provided through the measuring data. In addition, the authors would like to express their gratitude to the IPN for its financial support by the project SIP-20170588.

## References

1. Katz, A., Wood, J., Chokola, D.: The evolution of PA linearization: from classic feedforward and feedback through analog and digital predistortion. IEEE Microw. Mag. **17**(2), 32–40 (2016)
2. Roblin, P., Quindroit, C., Naraharisetti, N., Gheitanchi, S., Fitton, M.: Concurrent linearization: the state of the art for modeling and linearization of multiband power amplifiers. IEEE Microw. Mag. **14**(7), 75–91 (2013)
3. Liu, Y., Huang, C., Quan, X., Roblin, P., Pan, W., Tang, Y.: Novel linearization architecture with limited ADC dynamic range for green power amplifiers. IEEE J. Sel. Areas Commun. **34**(12), 3902–3914 (2016)
4. Nuñez-Pérez, J.C., Cárdenas-Valdez, J.R., Gontrand, C., Reynoso-Hernandez, J.A., Hirata-Flores, F.I., Jauregui-Duran, R., Perez-Pinal, F.J.: Flexible test bed for the behavioural modelling of power amplifiers. COMPEL - Int. J. Comput. Math. Electr. Electron. Eng. **33**(1/2), 355–375 (2013)
5. Rahati Belabad, A., Motamedi, S.A., Sharifian, S.: An adaptive digital predistortion for compensating nonlinear distortions in {RF} power amplifier with memory effects. Integr. {VLSI} J. **57**, 184–191 (2017)
6. Roblin, P., Myoung, S.K., Chaillot, D., Kim, Y.G., Fathimulla, A., Strahler, J., Bibyk, S.: Frequency-selective predistortion linearization of RF power amplifiers. IEEE Trans. Microw. Theory Tech. **56**(1), 65–76 (2008)
7. Li, H., Kwon, D.H., Chen, D., Chiu, Y.: A fast digital predistortion algorithm for radio-frequency power amplifier linearization with loop delay compensation. IEEE J. Sel. Top. Signal Process. **3**(3), 374–383 (2009)
8. Nuñez-Pérez, J.C., Cárdenas-Valdez, J.R., Montoya-Villegas, K., Reynoso-Hernandez, J.A., Loo-Yau, J.R., Gontrand, C., Tlelo-Cuautle, E.: FPGA-based test bed for measurement of AM/AM and AM/PM distortion and modeling memory effects in {RF} {PAs}. Integr. {VLSI} J. **52**, 291–300 (2016)
9. Naraharisetti, N., Roblin, P., Quindroit, C., Rawat, M., Gheitanchi, S.: Quasi-exact inverse PA model for digital predistorter linearization. In: 82nd ARFTG Microwave Measurement Conference, pp. 1–4 (2013)
10. Wu, X., Shi, J., Chen, H.: On the numerical stability of RF power amplifier's digital predistortion. In: 2009 15th Asia-Pacific Conference on Communications, pp. 430–433 (2009)

11. Dvorak, J., Marsalek, R., Blumenstein, J.: Adaptive-order polynomial methods for power amplifier model estimation. In: 2013 23rd International Conference Radioelektronika (RADIOELEKTRONIKA), pp. 389–392 (2013)
12. Ntoun, R.S.N., Bahoura, M., Park, C.W.: Power amplifier behavioral modeling by neural networks and their implementation on FPGA. In: 2012 IEEE Vehicular Technology Conference (VTC Fall), pp. 1–5 (2012)
13. Zhu, A., Brazil, T.J.: Behavioral modeling of RF power amplifiers based on pruned volterra series. IEEE Microw. Wirel. Compon. Lett. **14**(12), 563–565 (2004)
14. Ku, H., Kenney, J.S.: Behavioral modeling of nonlinear RF power amplifiers considering memory effects. IEEE Trans. Microw. Theory Tech. **51**(12), 2495–2504 (2003)
15. Cárdenas Valdez, J.R., Z-Flores, E., Núñez Pérez, J.C., Trujillo, L.: Local Search Approach to Genetic Programming for RF-PAs Modeling Implemented in FPGA, pp. 67–88. Springer International Publishing, Cham (2017)
16. Golovins, E., Ventura, N.: Modified order-recursive least squares estimator for the noisy OFDM channels. In: Fifth Annual Conference on Communication Networks and Services Research (CNSR'07), pp. 93–100 (2007)
17. Wang, Y., Ikeda, K., Nakayama, K.: A numerically stable fast newton-type adaptive filter based on order recursive least squares algorithm. IEEE Trans. Signal Process. **51**(9), 2357–2368 (2003)
18. HongWei, Z., Qiang, S., Wang, G., You, H.: System errors estimation of DOA and TDOA jointed locating system using sequential least squares. In: Proceedings of 2011 IEEE CIE International Conference on Radar, vol. 2, pp. 1025–1028 (2011)
19. Chen, Y., Zhang, D., Lin, Z., Lai, X.: A sequential weighted least squares procedure for design of IIR filters and two-channel IIR filter banks. In: 2014 IEEE International Symposium on Circuits and Systems (ISCAS), pp. 1195–1198 (2014)
20. Diniz, P.S.R.: Conventional RLS Adaptive Filter, pp. 209–247. Springer, Boston (2013)
21. Chani-Cahuana, J., Fager, C., Eriksson, T.: A new variant of the indirect learning architecture for the linearization of power amplifiers. In: 2015 10th European Microwave Integrated Circuits Conference (EuMIC), pp. 444–447 (2015)
22. Amin, S., Zenteno, E., Landin, P.N., Rnnow, D., Isaksson, M., Hndel, P.: Noise impact on the identification of digital predistorter parameters in the indirect learning architecture. In: 2012 Swedish Communication Technologies Workshop (Swe-CTW), pp. 36–39 (2012)
23. Dwivedi, N., Bohara, V.A., Hussein, M.A., Venard, O.: Fixed point digital predistortion system based on indirect learning architecture. In: 2014 International Conference on Advances in Computing, Communications and Informatics (ICACCI), pp. 1376–1380 (2014)
24. Chani-Cahuana, J., Landin, P.N., Fager, C., Eriksson, T.: Iterative learning control for RF power amplifier linearization. IEEE Trans. Microw. Theory Tech. **64**(9), 2778–2789 (2016)
25. Paaso, H., Mammela, A.: Comparison of direct learning and indirect learning predistortion architectures. In: 2008 IEEE International Symposium on Wireless Communication Systems, pp. 309–313 (2008)
26. Abd-Elrady, E., Gan, L., Kubin, G.: Direct and indirect learning methods for adaptive predistortion of IIR hammerstein systems. e & i Elektrotechnik und Informationstechnik **125**(4), 126–131 (2008)
27. Halstead, M.H.: Elements of Software Science. Operating and Programming Systems Series. Elsevier Science Inc., New York (1977)

# Optimal Sizing of Amplifiers by Evolutionary Algorithms with Integer Encoding and $g_m/I_D$ Design Method

**Adriana C. Sanabria-Borbón, Esteban Tlelo-Cuautle and Luis Gerardo de la Fraga**

**Abstract** The optimal sizing of analog integrated circuits (ICs) by evolutionary algorithms (EAs) has the challenge of reducing the search spaces of the design variables, guaranteeing the proper bias conditions and providing manufacturable feasible solutions. In this manner, this chapter applies two EAs, namely the non-dominated sorting genetic algorithm (NSGA-II) and differential evolution (DE) to optimize operational amplifiers designed with complementary metal-oxide-semiconductor (CMOS) IC fabrication technology. Those EAs link the simulation program with IC emphasis (SPICE) to evaluate performances characteristics, and apply the $g_m/I_D$ design method to guarantee bias conditions and to reduce the search spaces for the design variables of the MOS transistors, which are associated to the width (W) and length (L) of their channels. The W/L design variables are encoded with integer values that are converted to multiples of the IC fabrication technology within SPICE. That way, integer encoding of the design variables provides manufacturable transistor sizes, while the EAs are accelerated by using chromosomes with reduced search spaces provided by the $g_m/I_D$ design method. As examples, two CMOS operational transconductance amplifiers (OTAs) are sized with this optimization approach to highlight the EA's advantages when applying $g_m/I_D$ design method and integer encoding.

**Keywords** Multi-objective optimization · Circuit sizing · $g_m/I_D$ design method · NSGA-II · Differential evolution algorithm · Operational transconductance amplifier · MOS transistor · SPICE

A.C. Sanabria-Borbón (✉)
Department of Electrical and Computer Engineering, Texas A&M University,
College Station, TX 77843, USA
e-mail: adca.sanabria@gmail.com

E. Tlelo-Cuautle · L.G. de la Fraga
Computer Science Department, Cinvestav, Av. IPN 2508,
07360 Mexico City, Mexico
e-mail: fraga@cs.cinvestav.mx

E. Tlelo-Cuautle
Department of Electronics, INAOE, Luis Enrique Erro No. 1,
72840 Tonantzintla, Puebla, Mexico
e-mail: etlelo@inaoep.mx

© Springer International Publishing AG 2018

Y. Maldonado et al. (eds.), *NEO 2016*, Studies in Computational Intelligence 731,
https://doi.org/10.1007/978-3-319-64063-1_11

## 1 Introduction

The optimization of analog integrated circuits (ICs) consists of finding the correct topology and the circuit element values that satisfy a set of target specifications. In the traditional analog IC design process, a designer chooses an IC topology, and then the biasing conditions are determined by applying various techniques, as the operating point formulation approach introduced in [1]. In the majority of the cases, the most used technique consists of using explicit analytical equations to estimate the sizes of the circuit elements, and after that, a traditional IC designer uses a circuit simulator, e.g. SPICE, to accomplish the required specifications. This process is performed in a loop of trial and error cases until the design is tuned. In addition, because analog IC design involves many trade-offs among target specifications, the application of evolutionary algorithms (EAs) is quite useful for optimizing complementary metal-oxide-semiconductor (CMOS) operational amplifiers. It is also stressed in [2], that analog IC design is reluctant to be an automatic process. For this reason, optimizing analog ICs is currently one of the main research fields in the electronic design automation (EDA) industry [3].

The authors in [4] summarized current sizing approaches for analog ICs, and classified them into two main groups: knowledge-based and optimization-based approaches. As an example of the second group, the usefulness of EAs is highlighted in [5], where it can be inferred that the performance of an EA is different for different kinds of ICs. Nevertheless, those contributions as well as the approach reported in [2], do not consider the usefulness of exploiting biasing techniques into an optimization loop. In this context, this chapter shows the usefulness of the $g_m/I_D$ method [6–8], for estimating reduced search spaces for the width (W) and length (L) of the MOS transistors that guarantee appropriate bias level conditions. In addition, for the given IC fabrication technology, the W/L sizes must be multiples of lambda, which is provided by the manufacturer, therefore one can encode the design variables with integer numbers to provide feasible solutions that do not need a post-processing step like in [9], to round-off the sizes. In fact, rounding real W/L sizes values to be multiples of lambda, i.e. the IC fabrication technology, may degrade the performance of the objective functions or even violate the constraints. In this manner, integer encoding provides discrete W/L size values that are multiples of lambda, and reduces the computer execution time, and reduces the memory usage because in a 64-bit processor the integer variables employ 4 bytes, while real (double) variables employ 8 bytes [10].

EAs can be applied to single-objective and multi-objective problems. In the first case, the differential evolution (DE) algorithm is quite useful and for the second case, the non-dominated sorting genetic algorithm (NSGA-II) has proven to be the state-of-the-art in particular if two or three objectives are being considered. Multi-objective optimization algorithms are still under development [11–16]. Algorithms of that kind provide a population of feasible solutions that can be much better than the solutions provided by traditional circuit optimization approaches [17–19]. In this manner, EAs are quite useful for the optimization of operational amplifiers [9, 20, 21], and in this chapter, DE and NSGA-II are applied to optimize the sizes of CMOS operational transconductance amplifiers (OTAs), and those algorithms are improved by reducing the search spaces for the design variables W/L, and by applying integer encoding [10], to provide manufacturable feasible solutions for the given IC fabrication technology.

The remainder of this chapter is organized as follows: the chosen EAs are briefly described in Sect. 2. Second, the $g_m/I_D$ method is presented in Sect. 3. Our proposed optimization approach by applying the $g_m/I_D$ design method and integer encoding is described in Sect. 4. The examples on sizing OTAs with IC technology of 180 nm are given in Sect. 5. Finally, the conclusions and possible paths for future work are listed in Sect. 6.

## 2 Evolutionary Algorithms Solvers: DE and NSGA-II

The circuit optimization problem can be formulated by,

$$\text{Optimize} \quad \mathbf{f}(\mathbf{x})$$

$$\text{Subject to: } \mathbf{x} \in Q \subset \mathbb{N}^n_{>0} | \mathbf{g}(\mathbf{x}) \leq 0, \tag{1}$$

where $\mathbf{x}$ is a multidimensional vector $\mathbb{N}^n_{>0}$ of decision parameters delimited by $x^i_{\min} \leq x^i \leq x^i_{\max}$. $\mathbf{f}(\mathbf{x})$ represents a vector of $m$ objectives ($f_1(\mathbf{x})$,…, $f_m(\mathbf{x})$) that are being minimized/maximized, and $\mathbf{g}(\mathbf{x})$ is the vector of $p$ constraints that should be satisfied to guarantee feasible solutions. If $m = 1$, (1) is associated to a single-objective problem (SOP) that can be solved by DE, while for $m = \{2, 3\}$, (1) is associated to a multi-objective problem (MOP) that can be solved by NSGA-II. In both evolutionary algorithms, multiple constraints can be included in order to deal with a global and multi-dimensional optimization problem.

A SOP has only one solution. Contrary to a SOP, a MOP has a set of solutions, this is produced because their objective functions are in conflict, then when one is improved, the others could deteriorated. In a SOP it is easy to compare two solutions, one could be better or worse compared with the other, but in a MOP is a little more complicated.

In a MOP, two solutions are compared according to Pareto optimality: a decision vector $\mathbf{x} \in Q$ is Pareto optimal if there does not exist another decision vector $\mathbf{y} \in Q$ such that $f_i(\mathbf{y}) \leq f_i(\mathbf{x})$ for all $i = 1, \ldots, m$ and $f_j(\mathbf{y}) < f_j(\mathbf{x})$ for at least one index $j$. Therefore, in a MOP, the aim is to determine the Pareto optimal set from the set $Q$ of all the decision variable vectors that satisfy (1) with $m = \{2, 3\}$. The Pareto optimal front is the image, in the objective space, of the Pareto optimal set.

## 2.1 DE Algorithm

According to (1), the DE algorithm is a parallel direct search method that uses $n$-dimensional parameter vectors as a population for each generation $g$ [22]. In each generation, for every individual $\mathbf{x}_k$, a cost function $f(\mathbf{x}_k)$ is evaluated as the weighted sum of the error of the performance measurements with respect to the performance specifications, and is modeled by,

$$f(\mathbf{x}_k) = \alpha_1 C_A(\mathbf{x}_k) + \alpha_2 C_B(\mathbf{x}_k) + \ldots, \tag{2}$$

where $C_A$ given by (3), and $C_B$ given by (4), are the errors between a specification defined as $A_{\text{spc}}$ or $B_{\text{spc}}$ (upper and lower limits, respectively), and the value of that performance metric $A(\mathbf{x}_k)$ or $B(\mathbf{x}_k)$, respectively (evaluated for an individual $\mathbf{x}_k$). Each error is being multiplied by an associated factor $\alpha_n$, which represents the relevance of a determined specification in the whole optimization process. Therefore, the description of the cost function will determine which specifications will have prevalence in the optimization process. In (3) and (4), $\mathbf{x}_k$ corresponds to the vector of design variables (W/L) for each $k$ MOS transistor.

$$C_A(\mathbf{x}_k) = 1 - \frac{A(\mathbf{x}_k)}{A_{\text{spc}}}; \quad A(\mathbf{x}_k) < A_{\text{spc}}, \tag{3}$$

$$C_B(\mathbf{x}_k) = \frac{B(\mathbf{x}_k)}{B_{\text{spc}}} - 1; \quad B(\mathbf{x}_k) > B_{\text{spc}}. \tag{4}$$

Herein, DE algorithm is applied to minimize the cost function $f(\mathbf{x}_k)$ under the constraints related to keep all MOS transistors operating in the saturation region. The DE algorithm is based on an iterative procedure that employs genetic operators, namely: crossover and mutation, which modify the population in each generation. In addition, a selection operator is responsible for the choice of the new population's individuals, according to a cost function comparison. That way, only the most adapted individuals reach the next generation. The pseudocode of DE is sketched in Algorithm 1, it depends on the population size $N$, the number of generations $g_{\max}$, and the cost function $f$. The DE algorithm has three control parameters: crossover probability $CR$ [0, 1], difference constant $F$, and population size $N$. These parameters are usually kept as constants across the evolution process, and the population size is also constant for each generation $g$. The initial population is randomly selected and the gene values are distributed into the search space. Subsequently, the population is modified by the genetic operators in an iterative process until a threshold for the cost function is accomplished, or a maximum number of iterations is reached. The DE algorithm working principle is the generation of new individuals by adding the weighted difference between two population individuals to a third one, as given by,

$$\mathbf{v}_{g+1} = \mathbf{x}_{\text{best}} + F(\mathbf{x}_1 - \mathbf{x}_2), \tag{5}$$

where $\mathbf{x}_1$ and $\mathbf{x}_2$ are two random vectors selected from the population, $F$ determines the amount of change added to the best individual of the previous generation and in consequence determines the convergence rate. The individuals are randomly selected from the actual population with the constraint of being different among them. Then a *crossover* is performed between the created individual $\mathbf{v}_{g+1}$ and a target vector $\mathbf{x}$ as:

$$u_i = \begin{cases} v_{i,g+1} & \text{if rand}[0,1) \leq CR, \\ x_i, & \text{otherwise,} \end{cases} \tag{6}$$

where rand is a function that returns a real random number greater or equal to zero and lesser than 1, and $CR$ is the crossover probability. After crossover has been applied, a comparison is done between the generated individual and a predetermined population member cost. This work uses the cost function given in (2). Only the individual with the best cost value survives to the next iteration [22]. DE has some variants called strategies that are conventionally named $DE/x/y/z$, where $x$ is a string that denotes the base individual and can take values such as rand (a randomly selected individual) or best (the individual with the best cost value), $y$ refers to the amount of individuals implied on the operators, and $z$ refers to the crossover method, which can be binomial or exponential. Detailed information about crossover methods are given in [23, 24]. The most useful DE strategies are described in detail in [22, 25–28].

**Algorithm 1** Differential Evolution

```
1: procedure DE(N, g_max, f)
2:   Evaluate the initial population P (of size N) of random individuals.
3:   while stopping criterion is not met (g < g_max) do
4:     for each individual in Population do
5:       Create a candidate v from two randomly chosen parents using (5).
6:       Apply crossover
7:       Evaluate the cost function f(u) of the candidate.
8:       if the cost function of the candidate is less than the one of the parent then
9:         The candidate replaces the parent.
10:      else
11:        The candidate is discarded.
12:  return The best individual in the last population.
```

## 2.2 NSGA-II

The non-dominated sorting genetic algorithm (NSGA-II) is a non-domination based multi-objective optimization (MOO) algorithm, which has the following characteristics [29]: maintenance diversity, convergence and robustness of solutions in the Pareto front (PF) [30]. NSGA-II approximates the PF of a MOO problem by sorting and ranking all solutions in different Pareto sub-fronts, in order to choose the best solutions to create new offsprings [31], as sketched by Algorithm 2 [32], which has the population size $N$, the number of generations $g_{\max}$, and the objective functions $\mathbf{f}$ that depend on the circuit parameters $\mathbf{x}$.

---

**Algorithm 2** NSGA-II algorithm

1: **procedure** NSGA- II( $N$, $g_{\max}$, $\mathbf{f}$) ▷ $N$ members evolved $g_{\max}$ generations to solve $\mathbf{f}$
2: Initialize randomly the population
3: Calculate objective values
4: Assign rank (level) based on Pareto dominance - sort generated child population
5: Binary tournament selection
6: **for** $g = 1$ to $g_{\max}$ **do**
7: **for** each Parent and Child in Population **do**
8: Assign Rank (level) based on Pareto - sort
9: Generate sets of non-dominated vectors along PF
10: Loop (inside) by adding solutions to the next generation starting from the first front until $N$ individuals found determine crowding distance between points on each front
11: Select points (elitist) on the lower front (with lower rank) and are outside a crowding distance
12: Create next generation
13: Binary Tournament Selection
14: **return** The PF with highest rank and its associated PS.

---

NSGA-II is based on two main procedures: fast non-dominated sort (with $O(mN^2)$ computational complexity, where $m$ is the number of objectives and $N$ the population size) and crowding distance assignment. These two procedures ensure elitism and NSGA-II allows adding constraints to guarantee that the solutions are feasible. According to Algorithm 2, a population of competing individuals is created, and then the non-domination process ranks and sorts each individual. An offspring population is created by performing crossover, mutation and selection operations; and finally the parents and offsprings are combined before partitioning the new population into fronts.

## 3 $g_m/I_D$ Design Method and Integer Encoding

Metal-oxide-semiconductor field-effect-transistors (MOSFETs) are becoming the most widely used ones for integrated circuit technology within complementary (CMOS) integrated circuit (IC) fabrication technology [33]. Depending on the bias conditions, MOSFETs can operate within three regions known as linear, weak inversion and strong inversion. Also, it is possible to build two kinds of MOSFETs depending how silicon is doped: P-channel and N-channel built MOSFETs. For digital applications, CMOS ICs use one P-channel MOSFET for each N-channel, and the operation regions are mainly set to cut-off (OFF state) and saturation, which is the ON state. For analog applications, it is more common to operate the MOSFETs in the saturation region [34], while integrated resistors can be implemented with MOSFETs in the linear region and are used for linearization [35]. Also, integrated capacitors can be implemented using MOSFETs but for large capacitor values other techniques are used like capacitance multipliers or layout techniques. As one can infer, MOSFETs are quite useful for a great variety of today's modern applications. However, the optimization of the designs cannot be performed manually due to the very wide spectrum of target specification to accomplish and to the large quantity of variables involved, which can also have large ranges of feasible values.

The $g_m/I_D$ design approach was originally introduced in [6]. Its main advantage is the unification of all the MOS transistor operating regions. This method is based on the fact that the transconductance ($g_m$) and DC drain current ($I_D$) of a MOS transistor are proportional to its channel width ($W$) and length ($L$). That way, the ratio $g_m/I_D$ does not depend on the W/L sizes, but only on the biasing [7]. The ratio $g_m/I_D$ is standard for all MOS transistors in the same IC fabrication technology, and shows a relationship between small and large-signal models, and is helpful to describe other performance features like noise [36].

The $g_m/I_D$ design method begins by characterizing an IC fabrication technology through a simulation to measure MOS transistor parameters, such as: threshold voltage ($V_{th}$), transconductance efficiency $g_m/I_D$, intrinsic transistor gain ($g_m/g_o$), current density ($I_D/W$), and transit frequency ($f_T$). This is done for a set of values for the channel length $L$, while the gate-to-source voltage $V_{gs}$ is swept and $W$ remains constant. The characterization data is saved in look-up tables (LUTs) and it can be visualized in figures as already shown in [6–8]. That way, for different values of L, and for both kinds of MOS transistors N-channel and P-channel, the generated data is saved in different files to be used in the estimation of the $W/L$ sizes, which must be multiples of lambda, i.e. the IC fabrication technology. This issue can be guaranteed by applying integer encoding, as shown in [10], where from the SPICE netlist: all MOS transistors are parameterized by their design variables W/L using the sentence `.param`. In this manner, performing integer encoding to MOS transistor-based circuits, the EA assigns integers to the design variables W/L that are associated to $p$ and $q$ into SPICE as shown below in Fig. 1.

In the netlist above, the sizes $W/L$ are encoded by integer numbers ($p$ and $q$) and then they are scaled to be multiples of lambda, in this case 90 nm. Within the SPICE simulation, the MOS transistors take the real values. For example: into the circuit simulation $W$ and $L$ are multiplied by the IC fabrication technology using the

```
...
scale=90nm
.param p=100
.param q=2
...
MP1 drain-node gate-node source-node bulk-node PMODEL W='p' L='q'
...
```

**Fig. 1** SPICE netlist example for using integer W/L values

command *.scale*. In this manner, using an IC fabrication technology of 90 nm, the netlist for the MOS transistor MP1 is updated to W = 100*90 nm (9 μm), L = 2*90 nm (0.18 μm), and so on.

## 4 Proposed Sizing Optimization Approach

In EA solver, DE or NSGA-II, each member $\mathbf{x}_i$ in the population is a vector that contains the design variables (*W*s and *L*s of the MOS transistors and the values for passive circuit elements like resistors and capacitors). The cost function given in (2) is estimated by the sum of the errors of all the desired target specifications, and the constraints to guarantee the operating region of all N-type and P-type MOS transistors. From these conditions, the individuals (W, L and compensation capacitor) whose performances have the least cost error and keep the operating conditions, will pass to the next generation.

Figure 2 highlights our proposed circuit optimization approach by applying $g_m/I_D$ within the DE algorithm. The first step consists on characterizing the IC fabrication technology, where by simulation the voltage and channel length of the MOS transistor are swept in order to build the LUTs that save characteristics of both types of MOS transistors, P-type and N-type. Then, the operating regions, the biasing voltages, and the input and output swings are extracted and combined in order to establish a $V_{gs}$ range for each MOS transistor [8]. The operating region is a criteria extracted by the designer from the DC specifications, and they can be: weak, moderate and strong inversion, which offer different advantages in terms of performances.

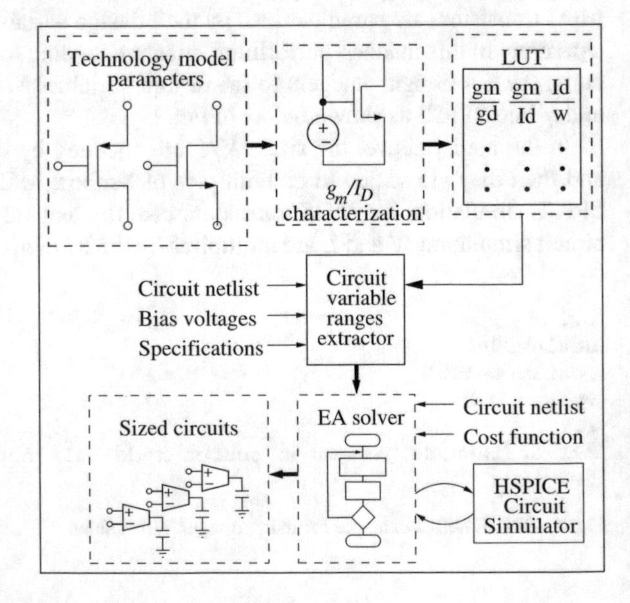


**Fig. 2** Proposed circuit sizing optimization approach

Defining a $\Delta V$ voltage, due to the overdrive variations tolerance as in [37], the MOS transistor saturation conditions can be expressed as:

$$V_{gs} \geq V_{th} + \Delta V \tag{7}$$

$$V_{ds} \geq V_{gs} - V_{th} + \Delta V. \tag{8}$$

Employing these simple inequalities, the biasing voltages $V_{dd}$ and $V_{ss}$ and the desired output voltage swing, the entire set of conditions that the nodal voltages must satisfy are found. After that, a sweep for each nodal voltage is performed from $V_{ss}$ to $V_{dd}$ in order to find all nodal voltages combinations that satisfy all the saturation biasing conditions.

The collected voltage ranges of nodal voltages are mapped into $V_{gs}$ of all MOS transistors. Then, the LUTs data and (9) are used to map the $V_{gs}$ in the corresponding current density $I_D/W$. Subsequently, the previous data is mapped into the corresponding $W$ ranges by (10).

$$I_D = \frac{g_m}{\left(\frac{g_m}{I_D}\right)^*} \tag{9}$$

$$W_x = \frac{I_D}{\frac{I_D}{W}} \tag{10}$$

Hence, there is a reduced search space for each design variable that accomplishes the biasing conditions and is feasible for the IC technology. The reduced search spaces are used to begin the optimization process by the EA solver. That is, now each design variable $W/L$ has its own range which is independent of the other variables. The ranges define the maximum and minimum values for the design variables during the optimization process. Then, EA solver is executed as detailed in Sect. 2, which uses the circuit simulator SPICE in the loop for measurements and the cost function calculation.

## 5 Examples on Sizing OTAs

Our proposed circuit optimization approach is tested for the sizing of two operational transconductance amplifiers (OTAs): the Miller OTA and the recycled folded-cascode OTA. The first one is shown in Fig. 3.

One important issue to solve before executing the selected EA solver is determining the values of their parameters. For instance, for DE algorithm, $F$ and $CR$ are problem dependent, so that one can apply the set of strategies presented by Storn [22]. Therefore, for the sizing of the OTA Miller shown in Fig. 3, the strategy with best convergence is $DE/best/1/exp$ shown in Eq. (5).

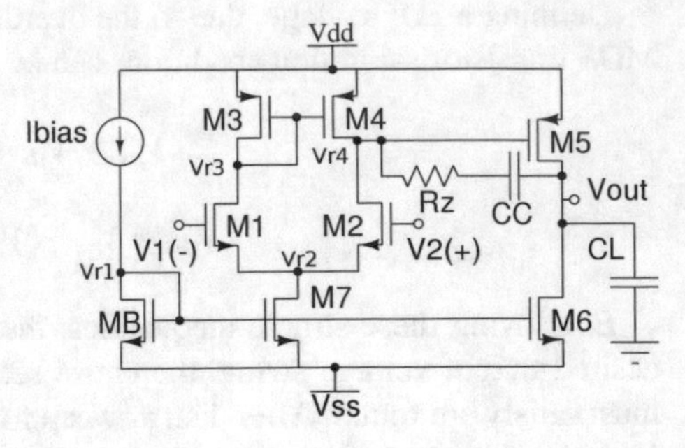


**Fig. 3** OTA Miller composed MOS transistors and the compensation $R_z C_c$ circuit elements

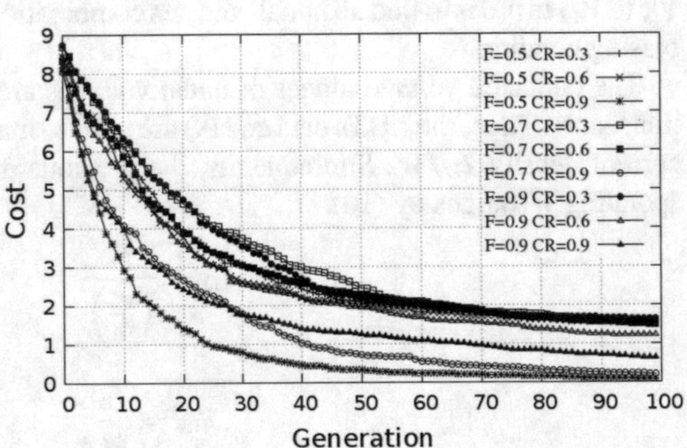


**Fig. 4** Comparison of the DE performances for different values of $F$ and $CR$

To find $F$ and $CR$ some simulations were executed on a benchmark problem taken from [22], for three combinations of $F$ [0.5, 0.7, 0.9], and also three combinations of $CR$ [0.3, 0.6, 0.9]. Figure 4 shows the comparison of the convergence of the strategy $DE/best/1/exp$ for these different parameters of $F$ and $CR$.

To apply the $g_m/I_D$ design method to estimate the design parameters ranges, the following inequalities resume the biasing conditions to keep all transistors operating in saturation region. In the next inequalities: $V_{r1}$, $V_{r2}$, and $V_{r3}$, are internal nodal voltages detailed in Fig. 3; $V_{dd}$ and $V_{ss}$ are the supply levels; $V_{icm}$ and $V_{ocm}$ are the input and output common-mode voltages, respectively; and $V_{tp}$ and $V_{tn}$ are the threshold voltages of N-type and P-type MOS transistors.

$$
\begin{aligned}
&V_{r1} \geq V_{ss} + V_{tn} + \Delta V \\
&V_{r1} \leq V_{ocm} + (V_{sw}/2) + V_{tn} + \Delta V \\
&V_{r2} \geq V_{r1} - V_{tn} + \Delta V \\
&V_{r2} \leq V_{icm} - V_{tn} - \Delta V \\
&V_{r3} \geq V_{icm} - V_{tn} + \Delta V \\
&V_{r3} \geq V_{r4} - |V_{tp}| + \Delta V \\
&V_{r3} \leq V_{dd} - |V_{tp}| - \Delta V \\
&V_{r4} \geq V_{icm} - V_{tn} - \Delta V \\
&V_{r4} \geq V_{ocm} + (V_{sw}/2) - |V_{tp}| + \Delta V \\
&V_{r4} \leq V_{dd} - |V_{tp}| + \Delta V
\end{aligned}
$$

**Table 1** Conditions for circuit simulation

| Parameter | Value |
|---|---|
| Technology | TSMC 180 nm |
| $V_{dd} = -V_{ss}$ | 0.9 V |
| $C_L$ | 3 pF |
| $I_{\text{bias}}$ | 100 μA |

**Table 2** Performance specifications for sizing the OTA Miller

| Parameter | Value |
|---|---|
| DC Gain | 60 dB |
| BW | 100 KHz |
| Phase Margin | 60 |
| Power | 1 mW |
| Vos | 100 μV |
| SR(+) | 20 V/μs |
| SR(−) | 20 V/μs |

From those equations, the $V_{gs}$ for each MOS transistor can be derived as:

$$V_{gs1} = V_{icm} - V_{r2}$$
$$V_{gs3} = V_{dd} - V_{r3}$$
$$V_{gs5} = V_{dd} - V_{r4}$$
$$V_{gs7} = V_{r1} - V_{ss}$$

The conditions employed in these simulations are listed in Table 1, where one can see that the biases $V_{dd}$, $V_{ss}$ and $I_{bias}$ are constants and also the load capacitance $C_L$. In this manner the optimization problem is devoted to improve the target specifications listed in Table 2, where the direct current (DC) gain, phase margin and slew rate (SR) must be as high as possible and the power and voltage offset (Vos) must be as low as possible.

In this MOS circuit sizing example, the cost function is calculated as a linear combination of the error of the specifications, as shown by (2). If the specifications overload a lower or upper threshold, Eqs. (3) or (4), can be used to estimate the error. Each value of $\alpha_i$ is the weight associated to each error component. Each constraint value is compared with a threshold of the specification. The cost of each specification using (3) and (4) is added to the global cost function only if the specification is not reached.

The DE algorithm was executed for $g = 150$ generations. The average cost was calculated for each generation and plotted for both cases with and without the $g_m/I_D$ design method with integer encoding, as shown in Fig. 5, where it becomes apparent that to reach the same minimal cost, DE needs 85 generations, while $DE + g_m/I_D$ takes only 40 generations. Using the same computer resources, the execution time for the first case was with the minimum value = 850.92s, while with our proposed approach it is 474.51s. It means that $DE + g_m/I_D$ is quite suitable for our purposes because it reduces the execution time for around 44%.

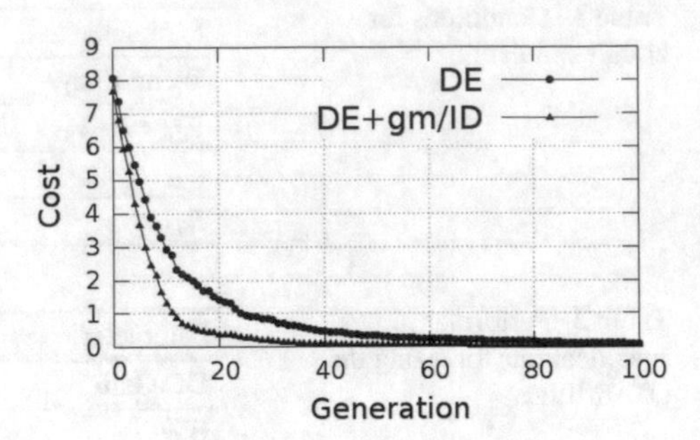


**Fig. 5** Cost evolution for both optimization cases by applying DE with and without the $g_m/I_D$ design method and integer encoding

**Table 3** Sizing statistical results by applying DE for 10 runs and 60 generations with the $g_m/I_D$ design method and integer encoding

| Performances | MAX | MIN | AVG | STD |
|---|---|---|---|---|
| Gain[dB]≥ 60 | 62.845 | 60.017 | 61.127 | 0.618 |
| BW[KHz]≥ 100 | 145.700 | 100.00 | 110.697 | 7.655 |
| PM≥ 60° | 67.607 | 60.045 | 61.912 | 1.252 |
| Power[mW]≤ 1 mW | 0.871 | 0.718 | 0.764 | 0.027 |
| Vos[μV]≤ 500 μV | 429 | 404 | 420 | 4.749 |
| SR(+)[V/μs]≥ 50 V/μs | 102.600 | 75.140 | 90.565 | 3.738 |
| SR(−)[V/μs]≥ 50 V/μs | 90.250 | 68.530 | 75.855 | 4.646 |

Table 3 lists the specification: minimum, maximum and average values of the solution for each specification. The best sizes for each target specification are listed in Table 4. Those best solutions are given by bold face numbers and below the target specifications, the Table list the corresponding W/L sizes. In this optimization case, the five feasible solutions (associated to each column) were selected among all optimal sizing results provided by DE algorithm.

The second example is for sizing the recycled folded-cascode OTA shown in Fig. 6. For this topology all MOS transistors are assumed to work in their saturation regions. NSGA-II was employed to perform the optimization process to accomplish the specifications listed in Table 5. NSGA-II was executed for a population of 100 individuals and for 100 generations. Among all the feasible solutions in the Pareto front shown in Fig. 7, Table 6 lists seven feasible ones, which accomplish seven target specifications, as it was done for the OTA Miller topology. The analog IC designer can choose the feasible solution according to the application, but from this table, one can conclude on the appropriateness of applying evolutionary algorithms with integer encoding and the $g_m/I_D$ design method.

**Table 4** Five optimal sizing results for the OTA Miller, which provide the best performance for each target specification (in bold face)

| Gain[dB]≥ 60 | **62.845** | 60.51 | 60.29 | 60.77 | 61.27 |
|---|---|---|---|---|---|
| BW[KHz]≥ 100 | 102.400 | **145.70** | 102.30 | 114.10 | 117.00 |
| PM≥ 60 ° | 62.76 | 60.24 | **67.60** | 60.93 | 60.82 |
| Power[mW]≤ 1 mW | 0.793 | 0.792 | 0.796 | 0.718 | 0.871 |
| Vos[μV]≤ 500 μV | **404** | 424 | 426 | 424 | 414 |
| SR(+)[V/μs]≥ 50 V/μs | 91.370 | 100.50 | 78.62 | 90.49 | **102.60** |
| SR(−)[V/μs]≥ 50 V/μs | 78.980 | 81.95 | 76.33 | 70.51 | **90.25** |
| W(M1,M2)[μm] | 37.35 | 37.44 | 37.62 | 37.62 | 37.26 |
| W(M3,M4)[μm] | 24.75 | 24.75 | 23.49 | 25.11 | 23.40 |
| W(M7,MB)[μm] | 14.58 | 11.07 | 12.15 | 12.42 | 13.14 |
| W(M5)[μm] | 21.24 | 21.87 | 22.14 | 19.89 | 21.60 |
| W(M6)[μm] | 67.68 | 51.66 | 59.40 | 58.32 | 63.45 |
| W(Mz)[μm] | 7.02 | 6.75 | 6.75 | 6.75 | 6.93 |
| L(M1,M2)[μm] | 0.63 | 0.63 | 0.81 | 0.90 | 0.99 |
| L(M3,M4)[μm] | 1.71 | 1.71 | 1.62 | 1.71 | 1.80 |
| L(M7,MB)[μm] | 0.99 | 1.08 | 1.08 | 0.81 | 0.81 |
| L(M5)[μm] | 0.18 | 0.18 | 0.18 | 0.18 | 0.18 |
| L(M6)[μm] | 1.53 | 1.62 | 1.71 | 1.53 | 1.17 |
| L(Mz)[μm] | 0.18 | 0.18 | 0.18 | 0.18 | 0.27 |
| C[fF] | 900 | 800 | 1100 | 900 | 800 |

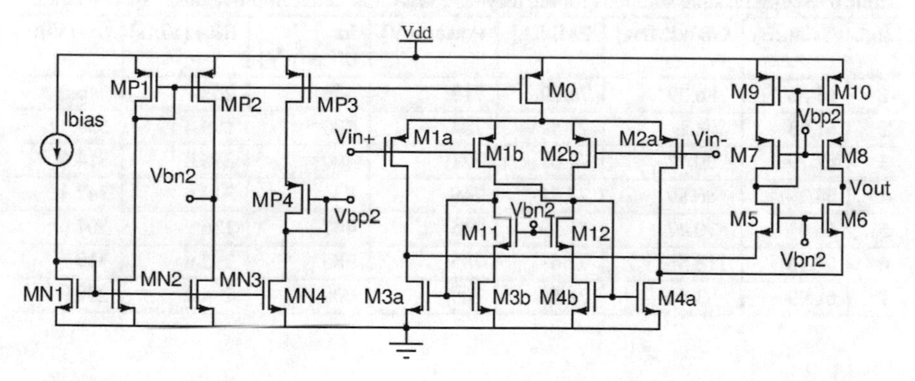


**Fig. 6** Recycled OTA topology

**Table 5** Recycling OTA target specifications

| Element | Value |
|---|---|
| Technology | TSMC 0.18 μm |
| Input offset[V] | ≤ 1 mV |
| Slew Rate[V/μs] | ≥ 90 |
| Open Loop PM | ≥ 70 |
| Capacitive Load | 5.6 pF |
| Power consumption | ≤ 800 μW |

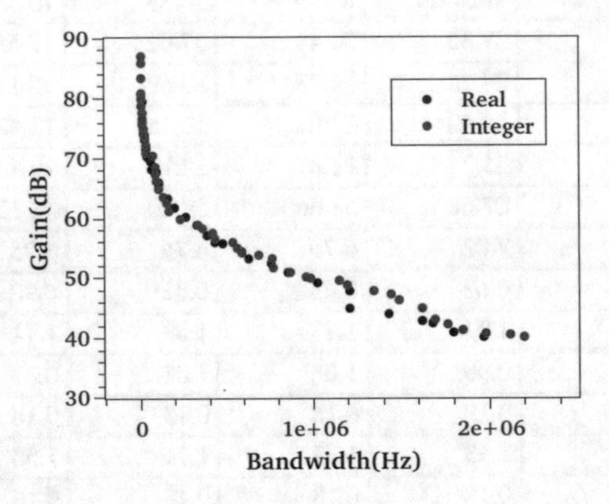


**Fig. 7** Pareto front for sizing the OTA Miller amplifier using integer and real encoding values

**Table 6** Sizing feasible solutions for the recycling OTA that better improve target specifications

| Ind. | Gain[dB] | GBW[MHz] | PM[deg] | Power[μV] | In. offset[μV] | SR+[V/μs] | SR-[V/μs] |
|---|---|---|---|---|---|---|---|
| 1 | 65.56 | 86.39 | 74.99 | 713 | 436 | 262 | 346.5 |
| 2 | 61.33 | 108.1 | 71.59 | 784 | 430 | 199.4 | 339 |
| 3 | 62.79 | 90.42 | 74.01 | 721 | 400 | 275.8 | 314.6 |
| 4 | 64.2 | 86.89 | 75.51 | 749 | 431 | 212.6 | 342.4 |
| 5 | 66.04 | 70.47 | 75.59 | 745 | 447 | 176 | 304 |
| 6 | 57.16 | 116.5 | 72.61 | 785 | 386 | 295.4 | 319 |
| 7 | 61.99 | 93.44 | 71.65 | 709 | 400 | 294.1 | 299.9 |

# 6 Conclusion

An optimal sizing approach based on EAs for CMOS amplifiers that applies the $g_m/I_d$ design method and integer encoding has been introduced. It was shown that the results provided by DE and NSGA-II converge in less generations when applying our proposed sizing optimization approach.

Another important advantage of applying the $g_m/I_d$ design method and integer encoding with evolutionary algorithms, is that the biasing conditions for the MOS transistors are guaranteed. As a result, we conclude that the $g_m/I_d$ design method and integer encoding are quite useful to reduce search spaces for the design variables, which at the end are multiples of the IC fabrication technology lambda, and then it also reduces time computing in the sizing optimization of analog ICs by applying evolutionary algorithms.

As a future work we will apply the proposed methodology to size more complex circuits and to incorporate a many objective optimization algorithm to solve problems with more than three objective functions.

**Acknowledgements** This work is partially supported by CONACyT-Mexico under grant 237991. The first author want to thank the Administrative Department of Science, Technology and Innovation of Colombia Colciencias for the scholarship.

## References

1. Guerra-Gómez, I., McConaghy, T., Tlelo-Cuautle, E.: Operating-point driven formulation for analog computer-aided design. Analog Integr. Circuits Signal Process. **74**(2), 345–353 (2013)
2. Graeb, H.E.: Analog Design Centering and Sizing. Springer, Berlin (2007)
3. McConaghy, T., Breen, K., Dyck, J., Gupta, A.: Variation-Aware Design of Custom Integrated Circuits: A Hands-on Field Guide. Springer, Berlin (2013)
4. Rocha, F.A., Martins, R.M., Lourenco, N.C., Horta, N.C.: State-of-the-art on automatic analog IC sizing. Electronic Design Automation of Analog ICs Combining Gradient Models with Multi-Objective Evolutionary Algorithms. SpringerBriefs in Applied Sciences and Technology, pp. 7–22. Springer, Berlin (2014)
5. Guerra-Gomez, I., Tlelo-Cuautle, E.: Sizing analog integrated circuits by current-branches-bias assignments with heuristics. Elektronika ir Elektrotechnika **19**(10), 81–86 (2013)
6. Silveira, F., Flandre, D., Jespers, P.G.A.: A $g_m/I_D$ based methodology for the design of CMOS analog circuits and its application to the synthesis of a silicon-on-insulator. Solid State Circuits IEEE **31**(9), 1314–1319 (1996)
7. Jespers, P.: The gm/ID Methodology, a Sizing Tool For Low-voltage Analog CMOS Circuits. Springer, Boston (2010)
8. Tlelo-Cuautle, E., Sanabria-Borbon, A.C.: Optimising operational amplifiers by evolutionary algorithms and gm/id method. Int. J. Electron. **103**(10), 1665–1684 (2016)
9. Kotti, M., Fakhfakh, M., Fino, M.H.: On the dynamic rounding-off in analogue and RF optimal circuit sizing. Int. J. Electron. **101**(4), 452–468 (2014)
10. Sanabria-Borbón, A.C., Tlelo-Cuautle, E.: Sizing analogue integrated circuits by integer encoding and NSGA-II. IETE Tech. Rev. 0(0):1–7, 0
11. Lara, A., Sanchez, G., Coello Coello, C.A., Schütze, O.: HCS: a new local search strategy for memetic multiobjective evolutionary algorithms. IEEE Trans. on Evol. Comput. **14**(1), 112–132 (2010)
12. Schütze, O., Martín, A., Lara, A., Alvarado, S., Salinas, E., Coello Coello, C.A.: The directed search method for multi-objective memetic algorithms. Comput. Optim. Appl. **63**(2), 305–332 (2016)
13. Xiong, F.-R., Qin, Z.-C., Xue, Y., Schütze, O., Ding, Q., Sun, J.-Q.: Multi-objective optimal design of feedback controls for dynamical systems with hybrid simple cell mapping algorithm. Commun. Nonlinear Sci. Numer. Simul. **19**(5), 1465–1473 (2014)
14. Schütze, O., Esquivel, X., Lara, A., Coello Coello, C.A.: Using the averaged Hausdorff distance as a performance measure in evolutionary multiobjective optimization. IEEE Trans. Evol. Comput. **16**(4), 504–522 (2012)

15. Luna, F., Zavala, G.R., Nebro, A.J., Durillo, J.J., Coello Coello, C.A.: Distributed multi-objective metaheuristics for real-world structural optimization problems. Comput. J. **59**(6), 777–792 (2016)
16. Zavala, G., Nebro, A.J., Luna, F., Coello Coello, C.A.: Structural design using multi-objective metaheuristics. comparative study and application to a real-world problem. Struct. Multidiscip. Optim. **53**(3), 545–566 (2016)
17. Sapatnekar, S.S., Rao, V.B., Vaidya, P.M., Sung-Mo Kang.: An exact solution to the transistor sizing problem for CMOS circuits using convex optimization. IEEE Trans. Comput. Aided Des. Integr. Circuits Syst. **12**(11), 1621–1634 (1993)
18. Mandal, P., Visvanathan, V.: CMOS op-amp sizing using a geometric programming formulation. IEEE Trans. Comput. Aided Des. Integr. Circuits Syst. **20**(1), 22–38 (2001)
19. Abbas, Z., Olivieri, M.: Optimal transistor sizing for maximum yield in variation-aware standard cell design. Int. J. Circuit Theory Appl. **44**(7), 1400–1424 (2015)
20. Shokouhifar, M., Jalali, A.: Evolutionary based simplified symbolic PSRR analysis of analog integrated circuits. Analog Integr. Circuits Signal Process. **86**(2), 189–205 (2016)
21. de la Fraga, L.G., Guerra-Gomez, I., Tlelo-Cuautle, E.: On the selection of solutions in multi-objective analog circuit design. NEO 2015, pp. 377–389. Springer, Berlin (2017)
22. Storn, R., Price, K.V.: Differential evolution - a simple and efficient heuristic for global optimization over continuos spaces. J. Glob. Optim. **11**(4), 341–359 (1997)
23. Mezura, E., Velázquez, J., Coello, C.A.: A comparative study of differential evolution variants for global optimization. In: CCO06: Proceedings of the 8th Annual Conference on Genetic and Evolutionary Computation, pp. 485–492. ACM Press, New York (2006)
24. Boussaïd, I., Lepagnot, J., Siarry, P.: A survey on optimization metaheuristics. Inf. Sci. **237**, 82–117 (2013)
25. Ao, Y., Chi, H.: Experimental study on differential evolution strategies. In: 2009 WRI Global Congress on Intelligent Systems, pp. 19–24 (2009)
26. Qing, A.: Differential Evolution: Fundamentals and Applications in Electrical Engineering. Wiley, New York (2009)
27. Onwubolu, G., Davendra, D.: Scheduling flow shops using differential evolution algorithm. Eur. J. Oper. Res. **171**(2), 674–692 (2006)
28. Kukkonen, S., Coello Coello, C.A.: Generalized differential evolution for numerical and evolutionary optimization. NEO 2015, pp. 253–279. Springer, Berlin (2017)
29. Deb, K., Pratap, A., Agarwal, S.: A fast and elitist multiobjective genetic algorithm: NSGA-II. IEEE Trans. Evol. Comput. **6**(2), 182–197 (2002)
30. Luo, B., Zheng, J., Xie, J., Wu, J.: Dynamic crowding distance-a new diversity maintenance strategy for MOEAs. In: 2008 Fourth International Conference on Natural Computation. pp. 580–585 (2008)
31. Guerra-Gomez, I., Tlelo-Cuautle, E., de la Fraga, L.G.: Richardson extrapolation-based sensitivity analysis in the multi-objective optimization of analog circuits. Appl. Math. Comput. **222**, 167–176 (2013)
32. Coello Coello, C.A., Lamont, G.B., Van Veldhuizen, D.A., Goldberg, D.E., Koza, J.R.: Evolutionary Algorithms for Solving Multi-Objective Problems. Genetic and Evolutionary Computation Series, 2nd edn. Springer, Berlin (2007)
33. Gray, P.R., Hurst, P.J., Lewis, S.H.: Analysis and Design of Analog Integrated Circuits, 5th edn. Wiley, New York (2009)
34. Arslan, E., Metin, B., Cicekoglu, O.: MOSFET-only multi-function biquad filter. AEU-Int. J. Electron. Commun. **69**(12), 7–10 (2015)

35. Zhang, H., Sanchez-Sinencio, E.: Linearization techniques for CMOS low noise amplifiers: a tutorial. In: IEEE Trans. Circuits Syst. I Regul. pap. 58(1), 22–36 (2011)
36. Jack, O., Ferreira, P.M.: A $g_m/I_D$-based noise optimization for CMOS folded-cascode operational amplifier. IEEE Trans. Circuits Syst. II Express Briefs **61**(10), 783–787 (2014)
37. Lin, C-W., Sue, P-D., Shyu, Y-T., Chang,S-J.: A bias-driven approach for automated design of operational amplifiers. In: International Symposium on VLSI Design, Automation and Test, 2009. VLSI-DAT '09. pp. 118–121 (2009)

# Index



© Springer International Publishing AG 2018
Y. Maldonado et al. (eds.), *NEO 2016*, Studies in Computational Intelligence 731,
https://doi.org/10.1007/978-3-319-64063-1

Printed in the United States
By Bookmasters